Math 3

God's Order In Numbers

Creators and Writers
Kevin Swanson (BS, Mechanical Engineering; MDiv.)
Elliott Best (BS, Mathematics; MS, Mathematics)

Editors
Tammy Sechrist (BS, Biology)
Amy Cool (MS, Secondary Education)

Generations
PASSING ON THE FAITH

Contents

To the Parent/Teacher

Welcome to the third level of "God's math"!

Math Demonstrated in Creation. Math shows God's wisdom, organization, and unchangeable nature—His patterns. We see God's attributes in all His creation. This course uses photos and realistic representations of God's design in creation whenever possible.

Use of the Concrete and the Abstract. Our goal is to start with the concrete, pointing out how math is demonstrated in God's creation first. Abstract math is then reinforced with concrete applications in science and economics. We want the student to make immediate application of math in the kitchen, the garden, business, transportation, and all of life.

Math Intersections with Music and Games. Since there is a strong connection between music and math, we include some of this. We also explore math and logic in games.

How to Use this Course

This course contains approximately 150 days of content for a 30 to 36-week school year. A suggested schedule follows this introduction but may be adapted to the student as necessary.

Typically, sequential math days will alternate between lessons of new content and the review of prior concepts. New lessons are varied and presented in an engaging, wonder-filled, enthusiastic way to immerse your child happily into the wonderful world of mathematics. "Practice" days use the time-tested idea of repetition to help the child retain a firm grasp on the old.

For the teacher's and student's convenience, all the content needed for the course is contained in a single book. Special information intended for the parent/teacher is provided in orange boxes. An answer key is also included in the back of the book. The simple, engaging text of the lessons is meant to be read aloud to the child so the parent/teacher does not need to take extra time to prepare. Also, the student may be able to read the lessons without the parent/teacher's assistance, as the text is set to a third-grade reading level.

Memorization is encouraged. While the parent/teacher may still wish to use flash cards, this course assumes that the students have learned their math facts. The student will have the opportunity to work on drills for multiplication facts in this course. Without getting the math facts down, it is extremely difficult to move on in math. God has given young children an amazing ability to memorize. We hope to set your student up for a lifetime of success in math. Just as drills are the fundamentals for athletes, the basic math facts must be memorized if the student will enjoy and further engage with the art and science of mathematics. For students having a hard time with memorization, parents may allow them more access to multiplication and division tables.

The most powerful form of learning comes by life application. Children who are not given the opportunity to life-apply math concepts will find the material irrelevant and boring. Chances are they will not retain much of what they have learned. Life application helps immensely with retention. Each chapter ends with suggestions of ways your children can apply math. Choose from several of these activities, or adapt them to provide the child with a tailor-made application of the material. Homeschoolers are much advantaged by bringing the math class into the kitchen, or including math with grocery shopping, or bringing math into any other context where real life is taking place.

We have also added "Everyday Math" exercises in which we use the second person singular, inviting the child into applying math to his or her own situation. Our goal is to make math as personally relevant as possible to the child.

Some children are naturally gifted in math and they may be capable of starting out earlier than the average. Others may not be able to finish this level in a single year. Some children may need more practice, and some less practice. We encourage respect for "the principle of individuality," especially in math. Allow the struggling child plenty of time to work with concrete manipulatives and don't worry about the schedule. For eager little mathematicians, we have included occasional "Extra Challenge" exercises. Most first and second level students will need one-on-one interaction from the parent/teacher. Once math facts are memorized and the student advances into this third level math, he or she should be able to work more independently.

We Need Your Feedback

Feel free to offer any comments and suggestions to us at *mail@generations.org*. We appreciate your input!

> And these words which I command you today shall be in your heart. You shall teach them diligently to your children, and shall talk of them when you sit in your house, when you walk by the way, when you lie down, and when you rise up. You shall bind them as a sign on your hand, and they shall be as frontlets between your eyes. You shall write them on the doorposts of your house and on your gates. (Deuteronomy 6:7-9)

Suggested Lesson Schedule

Date	Day	Lessons & Practice	✓	Progress Notes
First Semester—First Quarter				
Week 1	Monday	Chapter 1 Day 1 Lesson		
	Tuesday	Day 2 Lesson		
	Wednesday	Day 3 Practice		
	Thursday	Day 4 Practice		
	Friday			
Week 2	Monday	Day 5 Lesson		
	Tuesday	Day 6 Practice		
	Wednesday	Day 7 Lesson		
	Thursday	Day 8 Practice		
	Friday			
Week 3	Monday	Day 9 Lesson		
	Tuesday	Day 10 Practice		
	Wednesday	Day 11 Practice		
	Thursday	Day 12 Practice		
	Friday			
Week 4	Monday	Day 13 Lesson		
	Tuesday	Day 14 Practice		
	Wednesday	Day 15 Lesson		
	Thursday	Chapter 2 Day 16 Lesson		
	Friday	Day 17 Practice		
Week 5	Monday	Day 18 Lesson		
	Tuesday	Day 19 Practice		
	Wednesday	Day 20 Lesson		
	Thursday	Day 21 Practice		
	Friday			

**Weekly Memory Work: go over the lessons from that week and review the Memory Work sections.*

Date	Day	Lessons & Practice	✓	Progress Notes
Week 6	Monday	Day 22 Lesson		
	Tuesday	Day 23 Practice		
	Wednesday	Day 24 Lesson		
	Thursday	Day 25 Practice		
	Friday			
Week 7	Monday	Day 26 Practice		
	Tuesday	Chapter 3 Day 27 Lesson		
	Wednesday	Day 28 Practice		
	Thursday	Day 29 Practice		
	Friday			
Week 8	Monday	Day 30 Lesson		
	Tuesday	Day 31 Lesson		
	Wednesday	Day 32 Practice		
	Thursday	Day 33 Lesson		
	Friday			
Week 9	Monday	Day 34 Practice		
	Tuesday	Day 35 Lesson		
	Wednesday	Day 36 Practice		
	Thursday	Day 37 Lesson		
	Friday			

Date	Day	Lessons & Practice	✓	Progress Notes
First Semester—Second Quarter				
Week 1	Monday	Day 38 Lesson		
	Tuesday	Day 39 Practice		
	Wednesday	Day 40 Lesson		
	Thursday	Chapter 4 Day 41 Lesson		
	Friday			
Week 2	Monday	Day 42 Practice		
	Tuesday	Day 43 Lesson		
	Wednesday	Day 44 Lesson		
	Thursday	Day 45 Practice		
	Friday			
Week 3	Monday	Day 46 Lesson		
	Tuesday	Day 47 Practice		
	Wednesday	Day 48 Lesson		
	Thursday	Day 49 Practice		
	Friday			
Week 4	Monday	Day 50 Lesson		
	Tuesday	Day 51 Practice		
	Wednesday	Day 52 Lesson		
	Thursday	Chapter 5 Day 53 Lesson		
	Friday			
Week 5	Monday	Day 54 Practice		
	Tuesday	Day 55 Lesson		
	Wednesday	Day 56 Practice		
	Thursday	Day 57 Lesson		
	Friday			

Date	Day	Lessons & Practice	✓	Progress Notes
Week 6	Monday	Day 58 Practice		
	Tuesday	Day 59 Lesson		
	Wednesday	Day 60 Practice		
	Thursday	Day 61 Lesson		
	Friday			
Week 7	Monday	Day 62 Practice		
	Tuesday	Day 63 Lesson		
	Wednesday	Day 64 Practice		
	Thursday	Day 65 Lesson		
	Friday			
Week 8	Monday	Chapter 6 Day 66 Lesson		
	Tuesday	Day 67 Practice		
	Wednesday	Day 68 Lesson		
	Thursday	Day 69 Practice		
	Friday			
Week 9	Monday	Day 70 Lesson		
	Tuesday	Day 71 Practice		
	Wednesday	Day 72 Practice		
	Thursday	Day 73 Lesson		
	Friday			

Midterm Progress Notes	
Grade: _______	

Date	Day	Lessons & Practice	✓	Progress Notes
Second Semester—Third Quarter				
Week 1	Monday	Day 74 Practice		
	Tuesday	Day 75 Lesson		
	Wednesday	Day 76 Practice		
	Thursday	Day 77 Lesson		
	Friday			
Week 2	Monday	Chapter 7 Day 78 Lesson		
	Tuesday	Day 79 Practice		
	Wednesday	Day 80 Lesson		
	Thursday	Day 81 Practice		
	Friday			
Week 3	Monday	Day 82 Lesson		
	Tuesday	Day 83 Practice		
	Wednesday	Day 84 Lesson		
	Thursday	Day 85 Practice		
	Friday			
Week 4	Monday	Day 86 Lesson		
	Tuesday	Day 87 Practice		
	Wednesday	Day 88 Lesson		
	Thursday	Chapter 8 Day 89 Lesson		
	Friday			
Week 5	Monday	Day 90 Practice		
	Tuesday	Day 91 Lesson		
	Wednesday	Day 92 Practice		
	Thursday	Day 93 Lesson		
	Friday			

Date	Day	Lessons & Practice	✓	Progress Notes
Week 6	Monday	Day 94 Practice		
	Tuesday	Day 95 Lesson		
	Wednesday	Day 96 Practice		
	Thursday	Day 97 Lesson		
	Friday			
Week 7	Monday	Day 98 Practice		
	Tuesday	Day 99 Lesson		
	Wednesday	Day 100 Practice		
	Thursday	Day 101 Lesson		
	Friday			
Week 8	Monday	Chapter 9 Day 102 Lesson		
	Tuesday	Day 103 Practice		
	Wednesday	Day 104 Lesson		
	Thursday	Day 105 Lesson		
	Friday			
Week 9	Monday	Day 106 Practice		
	Tuesday	Day 107 Lesson		
	Wednesday	Day 108 Practice		
	Thursday	Day 109 Lesson		
	Friday			

Date	Day	Lessons & Practice	✓	Progress Notes
Second Semester—Fourth Quarter				
Week 1	Monday	Day 110 Practice		
	Tuesday	Day 111 Lesson		
	Wednesday	Chapter 10 Day 112 Lesson		
	Thursday	Day 113 Practice		
	Friday			
Week 2	Monday	Day 114 Lesson		
	Tuesday	Day 115 Practice		
	Wednesday	Day 116 Lesson		
	Thursday	Day 117 Practice		
	Friday			
Week 3	Monday	Day 118 Lesson		
	Tuesday	Day 119 Practice		
	Wednesday	Day 120 Lesson		
	Thursday	Day 121 Practice		
	Friday			
Week 4	Monday	Day 122 Lesson		
	Tuesday	Chapter 11 Day 123 Lesson		
	Wednesday	Day 124 Practice		
	Thursday	Day 125 Lesson		
	Friday			
Week 5	Monday	Day 126 Practice		
	Tuesday	Day 127 Lesson		
	Wednesday	Day 128 Practice		
	Thursday	Day 129 Lesson		
	Friday			

Date	Day	Lessons & Practice	✓	Progress Notes
Week 6	Monday	Day 130 Practice		
	Tuesday	Day 131 Practice		
	Wednesday	Day 132 Lesson		
	Thursday	Chapter 12 Day 133 Lesson		
	Friday			
Week 7	Monday	Day 134 Practice		
	Tuesday	Day 135 Lesson		
	Wednesday	Day 136 Practice		
	Thursday	Day 137 Lesson		
	Friday			
Week 8	Monday	Day 138 Practice		
	Tuesday	Day 139 Lesson		
	Wednesday	Day 140 Practice		
	Thursday	Day 141 Lesson		
	Friday			
Week 9	Monday	Day 142 Practice		
	Tuesday	Day 143 Lesson		
	Wednesday	Day 144 Practice		
	Thursday			
	Friday			
Progress Notes				
Final Grade: ______				

God multiplies animals when He gives babies to mama bears.

CHAPTER 1

More Numbers Than You Can Imagine!

Welcome to Math 3!

> "And Bezalel and Aholiab, and every gifted artisan in whom the Lord has put wisdom and understanding, to know how to do all manner of work for the service of the sanctuary, shall do according to all that the Lord has commanded." Then Moses called Bezalel and Aholiab, and every gifted artisan in whose heart the Lord had put wisdom, everyone whose heart was stirred, to come and do the work. (Exodus 36:1-2)

Bezalel and Aholiab were creative builders who helped build the tabernacle for the children of Israel. God appreciates skill. He gave special wisdom and understanding to these men so they would be skilled to do beautiful work on the house of worship. God is all wise, and He shares wisdom with us. He helps us to learn, and He encourages us to become skillful at our work. As you step through this next level of math, it is our hope and prayer that you will become more skillful at math. You will learn to add, subtract, multiply, and divide more skillfully and quickly.

You will learn more about fractions, shapes, and measurements. You will learn a little about using computers, calculators, spreadsheets, and coordinates. Math must be useful in the way that hammers and screwdrivers are useful. Remember that math is a tool you can use to study the world, to measure the world, to compare things, and to build things. Math will help you to cook and bake in the kitchen. Math gives you wisdom when you buy and sell things. Math helps you to understand travel, animal care, weather forecasts, and gardening.

This chapter will help you review some lessons you learned in Math 1 and 2.

Do these exercises, but watch the signs. Will you multiply (x)? Will you divide (÷)? Will you add (+)? Or will you subtract (–)?

Correct the wrong answers below. Cross out the wrong answers and write the correct answers. The first one is done for you.

1. 3 x 4 = ~~7~~ 12
2. 8 ÷ 1 = 1
3. 7 x 2 = 9
4. 6 ÷ 2 = 12
5. 10 – 1 = 11
6. 12 – 4 = 16
7. 4 x 2 = 6
8. 10 ÷ 1 = 11

Find the area of the shapes below.

9.
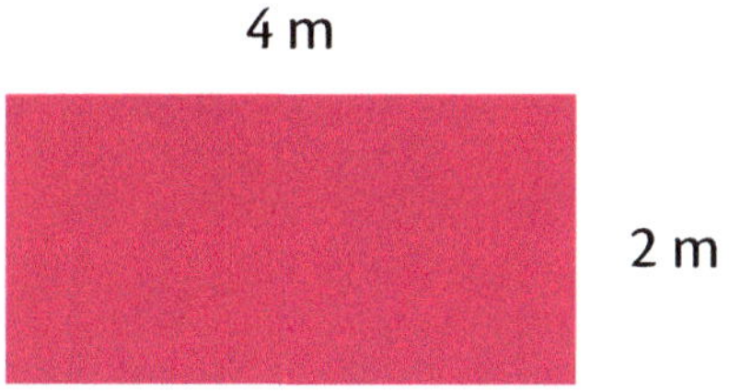

___ x ___ = ___ square m

11.
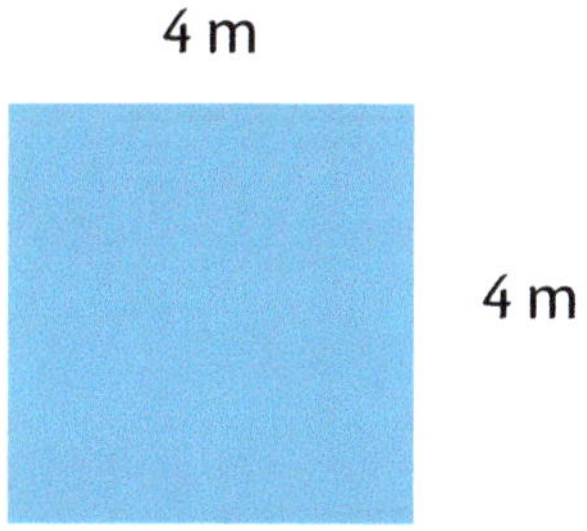

___ x ___ = ___ square m

10.
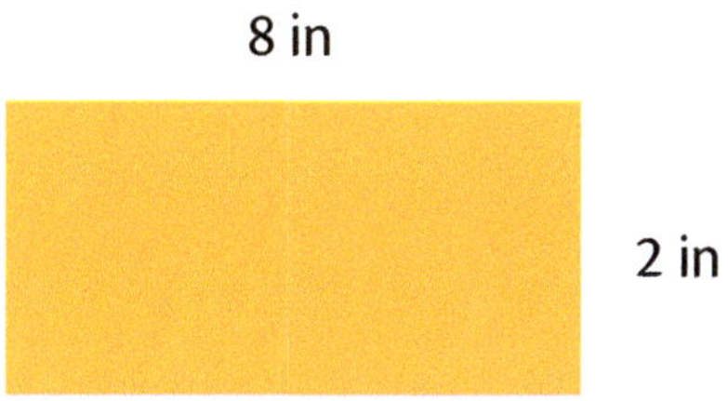

___ x ___ = ___ square in

12.
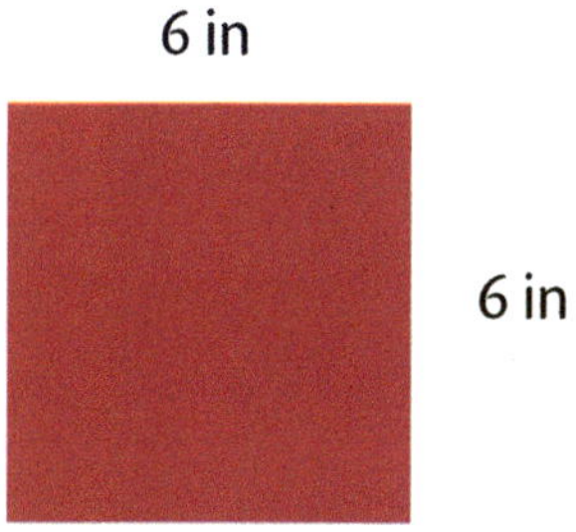

___ x ___ = ___ square in

For each block shape below, write the number of its columns and the number of its rows in the boxes. Then fill in its equation. See example at right.

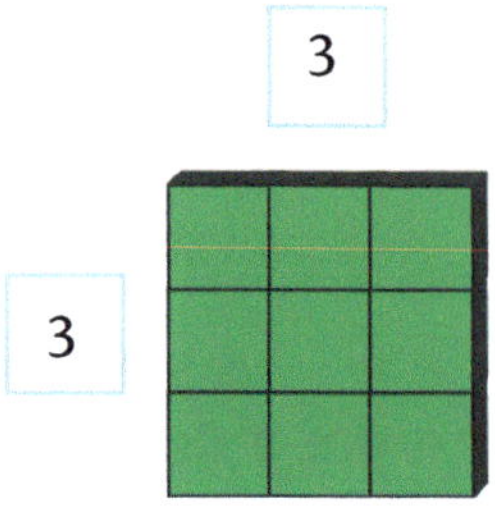

3 x 3 = 9

13.

___ X ___ = ___

14.

___ X ___ = ___

15.

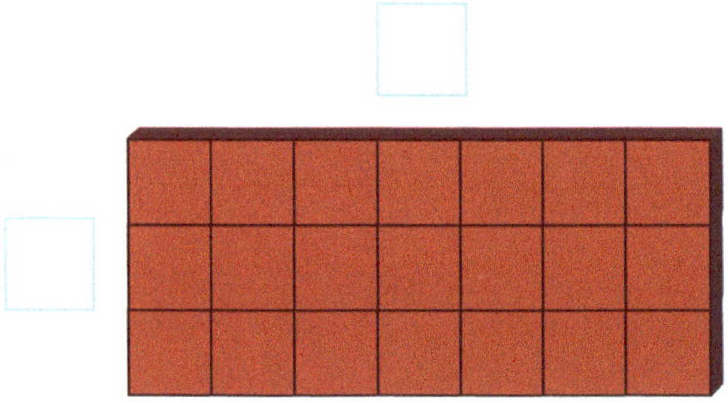

___ X ___ = ___

16.

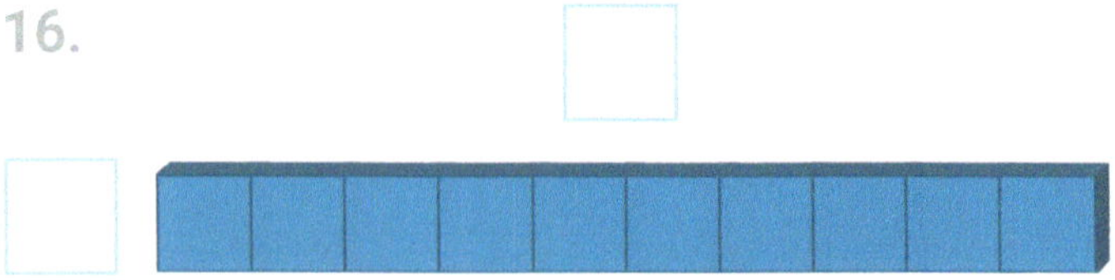

___ X ___ = ___

Counting Hundreds, Thousands, and Millions

DAY 2

> The LORD builds up Jerusalem;
> He gathers together the outcasts of Israel.
> He heals the brokenhearted
> And binds up their wounds.
> He counts the number of the stars;
> He calls them all by name.
> Great is our Lord, and mighty in power;
> His understanding is infinite.
> The LORD lifts up the humble;
> He casts the wicked down to the ground. (Psalm 147:2-6)

God counts the stars. There are more stars in the sky than we could ever count. But the Lord knows all of them by name. That's a lot of names!

There are so many stars that we would need very big numbers to count all of them! This lesson will introduce bigger and bigger numbers.

There is something inside boys and girls that wants to know huge numbers—you want to count the leaves on a tree and the stars in the sky. You want to tell others how many there are! Imagine telling your friend, "Can you believe there are almost 5,000 needles on that little tree? God must be powerful and wise to make such a tree!"

With big numbers, you need special patterns. You already know quite a lot about these patterns. Now you just need to take it to the next level. Look at the following exercises:

1. 19 + 1 =	3. 95 + 5 =	5. 298 + 2 =
2. 29 + 1 =	4. 99 + 1 =	6. 499 + 1 =

First, look at the big number in each equation. Can you see that 19 is almost 20? Can you see that 95 is almost as much as 100? Can you see that 298 is almost as high as 300?

Let's review place values. In the blocks below, the single block is the smallest chunk. It's all the way to the right, in the 1s place. Then comes the 10 chunk in the 10s place. Then, on the left, we see the 100 chunk in the 100s place.

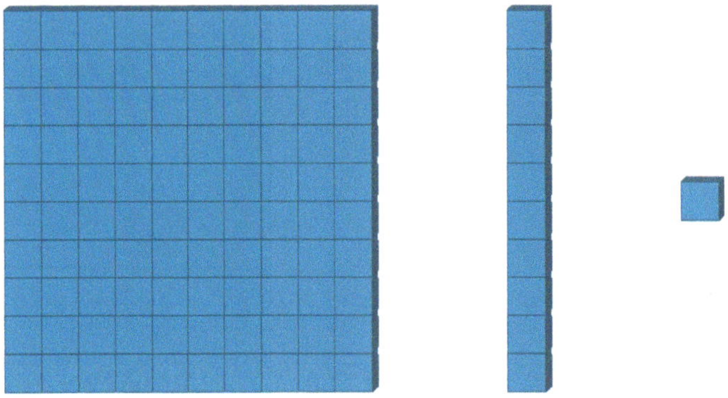

When we add 19 + 1, we first add everything in the 1s place together. That's 9 plus 1. We now have 10 1s. Then it's like trading the ten 1s for one chunk of 10. The 1s place goes to zero, and we have two chunks of 10. We have a 2 in the 10s place. With a 2 in the 10s place and a 0 in the 1s place, we see that our answer is 20.

To add 95 + 5, we first add the 5 + 5 because they are both in the 1s place. Just like before, the 1s place adds to 10. So we will write a zero in the 1s place and add 1 chunk of 10 to the 10s place. When we add that 1 chunk of 10 to the 9 chunks of 10 already in the 10s place, we get 10 chunks of 10. If we were using blocks, we could trade all 10 chunks of 10 for 1 chunk of 100. But if we are writing, we can just write a zero in the 10s place and a 1 in the 100s place. That will show the answer of 100.

When we add 2 to 298, the 1s place again becomes zero. We will again add a 10s chunk to the 10s place. But we already have 9 chunks of 10 in the 10s place. Another chunk of 10, will make 10 chunks of 10 in the 10s place. What do we do? We trade those 10 chunks of 10 for a chunk of 100. That means we would write a zero in the 10s place and add another chunk of 100 to the 100s place. We are adding 1 chunk of 100 to the 2 chunks of 100 that are already there. That makes an answer of 300! This is shown a different way below.

8 chunks of 1	+	2 chunks of 1	=	10 chunks of 1
9 chunks of 10	+	1 more chunk of 10	=	10 chunks of 10
2 chunks of 100	+	1 more chunk of 100	=	3 chunks of 100

Now let's do the exercises on the previous page together. Think of the numbers as blocks.

Now, here is a very big number. It's a gigantic number!

1234567

Here's the way we usually write large numbers: Use commas to break up groups of 3 digits, starting from the right. Our number should look like this:

1,234,567

⇧ ⇧

Do you see how we have started from the right and put a comma after each group of 3 digits? It just makes life easier! Can you follow that pattern and add commas to this ginormous number?

7. 3 4 5 3 4 6 2 3 4 5 7 2

Now, let's add 1 more to each of the numbers below. When you do, you'll find that you jump the number up so it has another place value on its left. Fill out the following chart. Recite these big numbers, with a little help from your parent/teacher.

	9 + 1 =	10	Ten
	99 + 1 =	100	One hundred
	999 + 1 =	1,000	One thousand
8.	9999 + 1 =		Ten thousand
9.	99999 + 1 =		One hundred thousand
10.	999999 + 1 =		One million

Do you see how hard it is to read the large numbers? Commas help!

Now, let's look at this big number: 1,234,567

How many 1s are there? How many 10s? How many 100s?

Imagine a box filled with 1,234,567 candies.

Place this number in the following table: 1,234,567

Million (1,000,000)	Hundred Thousand (100,000)	Ten-Thousand (10,000)	Thousand (1,000)	Hundred (100)	Ten (10)	One (1)

How many 1s of candies are there? How many 10s of candies are there? How many 100s of candies are there? How many 1000s of candies are there?

How do we read this gigantic number out loud? We read numbers from left to right. Here we go!

One million two hundred thirty-four thousand five hundred sixty-seven

Student Exercises

Can you fill in this place value chart for some more big numbers?

		1,000,000	100,000	10,000	1,000	100	10	1
1.	1,234,567	1	2	3	4	5	6	7
2.	32,431							
3.	145,943							
4.	95,703							
5.	999,999							
6.	1,999,991							
7.	458,392							
8.	1,010,010							
9.	9,999,999							
10.	8,275,384							

Add these double digit numbers. Remember to add the ones column first, and then the tens.

11. 21 + 11 = ____

12. 43 + 34 = ____

13. 33 + 44 = ____

14. 56 + 33 = ____

15. 12 + 34 = ____

16. 18 + 71 = ____

17. 21 + 45 = ____

18. 66 + 23 = ____

Let's explore God's patterns with sets. In these exercises, draw the new set that's formed when the members of the first sets are united. Remember, united sets (unions) are all of the members in the combined sets, without writing a repeated member twice. The symbol for union is U. In 24-26, draw brackets around your new sets.

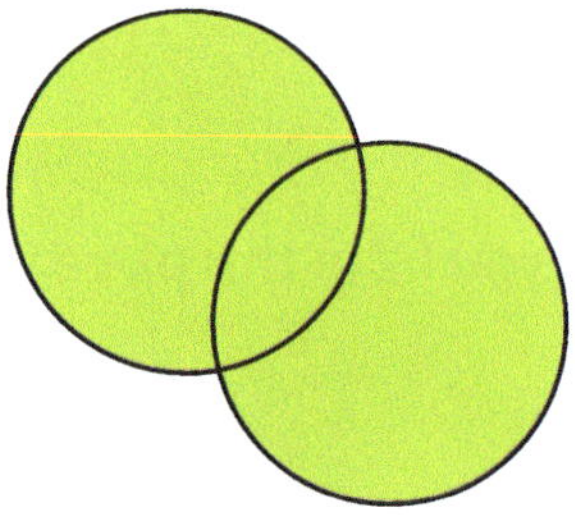

19.	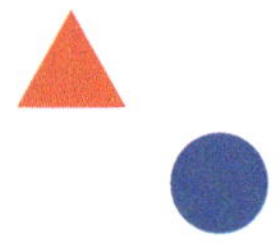	U		=	
20.		U	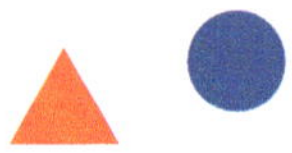	=	{ }
21.		U		=	{ }
22.		U		=	{ }
23.		U		=	{ }

24. {1,2} U {3,4} =

25. {1,2} U {2,3} =

26. {1,2} U {1,2} =

"Now the multitude of those who believed were of one heart and one soul; neither did anyone say that any of the things he possessed was his own, but they had all things in common." (Acts 4:32)

Practice DAY 3

Here is some practice with addition and subtraction!

1. $$\begin{array}{r} 5 \\ +\ 9 \\ \hline \end{array}$$

2. $$\begin{array}{r} 23 \\ +\ 46 \\ \hline \end{array}$$

3. $$\begin{array}{r} 321 \\ +\ 542 \\ \hline \end{array}$$

4. $$\begin{array}{r} 111 \\ +\ 181 \\ \hline \end{array}$$

5. $$\begin{array}{r} 4 \\ +\ 8 \\ \hline \end{array}$$

6. $$\begin{array}{r} 421 \\ +\ 511 \\ \hline \end{array}$$

7. $$\begin{array}{r} 217 \\ +\ 121 \\ \hline \end{array}$$

8. $$\begin{array}{r} 16 \\ -\ 8 \\ \hline \end{array}$$

9. $$\begin{array}{r} 45 \\ -\ 21 \\ \hline \end{array}$$

10. $$\begin{array}{r} 57 \\ -\ 14 \\ \hline \end{array}$$

11. $$\begin{array}{r} 397 \\ -\ 101 \\ \hline \end{array}$$

12. $$\begin{array}{r} 17 \\ -\ 5 \\ \hline \end{array}$$

13. $$\begin{array}{r} 45 \\ -\ 14 \\ \hline \end{array}$$

14. $$\begin{array}{r} 57 \\ -\ 17 \\ \hline \end{array}$$

15. $$\begin{array}{r} 852 \\ -\ 342 \\ \hline \end{array}$$

When you are learning multiplication facts, do you sometimes forget one of them? That's when you use a multiplication chart to find the answer. You do this by finding one of the numbers you are multiplying on the top of the chart and finding the other number on the side of the chart. Then you use your fingers to travel down the chart from the top number and sideways from the side number until your fingers meet. Where your fingers meet is your answer.

Did you know that you can also use a multiplication chart to help you find the answer to a division fact? Division facts have a large number that is being divided by a smaller number. Start by first putting your finger on the top of the chart, right on the smaller number you are dividing by. Now travel down until your finger hits the large number you are dividing. From that large number, travel sideways until you arrive at the number on the side of the chart. That number is your answer.

Here are some math questions. By now you know that they are called equations. Try to find the number to answer the question.

16. What plus 6 equals 15? $\square + 6 = 15$

17. What plus 8 equals 14? $\square + 8 = 14$

18. What minus 5 equals 9? $\square - 5 = 9$

19. What minus 7 equals 10? $\square - 7 = 10$

20. What times 3 equals 12? $\square \times 3 = 12$

21. What times 5 equals 10? $\square \times 5 = 10$

22. What divided by 2 equals 3? $\square \div 2 = 3$

23. What divided by 4 equals 1? $\square \div 4 = 1$

Practice DAY 4

Add these numbers as fast as you can. The first one shows how you can break the work into steps. There is room under each problem to write out your steps if you want to.

1. 5 + 2 + 3 + 2 =
   7 + 3 + 2 =
   10 + 2 = 12

2. 3 + 5 + 4 =

3. 5 + 3 + 4 + 5 =

4. 2 + 2 + 3 + 3 + 4 + 4 =

5. 10 + 10 + 10 + 3 =

6. 6 + 7 + 5 + 4 + 2 =

7. 2 + 3 + 4 + 5 + 2 + 1 + 1 =

8. 9 +8 + 2 + 3 + 6 =

9. 2 + 4 + 3 + 5 + 6 + 7 =

10. 1 + 1 +1 +1 + 2 + 2 +3 + 4 =

Let's practice division:

11. How many 3s can you find in these 6 blocks? Divide up these 6 blocks into groups of 3.

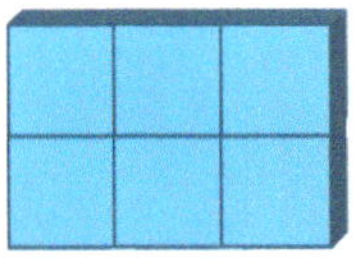

6 ÷ 3 = ☐

12. How many 3s can you see in these 9 blocks? Divide up these 9 blocks into 3s.

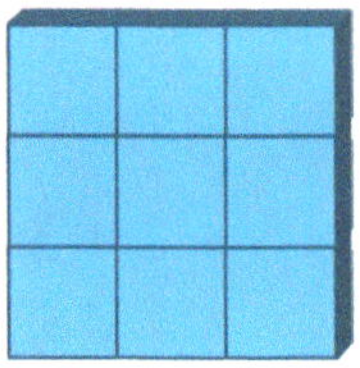

9 ÷ 3 = ☐

13. How many 2s can you see in these 20 blocks? Divide up these 20 blocks into 2s.

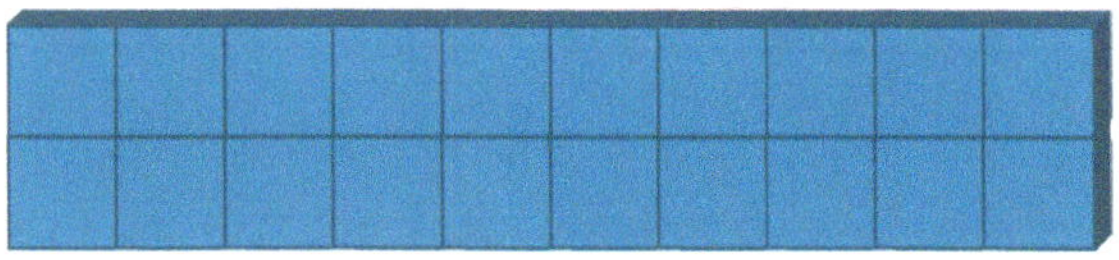

20 ÷ 2 = ☐

14. How many 4s can you see in these 16 blocks? Divide up these 16 blocks into 4s.

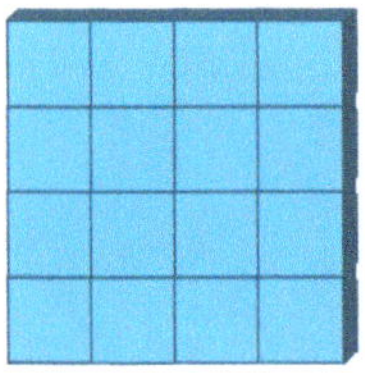

16 ÷ 4 = ☐

15. How many 10s can you see in these 30 blocks?

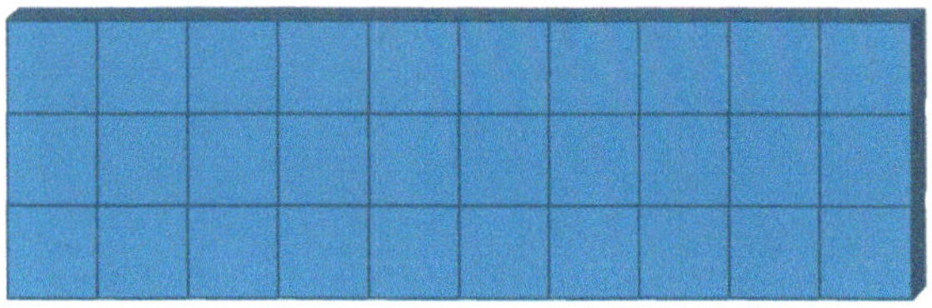

30 ÷ 10 = ☐

16. How many 3s can you find in these 15 blocks?

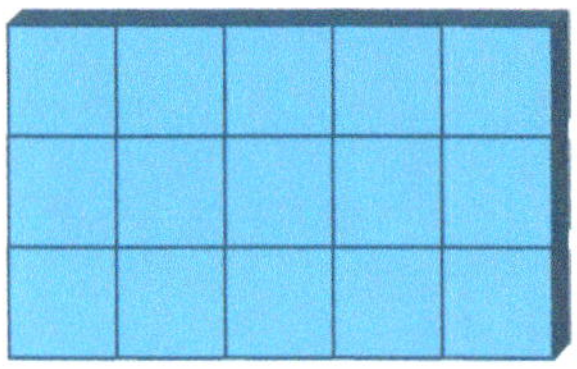

15 ÷ 3 = ☐

17. How many 5s can you find in these 15 blocks?

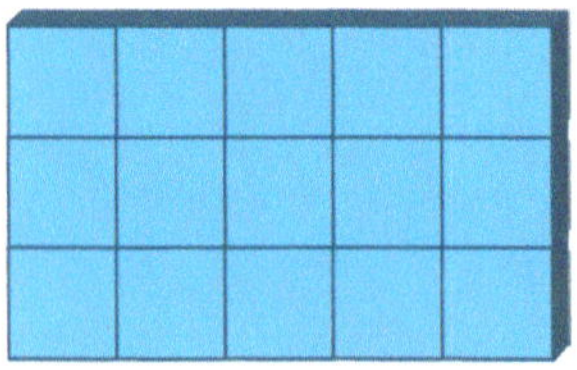

15 ÷ 5 = ☐

18. How many 6s can you find in these 18 blocks?

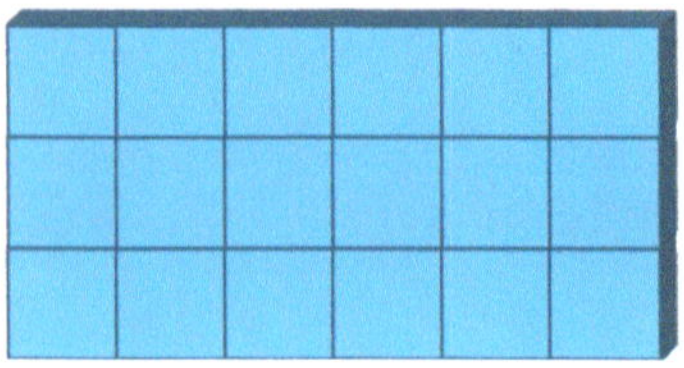

18 ÷ 6 = ☐

Adding More Big Numbers DAY 5

> Be diligent to know the state of your flocks,
> And attend to your herds; For riches are not forever,
> Nor does a crown endure to all generations. (Proverbs 27:23-24)

God has given each of us certain things to take care of. If you don't keep track of your sheep, goats, and cows, they will wander away. We are supposed to be good stewards. We are to take good care of the things God has put under our care.

Shepherds need to be able to count their sheep and add up their goats. Farmers keep track of their fruit and vegetable plants and their trees. Adding and multiplying help with this.

While some people will use calculators to add up things, this can be a slow way to do it. Sometimes we need to be able to add things quickly using our own minds. Often we will have to add up many numbers. So let's make sure that we can do this well.

How many chicks can you count in this brooding box?

Pretend that your family is going into raising chickens. You have three brooding boxes. You have counted 42, 35, and 48 chicks in the three boxes. Now, how many chicks do you have altogether?

Here is how to add these up. (See below) First add the 2 + 5 in the 1s column. That equals 7. Keep that in your mind, or write it down on the right side. Now you will need to add the 7 to the 8. That equals 15. Write down the 5 under the line in the 1s place, and carry the 1.

Now add the 10s column. Add 1+4+3+4. First add 1 and 4 together to get 5. Then add the 5 to the 3 to get 8. Add the 8 to the 4 to get 12. Write the 12 under the line, to the left of the 5. Your answer is 125. Notice that we did not have to carry the 1 into the hundreds place. There were no other hundreds to add, so we just wrote the 1 in the hundreds place of the answer. As you add the numbers in the 10s column, you can write the answers on the left side.

		1		
5		4	2	
8		3	5	7
12	+	4	8	15
	1	2	5	

Now try one for yourself! Add these chicks together! You can add them in your mind, or you can write the numbers in the boxes provided.

1.

Carry	1		
	3	5	
	2	2	
+	1	4	

2.

Carry	1		
	1	7	
	2	3	
+	5	7	

3.

Carry	1		
	6	8	
	2	1	
+	3	5	

Student Exercises

Try these addition exercises. Some of the numbers to carry have been provided as hints for you.

1.

	1	
	5	1
	3	5
+	9	9

2.

	1	
	2	5
	3	7
+	4	1

3.

	1	
	8	2
	3	3
+	2	6

4.

	1		
	1	3	
	6	7	
+	1	9	

5.

	1		
	1	4	
	2	6	
+	1	5	

6.

	2	4	
	4	1	
+		3	

7.

	7	2	
	5	2	
+	4	8	

8.

	6	5	
	3	6	
+	1	1	

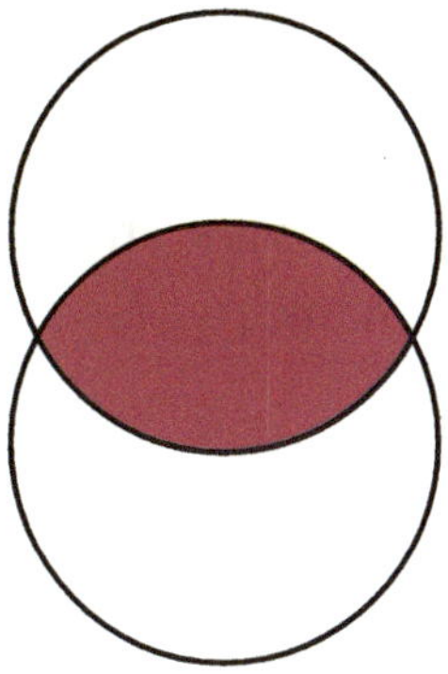

Set Intersection. Let's explore God's patterns with sets. Let's see how these sets come together to create new sets. Remember, intersections ask, "What do these two sets have in common?" The ∩ is the intersection symbol. If there is no intersection write "empty."

9.	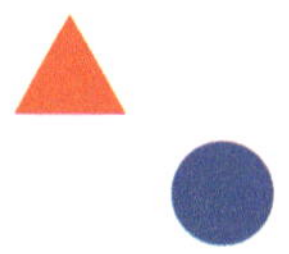	∩		=	
10.		∩		=	
11.		∩		=	
12.		∩		=	
13.		∩		=	

14. $\{1, 2\} \cap \{3, 4\} = \{ \qquad \}$

15. $\{1, 2\} \cap \{2, 3\} = \{ \qquad \}$

16. $\{1, 2\} \cap \{1, 2\} = \{ \qquad \}$

DAY 6 Practice

Check out each of these numbers and answer the following questions. This will help you see how big the numbers are, and pretty soon you will be ready to add, subtract, and do all sorts of things with them! Thank You, God! For each exercise, add commas to help you read the number more easily.

1. **26783**
   - How many 10s? ______
   - How many 1,000s? ______
   - How many 10,000s? ______

2. **3152345**
   - How many 1s? ______
   - How many 100s? ______
   - How many 100,000s? ______

3. **708310743**
   - How many 100s? ______
   - How many 1,000s? ______
   - How many 1,000,000s? ______

4. **35781**
   - How many 10s? ______
   - How many 1,000s? ______
   - How many 10,000s? ______

5. **56921**
   - How many 10s? ______
   - How many 100s? ______
   - How many 10,000s? ______

6. **125690**
   - How many 1s? ______
   - How many 1,000s? ______
   - How many 10,000s? ______

7. **347552**
   - How many 10s? ______
   - How many 10,000s? ______
   - How many 1,000,000s? ______

8. **1067254**
   - How many 1s? ______
   - How many 100s? ______
   - How many 10,000s? ______

Now for more practice with multiplication. Try answering these from memory.

9. 3 × 4 =	11. 2 × 5 =	13. 7 × 8 =	15. 8 × 5 =
10. 6 × 1 =	12. 7 × 3 =	14. 5 × 6 =	16. 8 × 9 =

Rounding DAY 7

You shall not steal, nor deal falsely, nor lie to one another. (Leviticus 19:11)

God wants honesty, especially when we are buying or selling things. When somebody asks for a large order of french fries at a restaurant, the servers should be sure the pouch is full. Restaurant servers don't count french fries. They don't weigh the french fries. They make a guess—an honest guess.

Today we are going to learn more about rounding numbers (making a good guess). You've already learned how to round numbers. But today's numbers will be a little bit harder.

First we will round to the nearest 100. Let's take the example of numbers between 200 and 300. If the number is lower than 250, we will round down to 200. If that boy on the steps doesn't quite make it to 250, he falls back down to 200. If the boy makes it to 250, or climbs up higher than 250, then we will round up to 300. The boy on the steps shoots all the way up to the top!

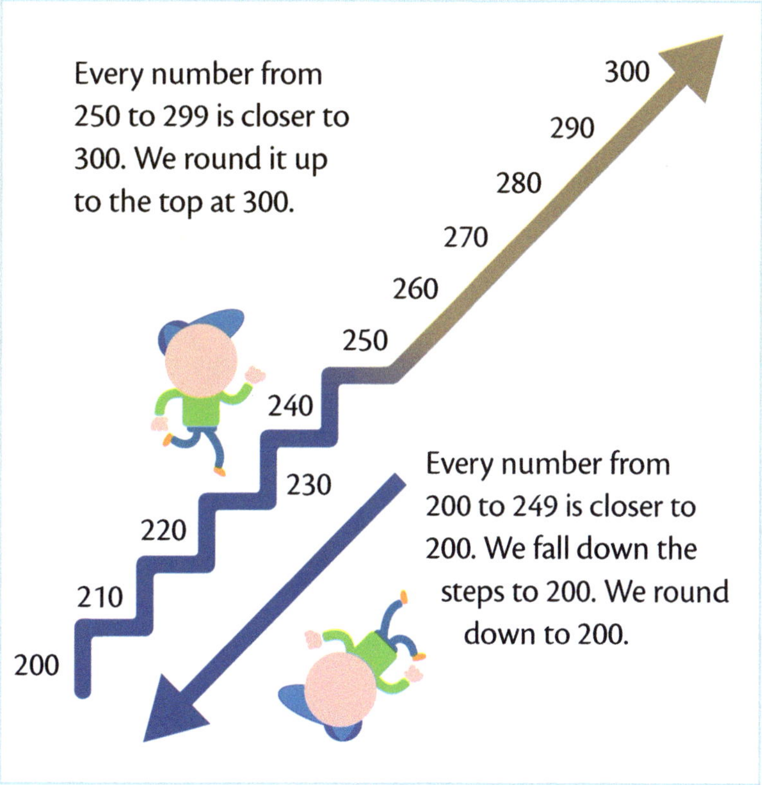

Round these numbers to the nearest 10:

1. 61 ⇨	3. 35 ⇨
2. 98 ⇨	4. 95 ⇨

Round these numbers to the nearest 100:

5. 480 ⇨	7. 632 ⇨
6. 960 ⇨	8. 981 ⇨

Rounding Bigger Numbers

Let's say the population of a city is 74,994. That's a lot of people! Let's use this number to practice rounding.

You can round 74,994 to the nearest 10. It rounds down to 74,990.

74,994 ⇨ 74,990

Round the number to the nearest 1,000. This means you will round 994 up to 1000.

74,994 ⇨ 75,000

Round to the nearest 100. This is means you will round 94 up to 100. Now if you round 94 up to 100, you will have to add 100 to 900. This means that 74,994 rounded to the nearest 100 is 75,000.

74,994 ⇨ 75,000

9. Round our number to the nearest 10,000. Can you do this on your own? Decide if 74,994 will round down to 70,000 or up to 80,000?

74,994 ⇨ ______

10. Round to the nearest 100,000. Is the 74,994 closer to 0 or to 100,000?

74,994 ⇨ ______

Now, when we make an estimate of how many people live in the city, we can say:

About 74,990 people (rounded to the nearest 10).
About 75,000 people (rounded to the nearest 100).
About 75,000 people (rounded to the nearest 1,000).
About 70,000 people (rounded to the nearest 10,000).
About 100,000 people (rounded to the nearest 100,000).

Or we can say there are exactly 74,994 people living in the city.

Extra Challenge

11. Round 950,715 to the nearest 10: ⇨

12. Round 950,715 to the nearest 100: ⇨

13. Round 950,715 to the nearest 1,000: ⇨

14. Round 950,715 to the nearest 10,000: ⇨

15. Round 950,715 to the nearest 100,000: ⇨

16. Round 950,715 to the nearest 1,000,000: ⇨

Before you finish the lesson today, talk to your parent/teacher about rounding. Why do we round numbers? Can you see how rounding in the examples above was helpful?

Student Exercises

Here are some big numbers! Well, they probably seem small to you now. Put them in the order that God made for them. Soon you will learn how to order the very large numbers you are learning about in this chapter. Remember, the alligator eats the bigger number: 45 < 78.

1. 123 ◯ 153
2. 824 ◯ 273
3. 899 ◯ 900
4. 349 ◯ 450
5. 348 ◯ 894
6. 398 ◯ 547
7. 537 ◯ 523
8. 54 ◯ 142
9. 345 ◯ 682
10. 400 ◯ 399
11. 34 ◯ 0
12. 980 ◯ 880

Practice DAY 8

Here is some practice with multiplication. Try answering the first set of exercises from memory. Then, you may use the multiplication chart to figure out the second set if you need to.

1. 3 x 2 = ☐
2. 6 x 2 = ☐
3. 4 x 4 = ☐
4. 5 x 3 = ☐
5. 2 x 4 = ☐
6. 6 x 3 = ☐
7. 1 x 11 = ☐
8. 3 x 1 = ☐
9. 4 x 3 = ☐

10. 6 x 8 = ☐
11. 5 x 7 = ☐
12. 6 x 6 = ☐
13. 10 x 9 = ☐
14. 7 x 5 = ☐
15. 8 x 8 = ☐
16. 4 x 9 = ☐
17. 6 x 4 = ☐

X	1	2	3	4	5	6	7	8	9	10
1	1	2	3	4	5	6	7	8	9	10
2	2	4	6	8	10	12	14	16	18	20
3	3	6	9	12	15	18	21	24	27	30
4	4	8	12	16	20	24	28	32	36	40
5	5	10	15	20	25	30	35	40	45	50
6	6	12	18	24	30	36	42	48	54	60
7	7	14	21	28	35	42	49	56	63	70
8	8	16	24	32	40	48	56	64	72	80
9	9	18	27	36	45	54	63	72	81	90
10	10	20	30	40	50	60	70	80	90	100

Everyday Math

18. How many feet are there in your family? How many people live in your house? How many feet does each person have? Multiply to find out how many feet are in your family.

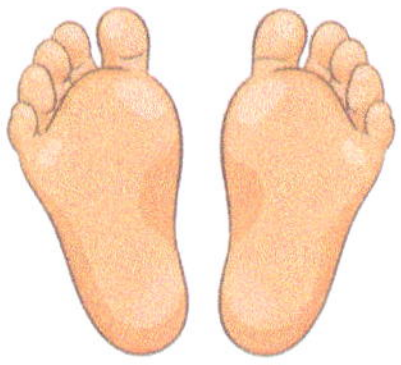

☐ x 2 = ☐

19. Sam took one bath every day last week. How many days are there in a week? How many baths did Sam take last week? How many baths would he take if he took one bath every day for 3 weeks?

1 x ☐ = ☐ 3 x ☐ = ☐

20. Juan has two octopuses in his fish tank. Each octopus has 8 legs. How many legs are there moving about in the fish tank?

2 x ☐ = ☐

21. How many little girls under 12 years of age go to your church? How many little boys under 12 go to your church? How many little children under 12 years of age go to your church altogether?

☐ + ☐ = ☐

22. Your parent, your sister, and you are doing messy work in the back yard. You will use gloves to keep your hands clean. How many gloves do you need for all three people?

☐ x ☐ = ☐

Let's practice measuring things today. Find the lengths of the lines, the angle sizes in degrees, and the perimeters of the rectangle and square. First, guess a measurement for each one. Then, measure. Measure using inches for this page instead of centimeters. You will need a protractor and a ruler for these exercises.

23. How big is this angle?

Your guess: _____

Your measurement: _____

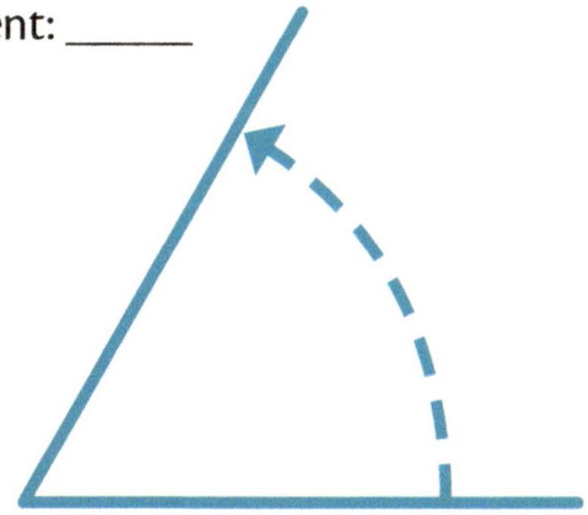

24. How big is this angle?

Your guess: _____

Your measurement: _____

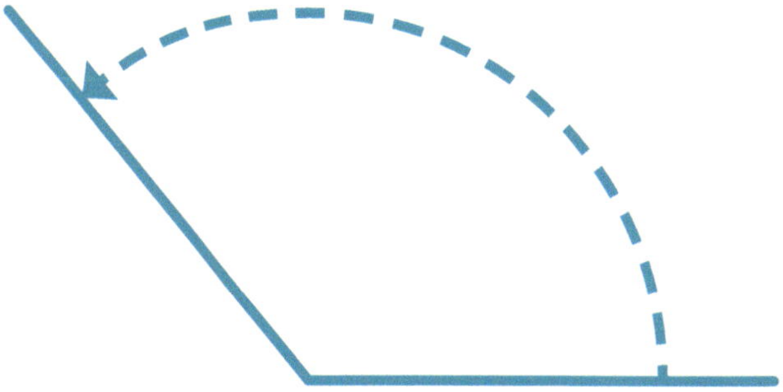

25. What is the length of this line?

Your guess: _____ Your measurement: _____

26. What is the perimeter of this rectangle?

Your guess: _____ Your measurement: _____
(Add it up!)

☐ + ☐ + ☐ + ☐ = ☐

27. What is the perimeter of this square?

Your guess: _____ Your measurement: _____
(Add it up!)

☐ + ☐ + ☐ + ☐ = ☐

DAY 9 Adding Bigger Numbers

> The lazy man says, "There is a lion in the road! A fierce lion is in the streets!" (Proverbs 26:13)

Sometimes math can get a little bit hard. But don't be like a lazy man who makes up an excuse about a lion! Don't be afraid of the challenge. Pray and ask God for help. Then, do your best to overcome the challenge. This lesson will be a little bit challenging.

Today we will learn to carry twice in the same addition exercise. Let's begin by adding these two big numbers together.

$$\begin{array}{r} 456 \\ +\ 147 \\ \hline \end{array}$$

First let's add 6 + 7 to get 13. That's 3 in the 1s place, and 1 in the 10s place. Carry the 1, writing it in the tens column above the 5. Now, we will add the carried 1, the 5, and the 4. That equals 10. Write 0 in the 10s place, and carry the 1. Now, we add 1 + 4 + 1, which equals 6. The answer is 603.

	1	1		
	4	5	6	
+	1	4	7	
	6	0	3	

Do you see that there are two carried 1s for this addition exercise? Now, try a few more exercises where you carry several times. The third one is a little bit harder. Some of the carries have been written in as hints.

1.

	1	1	
	5	2	4
+	3	9	8

2.

	1	1	
	6	5	8
+	2	8	5

3.

	4	5	3
+	5	4	7

In the last exercise, you added up 1+4+5 in the 100s column, and your answer should have been 10. That put a 1 in the 1000s place. You added these big 3-digit numbers together and got a much bigger 4 digit number!

Student Exercises

Here is some practice with addition and subtraction. You have been learning something new with large numbers. Today, remember the smaller numbers God has taught you so much about. Try to do these from memory!

1. 5 + 5 = ☐
2. 7 + 7 = ☐
3. 9 + 9 = ☐
4. 11 + 11 = ☐
5. 6 + 6 = ☐
6. 8 + 8 = ☐
7. 10 + 10 = ☐
8. 12 + 12 = ☐
9. 24 – 12 = ☐
10. 20 – 10 = ☐
11. 16 – 8 = ☐
12. 12 – 6 = ☐
13. 22 – 11 = ☐
14. 18 – 9 = ☐
15. 14 – 7 = ☐

Here are some math equations. Try to find the number to fill in the blank.

16. What divided by 3 equals 3?
☐ ÷ 3 = 3

17. What plus 9 equals 14?
☐ + 9 = 14

18. What minus 6 equals 9?
☐ – 6 = 9

19. What minus 2 equals 10?
☐ – 2 = 10

20. What times 3 equals 15?
☐ × 3 = 15

21. What times 6 equals 6?
☐ × 6 = 6

22. What divided by 4 equals 3?
☐ ÷ 4 = 3

23. What divided by 6 equals 2?
☐ ÷ 6 = 2

DAY 10 Practice

Let's add some big numbers today. Some of the carries have been written in for you.

1.

	1	
	1	7
	2	3
+	5	7

2.

	1	
	2	8
	6	1
+	5	5

3.

	1	
	3	5
	2	4
+	1	4

4.

	1	
	7	2
	4	7
+	1	2

5.

	1	
	6	8
	2	1
+	3	5

6.

	1	
	7	9
	1	3
+	3	2

Check out each of these numbers and answer the questions. These exercises will help you see how large the numbers are, and soon you will be ready to add, subtract, and do all sorts of things with them. Thank You, God! Add commas to each number to help you read it more easily.

7. 3 2 9 7 5 2 0 3 8

How many 10s? _______

How many 1,000s? _______

How many 10,000s? _______

8. 2 6 8 4

How many 1s? _______

How many 100s? _______

How many 1,000s? _______

9. 2 9 7 3 4 5 3

How many 100s? _______

How many 1,000s? _______

How many 1,000,000s? _______

Practice finding lengths, degrees, and area. First, guess a measurement for each of the shapes on the page. Then, measure. You should use inches for this page. You will need a protractor and a ruler for these exercises.

10. How big is this angle?

Your guess: ______

Your measurement: ______

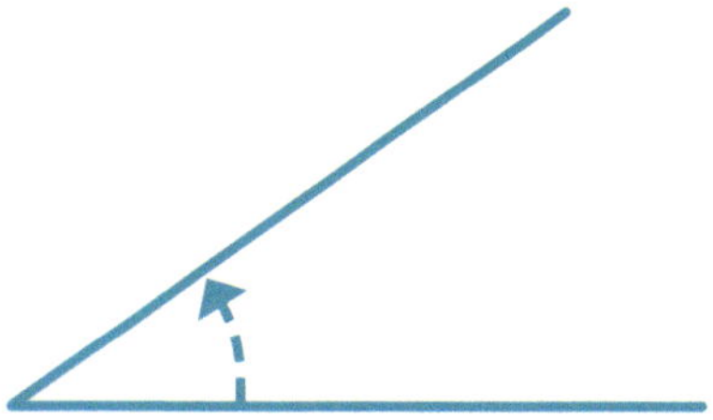

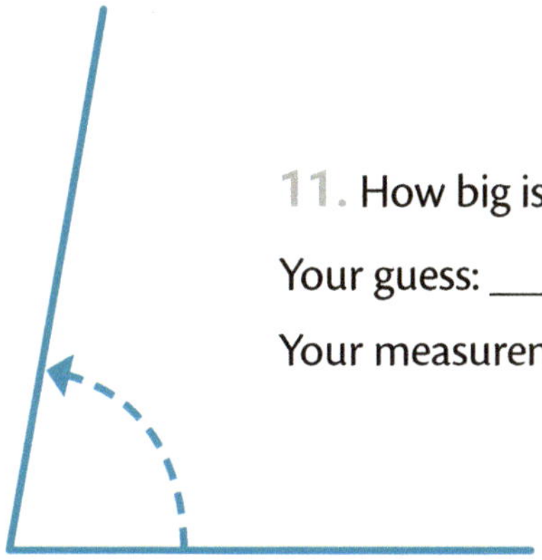

11. How big is this angle?

Your guess: ______

Your measurement: ______

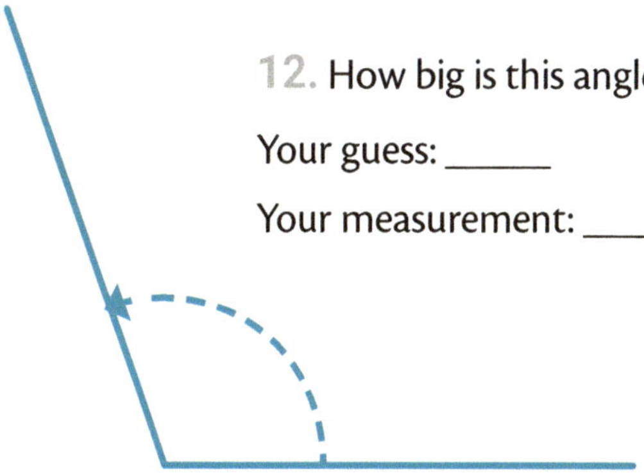

12. How big is this angle?

Your guess: ______

Your measurement: ______

13. How long is this line? Your guess: ______

Your measurement: ______

14. What is the area of this green rectangle?

Your guess: ______ in

Your measurement: ______ in

DAY 11 Activity and Practice

Let's use some math today for some real life investigations! The easiest way to do this is to ask your parent/teacher for use of the internet. You may also be able to find useful information in books or magazines at a library.

Of these two photos, which is the bigger city? Look up the population of your city along with four nearby cities. Record the numbers, and arrange the cities in order from largest to smallest. Does this help you understand the place God has given you to live?

Ask your parents about important years in your family history. Draw a timeline, and record those years in order from earliest to latest. You might want to use the following ideas, but you can be creative and choose other important events! In what year were your grandparents born? How about your parents? How about your siblings? How about you? What year did your parents get married?

Choose five companies whose stores you visit often or whose products you really enjoy. You may not know a lot of companies, and this may be a good time to learn about a few. For example, if you really like big trucks, you'd better know about Caterpillar. If you go shopping at Costco, you may not realize how many people are just like you all over the world! Look up each company's market capitalization (market cap). That basically means how much money the company is worth. The more the company is worth, the more people they can give jobs to and the more products and services they can offer. Arrange your companies from smallest to largest (the least market cap to the most market cap). Does anything surprise you about your list?

God has given you an amazing ability to see His world in a new way with big numbers. Thank You, God!

Student Exercises

Practice rounding these numbers!

1. Round 365,392 to the nearest 100. ____________
2. Round 365,392 to the nearest 10,000. ____________
3. Round 365,392 to the nearest 100,000. ____________
4. Round 1,345,765 to the nearest 1,000 ____________
5. Round 1,345,765 to the nearest 100,000 ____________
6. Round 1,345,765 to the nearest 1,000,000 ____________

Here is some practice with addition and subtraction. You are learning something new with bigger numbers. Today, let's remember how to add and subtract smaller numbers. Try these from memory!

7. 7 + 14 = ☐
8. 5 + 10 = ☐
9. 3 + 6 = ☐
10. 1 + 2 = ☐
11. 6 + 12 = ☐
12. 4 + 8 = ☐
13. 2 + 4 = ☐
14. 0 + 0 = ☐
15. 20 – 9 = ☐
16. 18 – 9 = ☐
17. 16 – 9 = ☐
18. 14 – 9 = ☐
19. 19 – 9 = ☐
20. 17 – 9 = ☐
21. 15 – 9 = ☐
22. 13 – 9 = ☐

DAY 12 Practice

Here is some practice with division. Try answering the first set of exercises from memory. Then, you may use the chart below to figure out the second set.

1. 6 ÷ 2 = ☐
2. 10 ÷ 2 = ☐
3. 6 ÷ 3 = ☐
4. 12 ÷ 3 = ☐
5. 8 ÷ 2 = ☐
6. 12 ÷ 2 = ☐
7. 9 ÷ 3 = ☐
8. 15 ÷ 3 = ☐
9. 16 ÷ 4 = ☐

10. 42 ÷ 6 = ☐
11. 50 ÷ 10 = ☐
12. 24 ÷ 6 = ☐
13. 24 ÷ 8 = ☐
14. 42 ÷ 7 = ☐
15. 50 ÷ 5 = ☐
16. 24 ÷ 4 = ☐

X	1	2	3	4	5	6	7	8	9	10
1	1	2	3	4	5	6	7	8	9	10
2	2	4	6	8	10	12	14	16	18	20
3	3	6	9	12	15	18	21	24	27	30
4	4	8	12	16	20	24	28	32	36	40
5	5	10	15	20	25	30	35	40	45	50
6	6	12	18	24	30	36	42	48	54	60
7	7	14	21	28	35	42	49	56	63	70
8	8	16	24	32	40	48	56	64	72	80
9	9	18	27	36	45	54	63	72	81	90
10	10	20	30	40	50	60	70	80	90	100

Check these numbers and answer the following questions. This will help you get ready to add, subtract, and do all sorts of things with large numbers. Add commas to help you read the numbers more easily.

17. 3 4 6 8 5 9
How many 10s? ________
How many 1,000s? ________
How many 10,000s? ________

18. 2 8 0 9 1 6 4 5
How many 1s? ________
How many 100s? ________
How many 100,000s? ________

Getting More Skillful with Subtraction

DAY 13

> Then Jerubbaal (Gideon) and all the people who were with him rose early and encamped beside the well of Harod, so that the camp of the Midianites was on the north side of them by the hill of Moreh in the valley. And the Lord said to Gideon, "The people who are with you are too many for Me to give the Midianites into their hands, lest Israel claim glory for itself against Me, saying, 'My own hand has saved me.' Now therefore, proclaim in the hearing of the people, saying, 'Whoever is fearful and afraid, let him turn and depart at once from Mount Gilead.'" And twenty-two thousand of the people returned, and ten thousand remained. (Judges 7:1-3)

When Gideon went out against the enemy, the Midianites, God wanted him to go out with a very small army, because God wanted everybody to know that He, God, would fight for His people. God wins our battles for us. Gideon had 32,000 men. But God told Gideon to take away the soldiers who were afraid to fight. Gideon took away 22,000 men. If we take away 22,000 from 32,000, there are only 10,000 left.

Today, we will learn more about subtracting numbers.

Subtracting large numbers is a little bit harder than adding. That's why we start with small numbers, and work up to bigger numbers. By the time this lesson is over, you'll be subtracting more than you ever thought you could! Praise God from Whom all blessings flow!

You've already learned how to borrow once in a subtraction exercise. Today, you will learn how to borrow twice. But first, let's try this exercise as a review:

$$\begin{array}{r} 465 \\ -\ 138 \\ \hline \end{array}$$

	100s	10s	1s
		5	15
	4	~~6~~	~~5~~
–	1	3	8
	3	2	7

Look in the 1s column. We can't subtract 8 from 5. We must borrow from the 10s column. See that we crossed out the 5 in the 1s column, and placed 15 above the 5. That's 15 instead of 5 because we borrowed a 10 from the 10s column.

Because we had to give up one 10 from the 10s column, we crossed out the 6 there and put a 5 just above it instead. Now we can subtract.

$15 - 8 = 7$

$5 - 3 = 2$

Before we do the double borrow, make sure you can do this single borrow by yourself. You will need to borrow a 10 from the 10s column.

	6	4	3
−	2	3	7

Now let's try the DOUBLE BORROW! Look at this subtraction exercise. We will need to borrow twice for this exercise. We'll borrow once for the 1s column and we'll borrow again for the 10s column because 4 is too big to subtract from 2.

	4	2	5
−	2	4	7

Here's how we double borrow. First we borrow a 10 for the 1s column. We need to cross out the 5 in the 1s column and place a 15 just above it. Then, we must change the 2 in the 10s column to a 1.

But we discover that, in the 10s column, we cannot subtract the 4 from the 1. We're going to have to borrow a 100 from the 100s column. So, just like when we borrow for the 1s column, we change the 1 in the 10s column to an 11 and write it above the crossed-out 2. Then, we must subtract the borrowed 1 hundred from the 4 hundreds in the 100s column. We write the 3 above the crossed-out 4.

	100s	10s	1s
	3	11	15
	~~4~~	~~2~~	~~5~~
−	2	4	7
	1	7	8

Now, we can finish the subtraction below:

$15 - 7 = 8$

$11 - 4 = 7$

$3 - 2 = 1$

Student Exercises

Try a few of these subtraction exercises on your own. The borrows have already been written in for you as hints.

1.

	10s	1s
	5	17
	~~6~~	~~7~~
–	1	9

2.

	100s	10s	1s
	8	13	15
	~~9~~	~~4~~	~~5~~
–	6	5	6

3.

	100s	10s	1s
	4	11	13
	~~5~~	~~2~~	~~3~~
–	3	5	8

Practice rounding these numbers! God's world is so organized. Rounding is an important way of enjoying that!

4. Round 3,748,583 to the nearest 100. ____________________

5. Round 3,748,583 to the nearest 10,000. ____________________

6. Round 3,748,583 to the nearest 100,000. ____________________

7. Round 56,132 to the nearest 1,000 ____________________

8. Round 56,132 to the nearest 100,000 ____________________

9. Round 56,132 to the nearest 1,000,000 ____________________

> Let him who stole steal no longer, but rather let him labor, working with his hands what is good, that he may have something to give him who has need. (Ephesians 4:28)

What are the answers to these multiplication exercises?

10. 7 x 4 =	11. 6 x 2 =	12. 3 x 9 =	13. 5 x 5 =

Everyday Math

14. Suppose that you were raising laying hens. Over three weeks, your chickens laid 45 eggs, 52 eggs, and 48 eggs. How many eggs did your chickens lay altogether over these three weeks?

+		

15. Suppose that you decide to plant tomatoes in your garden for three years in a row. For the first year's harvest, you pick 68 tomatoes. The second year's harvest yields 77 tomatoes. The third year yields 89 tomatoes. How many tomatoes did you pick over these three years?

+		

16. Suppose that your church decided to collect shoeboxes of gifts for needy children in other places around the world. During the first year of preparing these shoeboxes, they collected 20 boxes. They collected 35 boxes in the second year. Then, they collected 46 boxes in the third year, and 52 boxes in the fourth year. How many boxes did they collect all together?

+		

Practice DAY 14

Student Exercises

Below are a few sets. Circle the thing in each set that belongs in a set by itself.

1. 

2. 

3. 

4. 

5. 

6. 

Do these addition and subtraction exercises. Write the new numbers after borrowing or carrying in the top boxes as needed. Some have been provided for you.

7.

	1	1	
	1	7	6
+		8	9

10.

	1	1	
	1	6	2
+		5	8

13.

	1		
	3	9	0
+	1	9	1

8.

	1	1	
	2	4	5
+		7	7

11.

	1	15	16
	~~2~~	~~6~~	~~6~~
−	1	6	7

14.

	7	14	12
	~~8~~	~~5~~	~~2~~
−	3	6	6

9.

	4	5	7
+	1	4	8

12.

	5	5	7
−		4	3

15.

	2	4	5
+		7	7

Everyday Math

16. The grocery store is 12 miles from home. The post office is on the same road, 16 miles from home. How much farther is the post office from the grocery store?

____

17. Six friends were each given 3 pieces of candy. How many pieces of candy did they have altogether?

____

18. Kyle collects trains in a box that is 2 feet long and 1 foot wide. What is the perimeter of the box?

____

Go Find God's Numbers! DAY 15

This lesson integrates math into everyday life. With this activity, the student will apply God's patterns and wisdom in meaningful ways to home and community. Take a break from memory work and academic exercises to make math part of everyday life. The following are suggestions or examples. Other ideas may be added to the list.

There are many areas in which you can add, subtract, and compare numbers. Let's add up and compare numbers in your home.

1. Measure the perimeter of the rooms in your house. Round each measurement to the nearest foot (or meter). Compare the rooms. Which room is biggest? Which room is smallest?

Add up your room measurements. Most houses use baseboard trim (decorative wood strips) to hide the place where the walls and floor meet. If you were to put new baseboard trim into the house, how much would you need to buy?

2. How many minutes does it take to wash the breakfast dishes, the lunch dishes, and the supper dishes? Which task takes the shortest time? Which task takes the longest time? Use subtraction to find out the difference between the longest and the shortest times. "The family member that does the longest task serves the family the most.

3. Figure out how much paint you would need to paint a bedroom in your home. Suppose that one can of paint will cover 400 square feet. Let's measure to see how many square feet make up the four walls in your bedroom. Since walls are usually rectangles, measure the height and length of each wall and multiply to find their areas. Then add the areas of the walls together. How many cans of paint will you need? You may need a parent or teacher to help you to get an estimate. How much would it cost to paint one bedroom, if paint costs $20 per can? How much would it cost to paint 3 rooms? 4 rooms? Add it up! Use another sheet of paper if you need it to do the math.

Additional Activities

For extra activities, games, and helpful videos on these math concepts, refer to **generations.org/math3**.

Praise God for His beautiful shapes in creation!

CHAPTER 2

Patterns of Shapes

DAY 16 Discovering Amazing Patterns in God's Creation and God's Math

God's creation is full of amazing patterns, like the beautiful red flower on the previous page.

Someday you will learn how to use God's math to draw pictures. Computers can use math to create very complicated patterns like the one you see below. This design is a never-ending pattern called a **fractal**. We could have the computer use math to zoom in forever and still have the pattern repeat itself in smaller and smaller designs. Fractals give us a little bit of insight into the beauty, the creativity, and the wisdom of God. This is God's math displayed in a picture! Isn't it beautiful?

Today, let's look at some simple patterns made of shapes. This tile wall is made up of squares and rectangles.

These squares and rectangles never overlap, and all the shapes come together to fill the whole space. When this happens, it's called a tessellation. A **tessellation** is a way of using a shape or a few shapes to completely cover a space.

The wax cells in a bee hive might look like circles, but if you study them closely, they are actually six-sided shapes, called hexagons.

This is a tessellation God put in nature! The best designs begin with God.

Tessellations can get pretty complicated. We can use different colors, we can spin the shapes around, or we can add more than one shape or color to the pattern.

Check these out.

This tessellation uses the same shape in the same way, but it uses four different colors to add beauty.

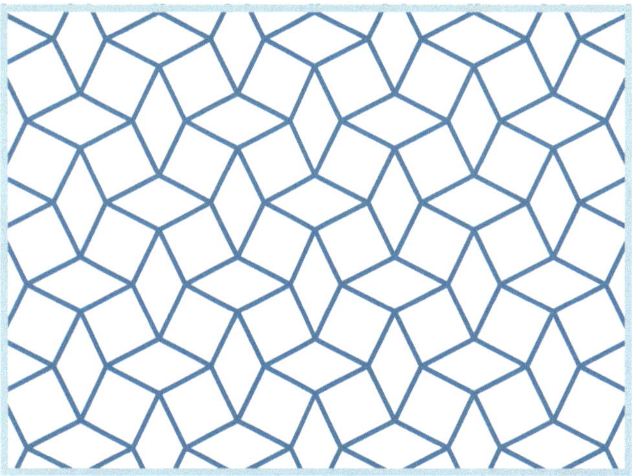

This tessellation uses two shapes, and some of those shapes are spun around in a different way. Can you find those two shapes? Does this pattern make your brain feel funny?

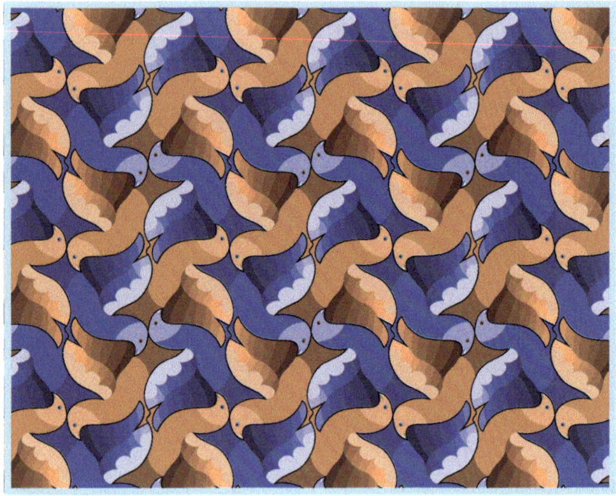

This tessellation uses only one shape. Can you see that these shapes are birds? The birds change colors and spin around. When you make patterns, they are often simpler than God's patterns. Most of God's patterns are so complicated that we can only just begin to understand them!

> Can you lift up your voice to the clouds,
> That an abundance of water may cover you?
> Can you send out lightnings, that they may go,
> And say to you, 'Here we are!'?
> Who has put wisdom in the mind?
> Or who has given understanding to the heart?
> Who can number the clouds by wisdom?
> Or who can pour out the bottles of heaven,
> When the dust hardens in clumps,
> And the clods cling together? (Job 38:34-38)

Here are some tessellations that God made. They don't exactly follow our patterns. That isn't because they are less beautiful and orderly. They are even better than what we can make. Can you see why we call these tessellations?

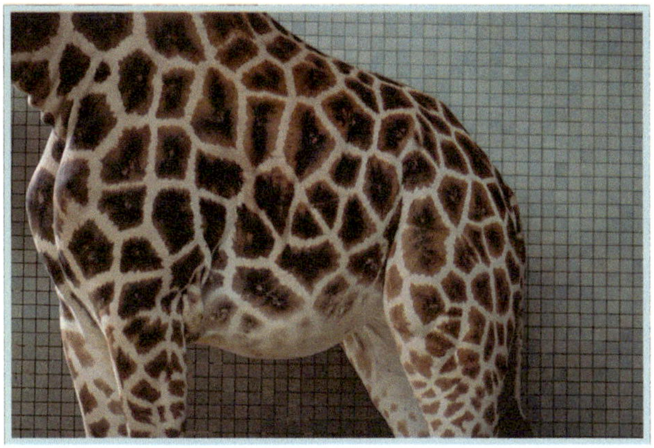

Do you recognize this animal?

Here is another of God's tessellations.

Who is wise like God? Who displays such spectacular beauty as this? Who is so worthy of praise? Our Father created the heavens and the earth. Through His Son Jesus Christ, and by the power of His Holy Spirit He has made all things good and beautiful! Amen.

Activity

Can you find 5 tessellations in your world? Check the kitchen or bathroom for tiles. Look for brickwork on the outside of a house. You might find a tessellation on a drinking glass or piece of furniture. You may find it on a pet snake. Take a look at a pineapple or study the bark of a tree.

When you are finished, pray: *Who is wise like God? Who works beauty so spectacularly? Who is so worthy of praise? Our Father created the heavens and the earth, and he did it all through His Son Jesus Christ. By the power of His Holy Spirit He has made it good, beautiful, and true. Amen.*

Student Exercises

Pretend to tile your own floor. Within the big box, draw some of these rectangle and square tiles to fill your floor.

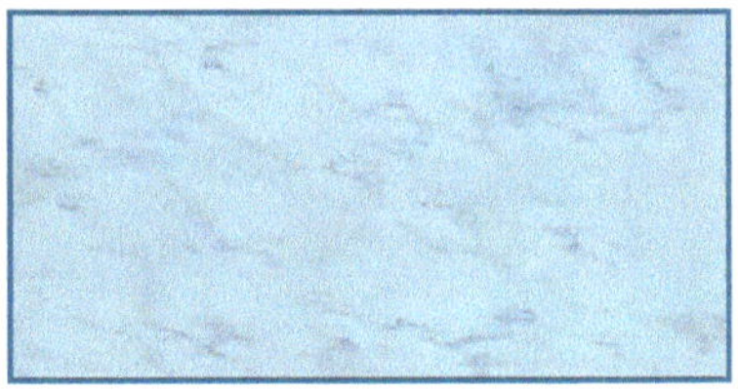

Spend some practice time today counting by chunks.

1. Count by 2s: 20, ______, ______, ______, ______, ______
2. Count by 3s: 30, ______, ______, ______, ______, ______
3. Count by 5s: 50, ______, ______, ______, ______, ______
4. Count by 8s: 40, ______, ______, ______, ______, ______
5. Count by 10s: 100, ______, ______, ______, ______, ______

Let's think about how many 1s are in these big numbers. How many 10s? How many 100s? . . . Fill in the place-value chart below.

	NUMBER	1,000,000	100,000	10,000	1,000	100	10	1
6.	123,456		1	2	3	4	5	6
7.	7,654,321							
8.	234,987							
9.	99,999							
10.	100,004							
11.	404,040							

What are the answers to these multiplication exercises?

12. 3 x 5 =	14. 4 x 7 =	16. 4 x 8 =	18. 3 x 8 =
13. 6 x 3 =	15. 9 x 2 =	17. 5 x 7 =	19. 7 x 4 =

DAY 17 Practice

Here is some practice with multiplication. Try answering the first set of exercises from memory. Then, you may use the chart to figure out the second set.

1. 3 × 4 = ☐
2. 6 × 1 = ☐
3. 5 × 4 = ☐
4. 1 × 5 = ☐
5. 2 × 5 = ☐
6. 7 × 3 = ☐
7. 1 × 12 = ☐
8. 2 × 10 = ☐

9. 7 × 8= ☐
10. 5 × 6 = ☐
11. 7 × 6 = ☐
12. 10 × 7 = ☐
13. 8 × 5 = ☐
14. 8 × 9 = ☐
15. 3 × 9 = ☐
16. 6 × 6 = ☐

X	1	2	3	4	5	6	7	8	9	10
1	1	2	3	4	5	6	7	8	9	10
2	2	4	6	8	10	12	14	16	18	20
3	3	6	9	12	15	18	21	24	27	30
4	4	8	12	16	20	24	28	32	36	40
5	5	10	15	20	25	30	35	40	45	50
6	6	12	18	24	30	36	42	48	54	60
7	7	14	21	28	35	42	49	56	63	70
8	8	16	24	32	40	48	56	64	72	80
9	9	18	27	36	45	54	63	72	81	90
10	10	20	30	40	50	60	70	80	90	100

Check out these numbers and answer the following questions. Get used to large numbers because soon you will learn to add and subtract them. Write in commas to help you read the numbers more easily.

17.
7 6 3 8 0 5 1
How many 10s? ______
How many 1,000s? ______
How many 10,000s? ______

18.
9 0 4 7 3 5 2 3 4
How many 100s? ______
How many 1,000s? ______
How many 1,000,000s? ______

19.
4 7 2 9
How many 1s? ______
How many 100s? ______
How many 1,000s? ______

20.
5 6 7 2 0 8
How many 1s? ______
How many 10s? ______
How many 1,000s? ______

Activity

Create a Place Value Bracelet!

You will need:

- Froot Loops cereal
- A pipe cleaner or chenille wire
- 6 small bowls

Separate the pieces of cereal by color into the six different bowls. You will need no more than nine pieces of each color. Arrange the bowls in a line so that each bowl will stand for a certain place value. The bowl on the far right will contain cereal pieces worth 1. The bowl on the far left will contain cereal pieces worth 100,000 . The bowls between will stand for the other place values. Now string pieces of cereal onto your pipe cleaner or chenille wire. You will put between 0 and 9 of each color on your wire. Be sure to group pieces of the same color together. Be sure to put them on the wire in the same order as your place value bowls.

When you are finished, say the number you have made. Twist the ends of the wire together around your wrist to make a bracelet.

DAY 18 Making Tessellations

Activity

Today you will make your own tessellation. Anybody can make one with squares, but you are going to learn to make fancy, fun ones! Read through these steps, and then follow them to create your own!

Step 1: Start with a shape that already makes a tessellation. How about a rectangle?

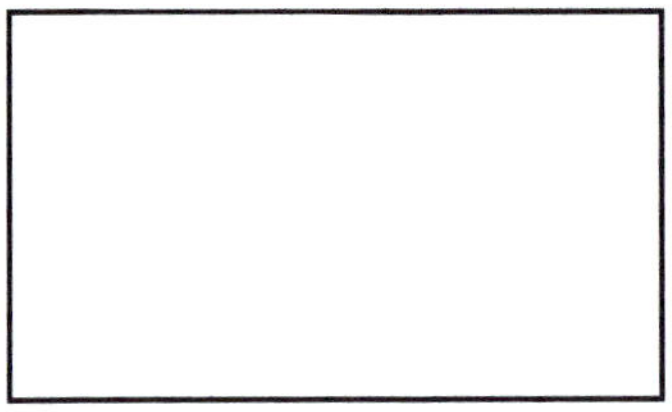

Step 2: Draw a line that starts on one side of the rectangle and ends on the opposite side. Cut the shape along that line. You can make it straight or curved or a little bit of both!

Step 3: Take the two uncut sides of the rectangle and tape their straight edges together so they match up perfectly.

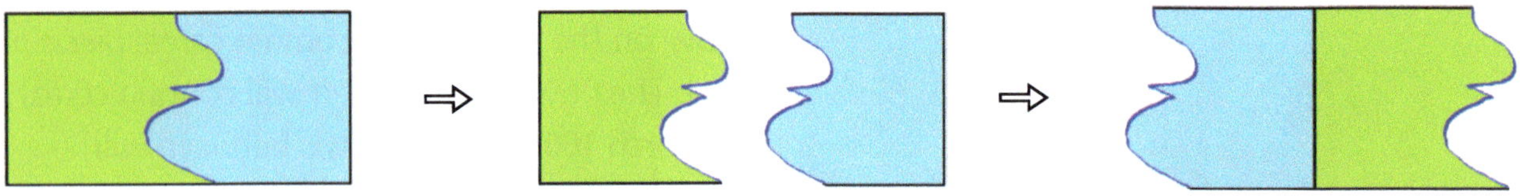

Step 4: Draw another line that starts on one of the sides that you already cut and ends on the opposite side. Cut the shape along that line. Then, tape their two straight sides together so that they match up perfectly.

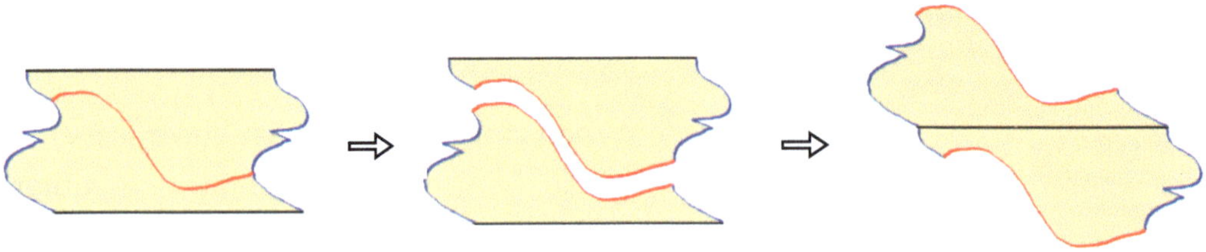

Step 5: Trace your shape onto a piece of paper. You may want to use colored paper or construction paper to make your tessellation even more creative! Cut out your shape. Do this over and over for more shapes. Then, once you have made a bunch of copies of your shape, piece them together and glue them to another sheet of paper. Here is an example:

You could keep going and fill the entire page, or you could just do a spot like this. After seeing this one, you might think more simple lines would create a more interesting pattern. Or you might want to make a very complicated tessellation. As long as you follow the steps, it's all up to you.

DAY 19 Practice

Let's practice finding lengths and degrees. First, go through and guess a measurement for each. Then, measure to see how close you were. Measure the lengths in inches.

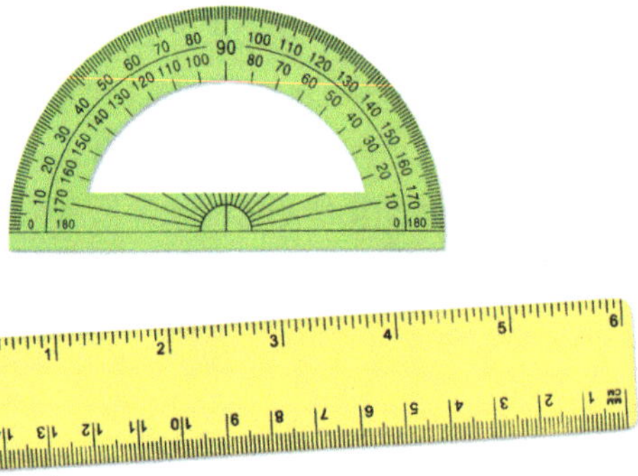

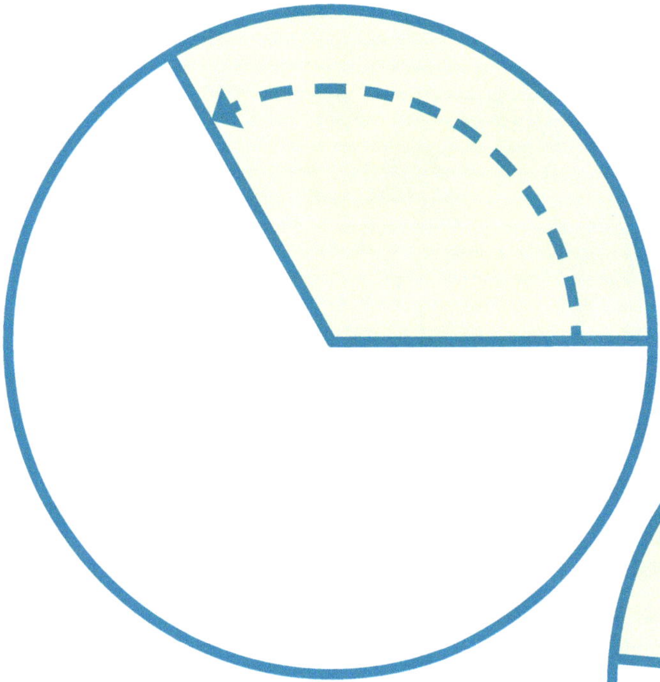

1. Guess the size of this angle in degrees.

____

2. Measure the angle using a protractor.

____

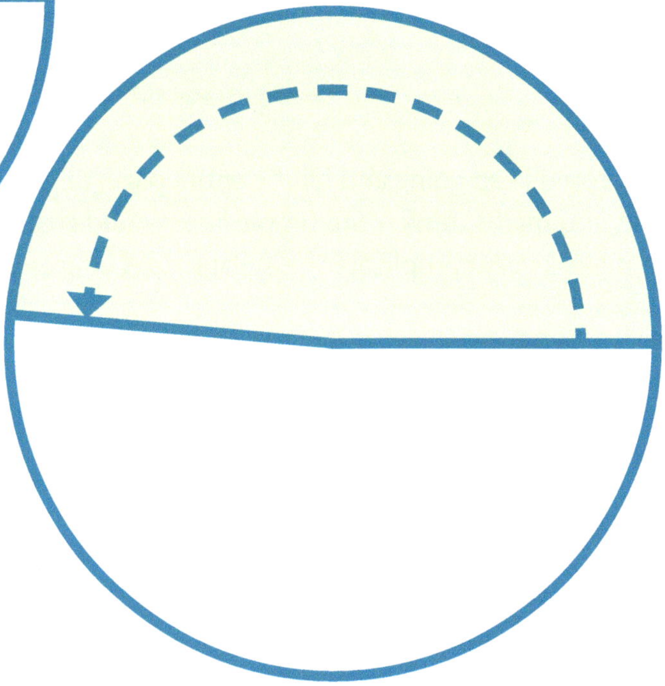

3. Guess the size of this angle in degrees.

____

4. Measure the angle using a protractor.

____

5. Guess the length of this line in inches ______ Measure the line using a ruler ______

6. Guess the length of this line in inches ______ Measure the line using a ruler ______

Measure the length and width of these rectangles in inches. Then use multiplication to find the area of the rectangles. Label your answers with "square inches" at the end.

7. What is the area of this rectangle?

________ x ________ = ________ sq. in.

8. What is the area of this rectangle?

________ x ________ = ________

9. What is the area of this rectangle?

_________ x _________ = _________

10. What is the area of this rectangle?

_________ x _________ = _________

11. What is the area of this rectangle?

_________ x _________ = _________

Check out each of these numbers and answer the following questions. This will help you see how large the numbers are, and soon you will be ready to add, subtract, and do all sorts of things with them. Thank You, God! Write in commas to help you read the numbers more easily.

12. 3 1 5 2 3 4 5

How many 10s? _______

How many 1,000s? _______

How many 10,000s? _______

13. 7 0 8 3 1 0 7 4 3

How many 1s? _______

How many 100s? _______

How many 1,000,000s? _______

14. 6 8 4 2 5

How many 100s? _______

How many 1,000s? _______

How many 100,000s? _______

Add or subtract. Some of the borrows and carries have already been provided for you as hints.

15.

		1	
	3	2	4
+	1	4	6

17.

	1	1	
	5	3	5
+	1	6	6

19.

	6	1	6
+	1	8	7

16.

		4	12
	4	~~5~~	~~2~~
−	1	3	4

18.

	5	12	11
	~~6~~	~~3~~	~~1~~
−	2	5	3

20.

		4	14
	1	~~5~~	~~4~~
−		3	5

DAY 20 Constant & Moving Patterns

Have you noticed that, throughout each month, the moon goes through phases? Sometimes the moon is shaped like a crescent. Sometimes it is full. These are God's patterns that have repeated over and over again for thousands of years! For the rest of this chapter you are going to explore patterns. Tessellations repeat the same shape over and over. That is called a constant pattern. The moon phases are another kind of constant pattern. Here are some more constant patterns: The sun is constantly rising. You are constantly breathing. Mom is constantly loving! Over and over the same thing happens. Here are two more constant patterns:

Read these patterns from left to right. Circle, circle, circle, all the same, one after another.

Here are the beginnings of two constant patterns. What comes next? Can you continue these patterns?

1. 

2. 3 2 3

Okay, that's too easy. God has gifted you too much to stop there today! The next pattern you will learn about is a moving pattern. Check out these examples:

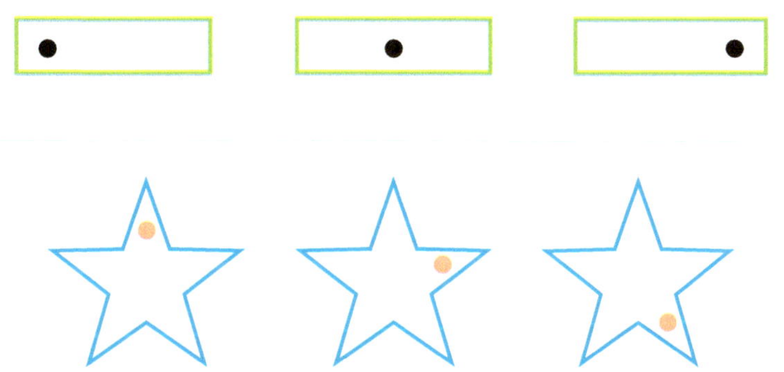

Do you see the constant patterns? The rectangles are constantly the same shape and in the same position. The stars are too. You need a constant pattern so that you can see the moving pattern. You couldn't see the dot move if the rectangle moved too. Now, check out that dot. In the rectangles, the dot moves from the left to the right. In the stars, the dot moves around the points. Get it?

Continue these moving patterns. For exercises 3 and 4, fill in the drawing on the third image. For exercise 5, complete the fifth image. For exercise 6, complete the third image.

3. 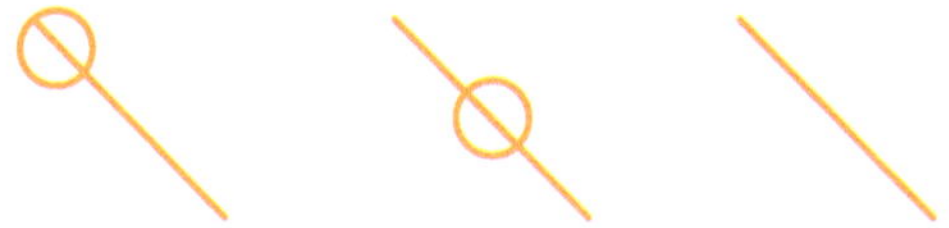

4. 

5. 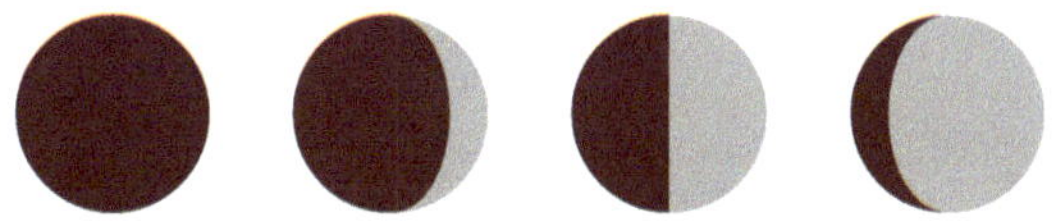

6. 

We use a moving pattern when we count by chunks. To get the next number in a list, we add the same number each time. It's like moving the dot the same distance around the star each time. Continue the moving patterns below by listing the next three numbers.

7.	6 9
8.	10 15

Constant patterns describe who God is.

> Jesus Christ is the same yesterday, today, and forever. (Hebrews 13:8)

Moving patterns have something to say about Him too! Jesus set his mind on the cross, and he marched right to it, step-by-step.

> Now it came to pass, when the time had come for Jesus to be received up, that He steadfastly set His face to go to Jerusalem… "For the Son of Man did not come to destroy men's lives but to save them." (Luke 9:51, 56a)

Take a moment and give God some praise! Admire Him with constant patterns and moving patterns. Remember, this is not an extra thought. Looking for God's glory is the most important part of the lesson. This is where you will make it or break it as a mathematician.

Student Exercises

Practice counting by chunks. This time you are counting backwards.

1. Count by 2s backwards: 20, 18, ______, ______, ______, ______
2. Count by 3s backwards: 30, 27, ______, ______, ______, ______
3. Count by 5s backwards: 50, 45, ______, ______, ______, ______
4. Count by 8s backwards: 72, 64, ______, ______, ______, ______
5. Count by 10s backwards: 100, 90, ______, ______, ______, ______

Can you finish today by solving these equations? Write the answer in the blank.

6. What plus 5 equals 13? ________ + 5 = 13
7. What plus 9 equals 17? ________ + 9 = 17
8. What minus 6 equals 5? ________ − 6 = 5
9. What minus 2 equals 15? ________ − 2 = 15
10. What times 3 equals 9? ________ × 3 = 9
11. What times 6 equals 18? ________ × 6 = 18
12. What divided by 3 equals 4? ________ ÷ 3 = 4
13. What divided by 6 equals 3? ________ ÷ 6 = 3

DAY 21 Practice

Here is some practice with addition and subtraction. Watch out for the double carries (with addition) and the double borrows (with subtraction). Some of the carries and borrows have been written in for you.

1.
$$\begin{array}{cccc} & 1 & 3 & 2 \\ + & 5 & 4 & 7 \\ \hline \end{array}$$

2.
$$\begin{array}{cccc} & & 1 & \\ & 3 & 1 & 7 \\ + & 1 & 5 & 8 \\ \hline \end{array}$$

3.
$$\begin{array}{cccc} & 2 & 5 & 7 \\ + & 4 & 3 & 4 \\ \hline \end{array}$$

4.
$$\begin{array}{cccc} & & 5 & 13 \\ & 7 & \cancel{6} & \cancel{3} \\ - & 3 & 5 & 4 \\ \hline \end{array}$$

5.
$$\begin{array}{cccc} & 4 & 12 & 12 \\ & \cancel{5} & \cancel{3} & \cancel{2} \\ - & 1 & 5 & 9 \\ \hline \end{array}$$

6.
$$\begin{array}{cccc} & 1 & 1 & \\ & 4 & 8 & 4 \\ + & 4 & 4 & 6 \\ \hline \end{array}$$

7.
$$\begin{array}{cccc} & 1 & 1 & \\ & 3 & 9 & 4 \\ + & 1 & 9 & 9 \\ \hline \end{array}$$

8.
$$\begin{array}{cccc} & 1 & 5 & 4 \\ + & & 5 & 4 \\ \hline \end{array}$$

9.
$$\begin{array}{cccc} & 0 & 16 & 15 \\ & \cancel{1} & \cancel{7} & \cancel{5} \\ - & & 8 & 6 \\ \hline \end{array}$$

10.
$$\begin{array}{cccc} & 8 & 5 & 2 \\ - & 1 & 4 & 6 \\ \hline \end{array}$$

11.
$$\begin{array}{cccc} & 4 & 0 & 0 \\ + & 5 & 9 & 9 \\ \hline \end{array}$$

12.
$$\begin{array}{cccc} & 1 & 1 & \\ & 8 & 3 & 5 \\ + & & 6 & 7 \\ \hline \end{array}$$

13.
$$\begin{array}{cccc} & 2 & 4 & 9 \\ + & & 3 & 1 \\ \hline \end{array}$$

14.
$$\begin{array}{cccc} & 3 & 12 & \\ & \cancel{4} & \cancel{2} & 6 \\ - & & 3 & 1 \\ \hline \end{array}$$

15.
$$\begin{array}{cccc} & 6 & 4 & 5 \\ - & 2 & 5 & 6 \\ \hline \end{array}$$

Everyday Math

16. Your father's motorized bicycle can only carry 300 pounds. Let's pretend that you weigh 72 pounds, and your father weighs 212 pounds. How much do you weigh altogether? Can you ride the bike with your father, or would that be too much weight?

$$\begin{array}{r} 212 \\ +\ 72 \\ \hline \square \end{array}$$

17. You are planning to go Christmas shopping for four family members and friends. You have saved $165. Here are the gifts you would like to purchase: A Bible for $35, a wagon for $46, a pair of shoes for $39, and a skateboard for $28. How much does this add up to? How much money will you have left over after buying the presents?

$$\begin{array}{r} 35 \\ 46 \\ 39 \\ +\ 28 \\ \hline \square \end{array} \qquad \begin{array}{r} 165 \\ -\ \square \\ \hline \square \end{array}$$

18. Your family has 26 worship songs downloaded on the computer, 37 different worship songs on a tablet, and 55 other worship songs on phones. How many God-praising songs can your family enjoy listening to all together?

$$\begin{array}{r} \square \\ \square \\ +\ \square \\ \hline \square \end{array}$$

19. A winning basketball team scored 106 points in the game. The losing team scored 79 points. How many more points did the winning team have?

−

20. Many people had to leave their homes and find a safe place to shelter during a storm. Your church opened its doors as a shelter for these people. 135 people came to take refuge. If 86 of these were adults, how many were children? How many of the children were boys if there were 22 girls?

$$\begin{array}{r} 135 \\ -\ 86 \\ \hline \end{array}$$

−

Pretend that the shapes below are pieces of paper that have been folded and opened up again. The light blue lines are their creases. There is a dark blue smaller shape inside each larger shape. Figure out how many small blue shapes fit in the big shape, and write the number next to the shape. A nice way to do this is to color in copies of the small shape (counting as you go) until you fill the whole space.

21.

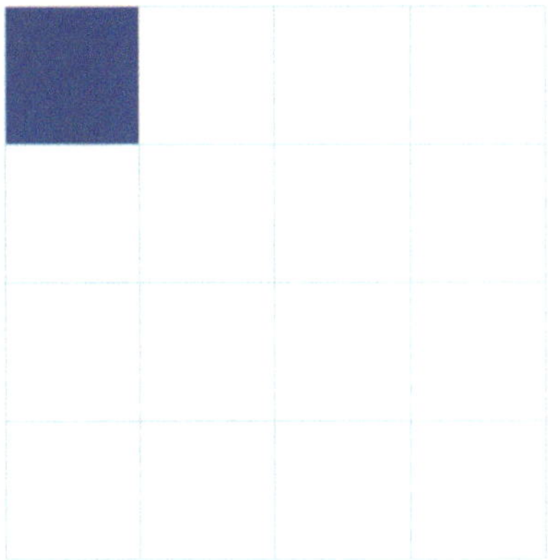

22.

23.

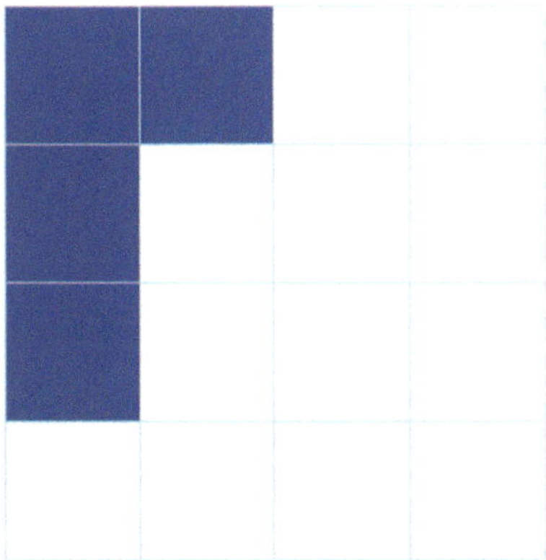

24.

25.

26.

DAY 22 Rotating Shape Patterns

You have learned about the constant pattern. You have also seen the moving pattern. Today, we will look at the spinning pattern. Spinning is a fine word, but you are ready for some new vocabulary. We call this new pattern the *rotating pattern*. Since you've learned the moving pattern, you can probably guess what the rotating pattern looks like. Before we look at the rotating pattern, let's learn how to say which way something is turning.

Life is full of rotations! Usually, when we turn a screw to tighten it, we say righty tighty! When we loosen a screw we say "lefty loosey." That's because when we turn a screwdriver towards the right, we are tightening the screw. When we tighten the screw, we are turning it **clockwise**. That means we turn it in the same direction that the hands of a clock move. Which way does the second hand move on a clock? Take a moment and watch it move. Which way is it moving? We call this "clockwise." If something moves around the other direction, we call its direction **counterclockwise**.

Activity

Using a screwdriver, tighten a screw and loosen it. Which way are you twisting the screwdriver when you tighten the screw? Which way are you twisting the screwdriver if you are loosening the screw?

Here is a picture to help you remember clockwise and counterclockwise. Run your finger along the arrow to get the idea.

Clockwise

Counterclockwise

Now check out some rotating patterns. Look at the following pictures from left to right. Try to imagine how the shapes are spinning (or rotating). Are they going clockwise or counterclockwise? Draw a few more pictures in each exercise to keep the pattern going.

1. 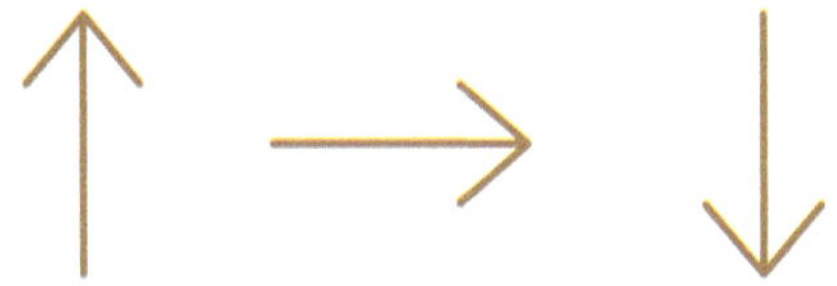

2. 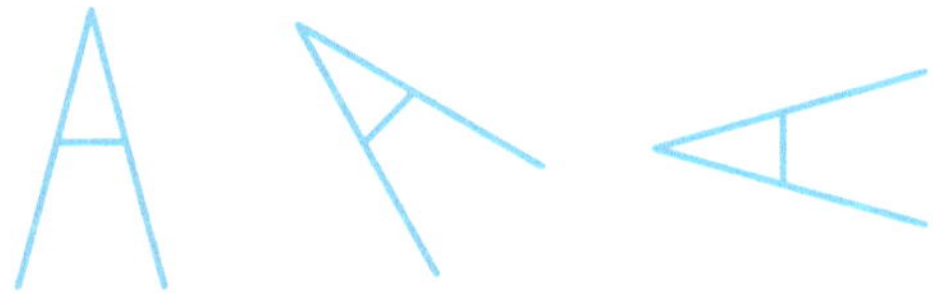

3. 

Direction is important for seeing the rotating patterns that God has made. But there is a little more to it than that. We also need to ask, "How much are they rotating?" Sometimes the best answer to a question is another question. In this case we should ask, "How many positions will this shape rotate to before it gets back to where it started?"

With that question in mind, look at the patterns above again. The first will rotate to 4 positions before it gets back to where it started. The other two will rotate to 8 positions. A better way to see this is to get out a piece of paper, draw the shape, and rotate it yourself!

Check out these rotating shapes, and draw their next positions until the shape is back to where it started its rotation.

4. 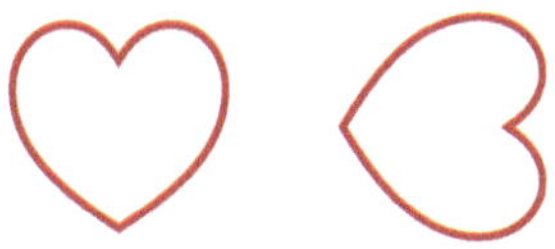

5.

Student Exercises

Here are some spinning patterns you will see everyday if you watch the clock. Let's see how many positions there will be as the minute hand rotates once around the clock, if you count a new position every ten minutes. Finish sketching the hands on the clocks.

Count the positions!

			1.	2.	3.
1:10	1:20	1:30	1:40		

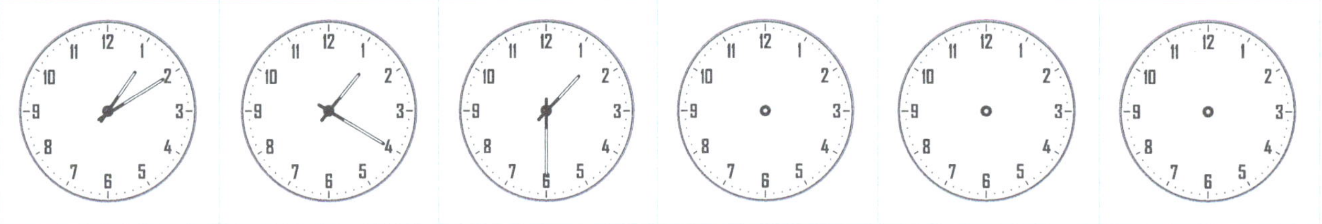

How many positions will the minute hand of the clock show as it rotates once around the clock using 15 minute intervals? Finish sketching the hands on the clocks and write in the times.

Count the positions!

		4.	5.
1:15	1:30		

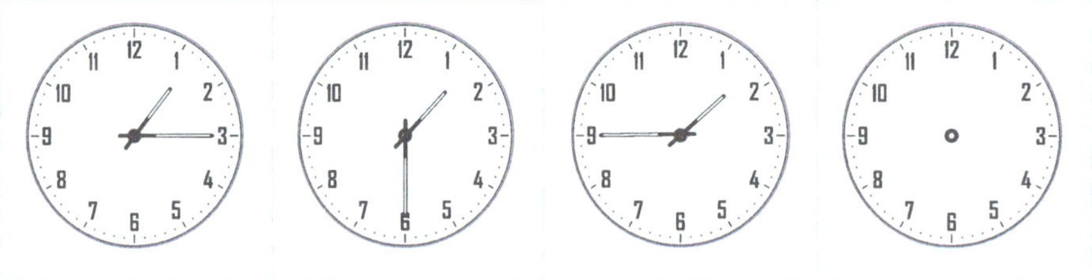

How many positions will the minute hand show as it rotates once around the clock using 5 minute intervals? Finish sketching the hands on the clocks and write in the times.

Count the positions!

				6.	7.	8.	9.	10.	11.	12.	13.
1:05	1:10	1:15	1:20	1:25	1:30	1:35	1:40				

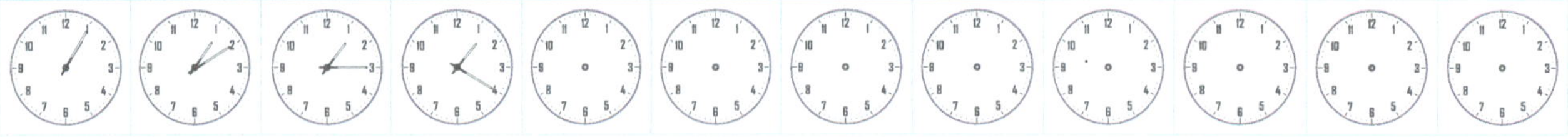

Practice division. Answer the first set of exercises from memory. You may use the chart for the second set if you need to.

14. 8 ÷ 4 = ☐
15. 16 ÷ 4 = ☐
16. 5 ÷ 5 = ☐
17. 15 ÷ 5 = ☐
18. 12 ÷ 4 = ☐
19. 20 ÷ 4 = ☐
20. 10 ÷ 5 = ☐
21. 20 ÷ 5 = ☐

22. 36 ÷ 6 = ☐
23. 36 ÷ 1 = ☐
24. 45 ÷ 5 = ☐
25. 32 ÷ 4 = ☐
26. 36 ÷ 4 = ☐
27. 36 ÷ 9 = ☐
28. 45 ÷ 9 = ☐
29. 32 ÷ 8 = ☐

X	1	2	3	4	5	6	7	8	9	10
1	1	2	3	4	5	6	7	8	9	10
2	2	4	6	8	10	12	14	16	18	20
3	3	6	9	12	15	18	21	24	27	30
4	4	8	12	16	20	24	28	32	36	40
5	5	10	15	20	25	30	35	40	45	50
6	6	12	18	24	30	36	42	48	54	60
7	7	14	21	28	35	42	49	56	63	70
8	8	16	24	32	40	48	56	64	72	80
9	9	18	27	36	45	54	63	72	81	90
10	10	20	30	40	50	60	70	80	90	100

Can you finish this sudoku puzzle? Each line needs to have numbers 1 through 9. Each column also must have numbers 1 through 9. And each 9-number square must have numbers 1 through 9. If you can't figure out one number, try another. Pretty soon, the whole puzzle will be filled up. Have fun!

5		4	6	7	8	9	1	2
6	7	2	1	9	5	3		8
1	9	8	3	4	2	5	6	7
8	5	9	7	6	1	4	2	3
	2	6	8		3	7	9	1
7	1	3	9	2	4	8	5	6
9	6	1	5	3	7	2	8	
2		7	4	1	9	6	3	5
	4	5	2	8	6		7	9

If this puzzle was really easy, go find more challenging sudoku puzzles!

Practice DAY 23

Practice addition and subtraction. Some of the carries and borrows have been written in for you.

1.

		1	
	8	3	5
+	2	2	7

2.

	1	1	
	6	4	7
+	1	5	3

3.

	2	8	0
+	5	3	0

4.

	1	1	
	9	1	2
+	1	8	9

5.

	7	11	14
	~~8~~	~~2~~	~~4~~
−	5	3	9

6.

	4	3	1
−		5	7

7.

		7	13
	7	~~8~~	~~3~~
−	3	2	5

8.

	5	7	6
−	1	7	2

Solve these multiplication problems.

9. 8 x 4 = ☐

10. 12 x 2 = ☐

11. 4 x 4 = ☐

12. 2 x 9 = ☐

13. 5 x 5 = ☐

14. 0 x 3 = ☐

15. 5 x 3 = ☐

16. 8 x 6 = ☐

17. A barn contains 108 sheep and 35 goats. How many more sheep than goats does the barn contain?

18. Your family needs to bring 48 cookies for a meal at church. You have already decorated 26 cookies. How many more cookies do you still need to decorate?

19. A bookcase contains 3 shelves. The first shelf has 33 books. The second shelf has 30 books. The last shelf has 31 books. How many books are there altogether?

20. There were 115 jellybeans in a jar. You and your brother ate 40 jellybeans. How many jellybeans were left?

Cancellation Patterns

DAY 24

Today's patterns will be a little harder. God is very wise. He is much wiser and smarter than any of us. But God reveals some of His patterns to us. How do we respond when we learn about His patterns? We are in awe of God's wisdom, His order, and His amazing creativity!

Math shows us a little bit of God's patterns. We have to work hard to understand them. The more we work at it, the more we will enjoy these patterns.

Take a minute and try to figure out what is going on in the pattern below. The blue dots are shapes. The gray circles are there to show you an empty space. Move from left to right. Here is a hint: The first two pictures are coming together to create the third. Spend a few minutes thinking about this.

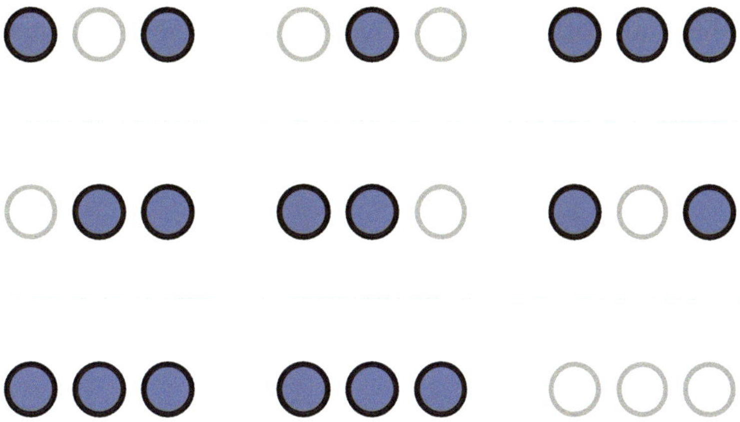

Okay, here is the secret.

> It is the glory of God to conceal a matter,
> But the glory of kings is to search out a matter. (Proverbs 25:2)

We add the first two sets to arrive at the third. Put the first two together like this, to get the answer in the third set:

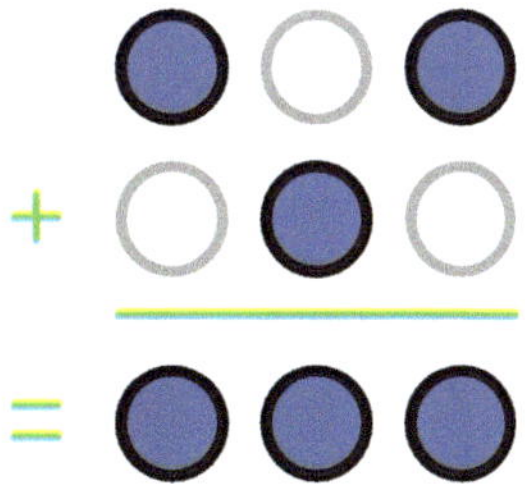

Let's try the second pattern. Just as before, when you combine the first set of dots with the second, if a dot from one set finds an empty spot in the other, it fills it. But now we notice something else happening. If two dots come together and want the same space, they both disappear! When this happens in math we call it **canceling**. Can you see the canceling pattern when we combine these together?

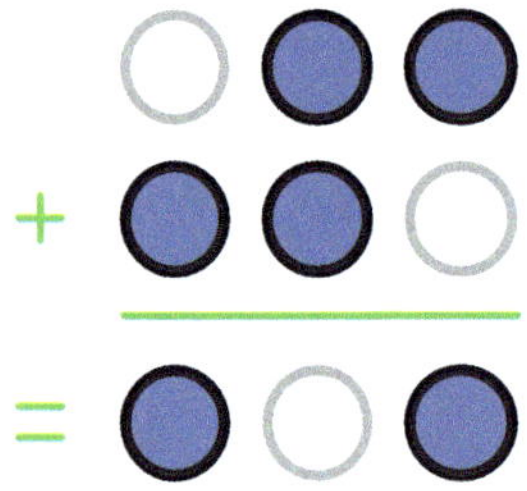

Now look at the third blue-dot pattern on the previous page. Do you see that all the dots are canceled?

Can you see how the first two triangles come together to make the third?

Did you figure them out on your own? Were you able to explain how the first two triangles in each set combined to make the the third triangle?

Finish the canceling patterns below by drawing the third part.

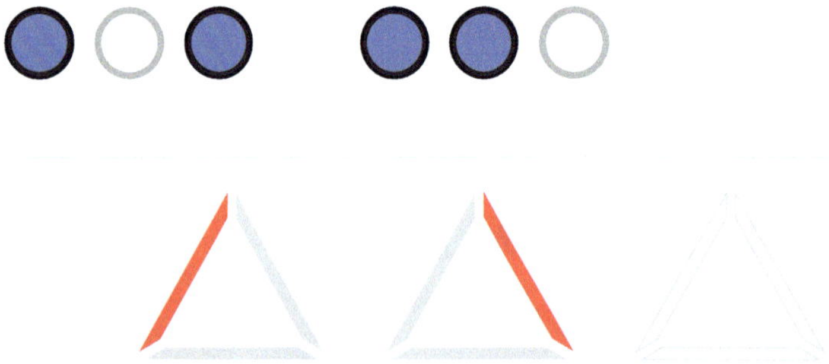

Student Exercises

Find the correct number to answer these math questions.

1. What plus 4 equals 13? ________ + 4 = 13
2. What plus 10 equals 13? ________ + 10 = 13
3. What minus 8 equals 8? ________ − 8 = 8
4. What minus 7 equals 8? ________ − 7 = 8
5. What times 2 equals 10? ________ × 2 = 10
6. What times 4 equals 8? ________ × 4 = 8
7. What divided by 2 equals 10? ________ ÷ 2 = 10
8. What divided by 5 equals 4? ________ ÷ 5 = 4

9. Guess the size of this angle in degrees.

___

Measure the angle using a protractor.

___

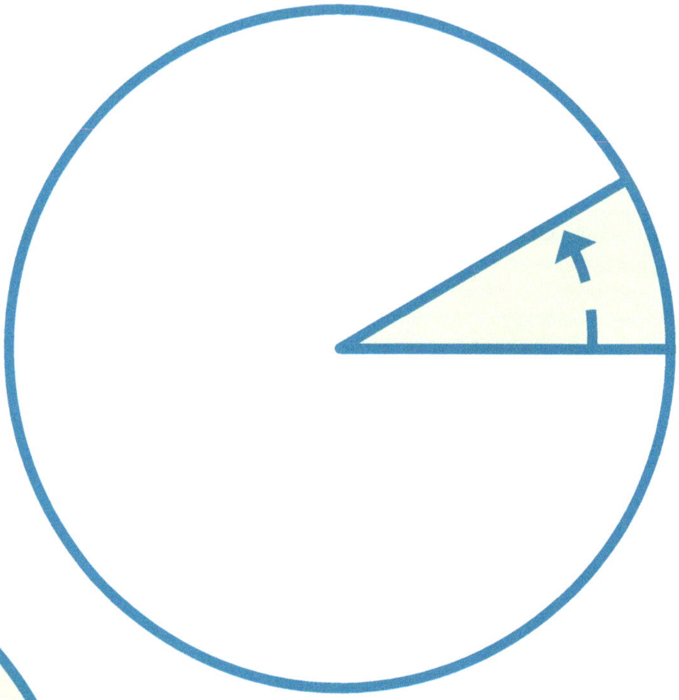

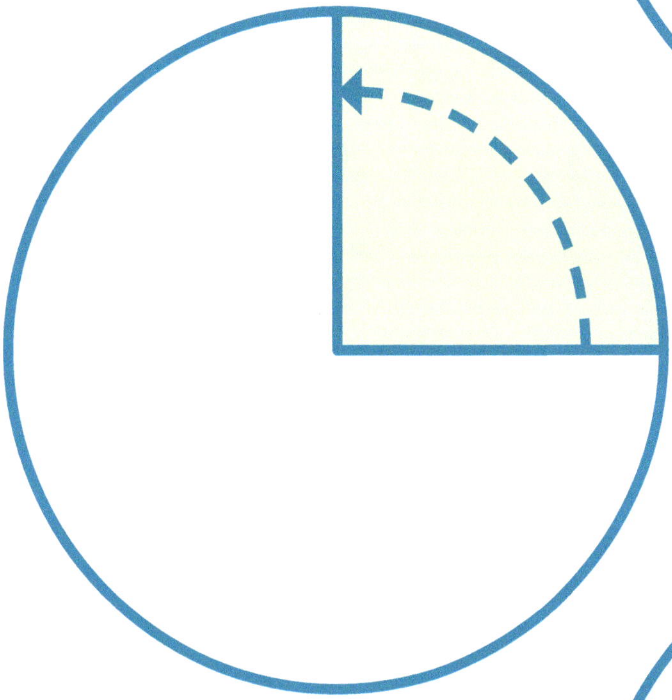

10. Guess the size of this angle in degrees.

___

Measure the angle using a protractor.

___

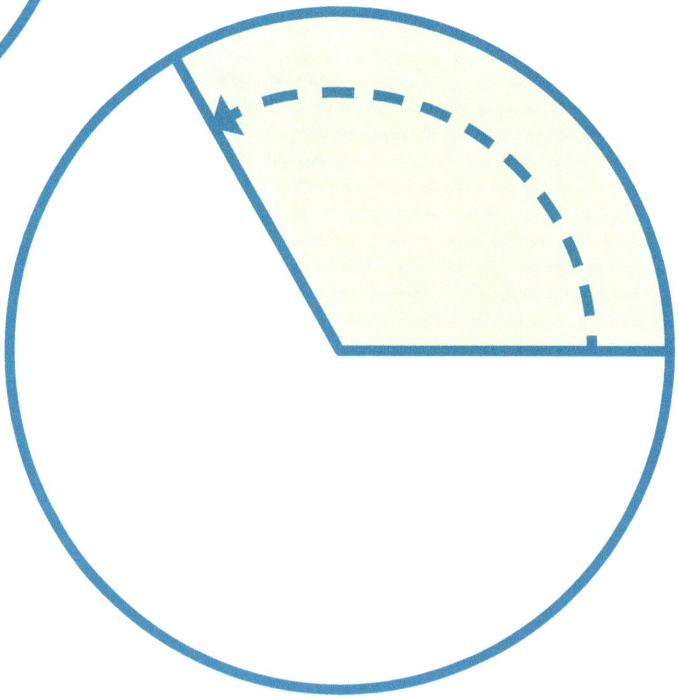

11. Guess the size of this angle in degrees.

___

Measure the angle using a protractor.

___

12. Guess the length of this line in centimeters _______ Measure the line using a ruler _______

13. Guess the length of this line in centimeters _______ Measure the line using a ruler _______

14. What is the area of this rectangle in square centimeters?

_________ x _________ = _________

15. What is the area of this square in square inches?

_________ x _________ = _________

16. What is the area of this rectangle in square centimeters?

_________ x _________ = _________

DAY 25 Practice

Practice counting by chunks. These counting patterns will help you remember your multiplication fact patterns. God's patterns are often connected!

1. Count by 3s: 15, ______, ______, ______, ______, ______, ______, ______

2. Count by 6s: 12, ______, ______, ______, ______, ______, ______, ______

3. Count by 9s: 0, ______, ______, ______, ______, ______, ______, ______

Here is some practice with addition and subtraction.

4. 6 + 8 = ☐

5. 8 + 6 = ☐

6. 7 + 7 = ☐

7. 8 + 8 = ☐

8. 8 + 3 = ☐

9. 3 + 8 = ☐

10.

	1	5	3	3	4
+	2	2	7	4	9

11. 14 – 7 = ☐

12. 16 – 7 = ☐

13. 18 – 7 = ☐

14. 19 – 7 = ☐

15. 20 – 7 = ☐

16. 35 – 12 = ☐

17.

	3	4	5	6	5
–	1	3	4	3	1

Go Use God's Patterns God's Way! DAY 26

This lesson integrates math into everyday life. With this activity, the student will apply God's patterns and wisdom in meaningful ways to home and community. Take a break from memory work and academic exercises to make math part of everyday life. The following are suggestions or examples. Other ideas may be added to the list.

God has created many patterns in nature and in math. He has given you creativity too!

Today, let's put together shapes to create beautiful patterns. Which kind of bathroom floor would you like? God's squares and triangles makes everything look beautiful!

1. Design a pattern for bathroom tiles.
2. Design a floor pattern for a porch, patio, or deck.
3. Design a stained glass window. Create your own stained glass window using black construction paper, tissue paper, and contact paper.

Note: Patterns for stained glass windows may be found on the internet. The parent/teacher may assist the student with this project.

Additional Activities

For extra activities, games, and helpful videos on these math concepts, refer to **generations.org/math3**.

God created the orange, made up of separate segments perfect for eating!

CHAPTER 3

Finding Fractions

DAY 27 A World Full of Fractions

Memory Drill

These memory drills may be completed either in writing or orally with a parent/teacher. Your goal is to complete this drill of twelve exercises in 60 seconds.

1. 9 + 4 =	5. 6 – 2 =	9. 8 + 6 =
2. 13 + 7 =	6. 11 – 4 =	10. 12 + 3 =
3. 8 x 4 =	7. 10 ÷ 2 =	11. 16 ÷ 4 =
4. 10 x 5 =	8. 12 ÷ 4 =	12. 25 ÷ 5 =

What part of your day do you spend learning? How about playing and sleeping? Can you eat a whole watermelon? Did God make the whole world or just a part of it?

> God, who made the world and everything in it (Acts 17:24)

Let's learn about parts and wholes.

How many apples do you see here? That's easy! There's only 1 apple here.

How many apples do we have now? This is **less than** 1 apple. This is half an apple. We write one-half like this:

$$\frac{1}{2}$$

What if we cut an apple in half and you eat both halves? You have eaten the amount of a whole apple. We can show this with numbers:

$$\frac{2}{2} = 1$$

Now how many apples do you see here? That's right. This is one and a half apples, We can say this as a number two ways: One and a half or three-halves. We can write it like this:

$$1\frac{1}{2} \qquad \frac{3}{2}$$

We call these parts of whole things *fractions*. A **fraction** is a part of a whole. Let's try a few more. How many apples do you see here? Finish writing the number in the box. The fraction bar is provided for you.

1. 2 —

Now how many half apples can you count here?

That's right! There are seven-halves. This is how we write it as a fraction: $\frac{7}{2}$

Now, suppose you glued 6 of these half apples back together. This is what it would look like:

How many apples do we have now? Finish writing the number in the box. Now do you see that 7 half apples is the same as 3 and a half apples?

2. 3 —

Fractions have a top number and a bottom number. The top number is called the **numerator**. The bottom number is called the **denominator** The denominator tells us how many equal parts a whole thing is divided into. The numerator tells us how many of those parts we are talking about.

$\frac{1}{2}$	$\frac{2}{5}$	$\frac{8}{3}$
You can say: One over two Or you can say: One-half,	Two over five Two-fifths,	Eight over three Eight-thirds.

Say these fractions both ways.

3. $\frac{8}{5}$

4. $\frac{3}{8}$

5. $\frac{9}{4}$

Activity

There are so many fractions to discover all around us. Write the fraction for pictures 1 through 3.

Then make up three of your own fractions from things you find around the house. Draw pictures in boxes 4 and 5 and write the fractions.

1. 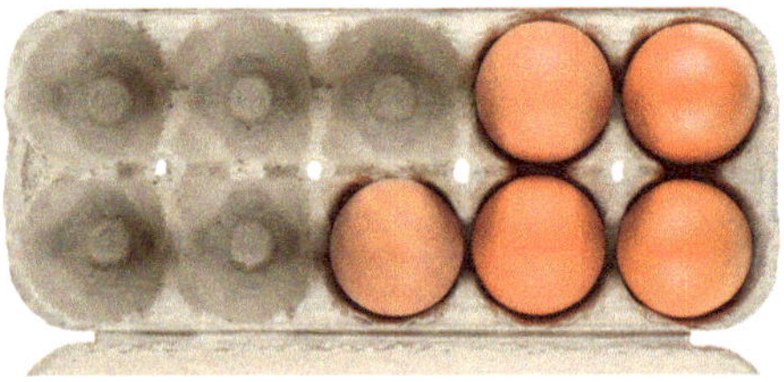 $\frac{\quad}{10}$

2.  $\frac{\quad}{2}$

3.  $\frac{4}{\quad}$

4. $\frac{\quad}{\quad}$

5. $\frac{\quad}{\quad}$

6. Write your own fraction where the top number is the same as the bottom number. Remember, that means you have all the parts you need to make up a whole.

7. Write a fraction where the top number is larger than the bottom number. That means you have more parts than you need to make up a whole.

8. Write a fraction where the bottom number is larger than the top number. That means you don't have enough parts to make up a whole.

Let heaven and earth praise Him,
The seas and everything that
moves in them. (Psalm 69:34)

Let all the parts of the earth worship God!
Let all the earth worship God!
All the parts and the whole is full of God's glory!

The more fractions you find, the more you'll appreciate how God's fractions fill this wonderful world He's made.

Student Exercises

Let's explore God's patterns with sets. Think about how these sets come together to create new sets.

You may enjoy talking through each exercise with your parent/teacher. For unions (∪), ask yourself, "Do these sets come together to make a bigger set?" For intersections (∩), ask yourself, "Do the sets contain members in common with each other?"

1.	{Men}	∪	{Women}	=	{ }
2.	{Shirts}	∪	{Clothing}	=	{ }
3.	{Heaven}	∪	{Earth}	=	{ }
4.	{Time in the Day}	∩	{Time to Work}	=	{ }
5.	{Sea Creatures}	∩	{Mammals}	=	{ }
6.	{Fruits}	∩	{Meats}	=	{ }

Practice DAY 28

Memory Drill

These memory drills may be completed either in writing or orally with a parent/teacher. Your goal is to complete this drill of twelve exercises in 60 seconds.

1. 5 + 3 = ☐	5. 12 + 7 = ☐	9. 4 + 13 = ☐
2. 13 – 5 = ☐	6. 16 – 4 = ☐	10. 10 – 3 = ☐
3. 5 x 4 = ☐	7. 6 x 7 = ☐	11. 10 x 10 = ☐
4. 56 ÷ 8 = ☐	8. 12 ÷ 4 = ☐	12. 50 ÷ 5 = ☐

Think about how many parts of these shapes are shaded. Write the fraction.

13. 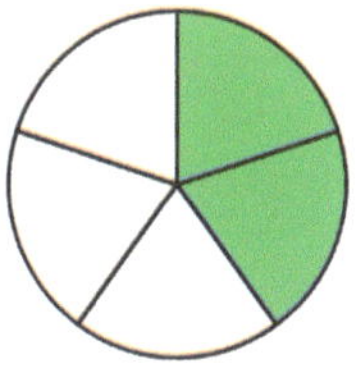 —

16. 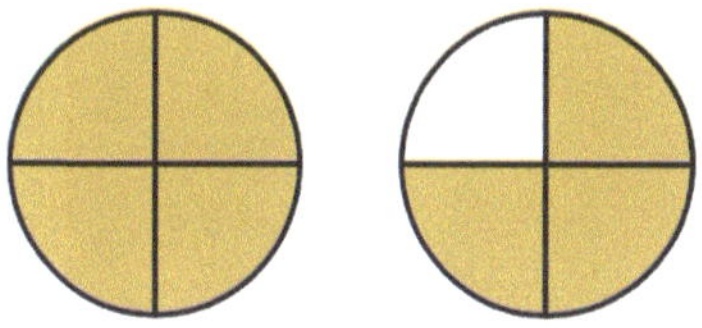 1 —

14. 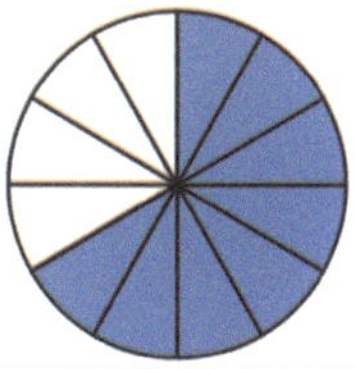 —

17. 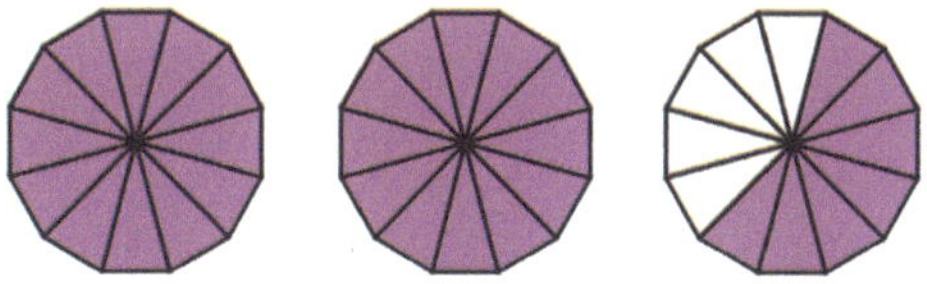 $2\frac{\quad}{12}$

15. 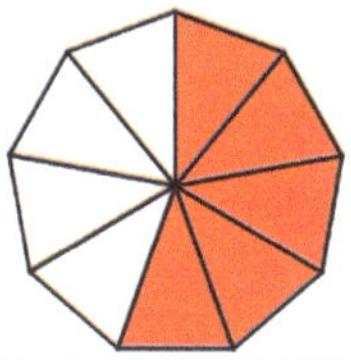 —

18. 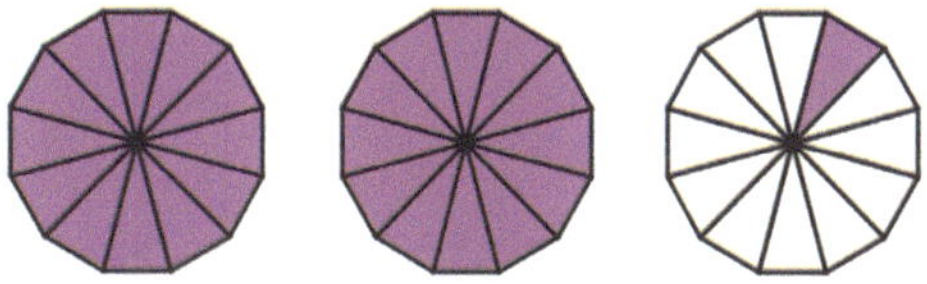 $\frac{\quad}{12}$

Put these numbers in the order that God made for them. Think about which number is bigger than the other. Fill in the < or > signs for each of these exercises.

3 < 5   5 > 3   5 = 5

19.	1345		1323	24.	8264		6912
20.	4681		4687	25.	7425		7369
21.	1,000,000		10,000	26.	100		100,000
22.	47,234		47,251	27.	47,980		47,380
23.	1,999		2,001	28.	4,564,234,534		0

Practice DAY 29

Memory Drill

These memory drills may be completed either in writing or orally with a parent/teacher. Your goal is to complete this drill of twelve exercises in 60 seconds.

1. 7 + 4 = ☐
2. 12 – 5 = ☐
3. 5 x 6 = ☐
4. 42 ÷ 7 = ☐
5. 15 + 2 = ☐
6. 10 – 3 = ☐
7. 7 x 8 = ☐
8. 16 ÷ 4 = ☐
9. 6 + 12 = ☐
10. 9 – 4 = ☐
11. 7 x 10 = ☐
12. 30 ÷ 10 = ☐

13. Count by 2s:

50, ______, ______, ______, ______, ______

14. Count by 4s:

40, ______, ______, ______, ______, ______

15. Count by 6s:

0, ______, ______, ______, ______, ______

16. Count by 8s:

0, ______, ______, ______, ______, ______

17. Count by 8s:

40, ______, ______, ______, ______, ______

For each exercise, draw in the missing triangles all the way to the edge of the page. You will be making a tessellation pattern (no gaps) for each color.

18. 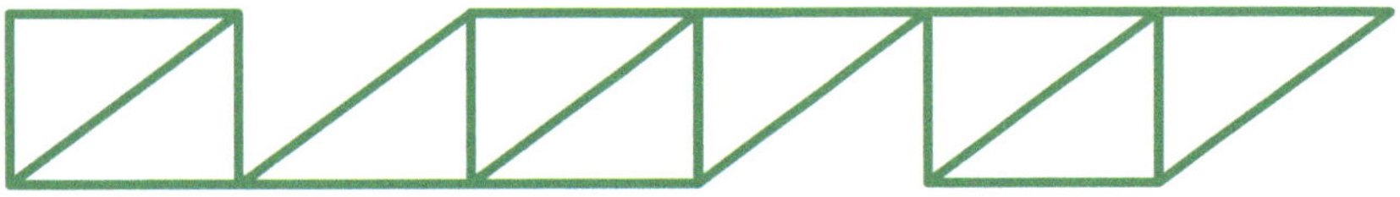

19. 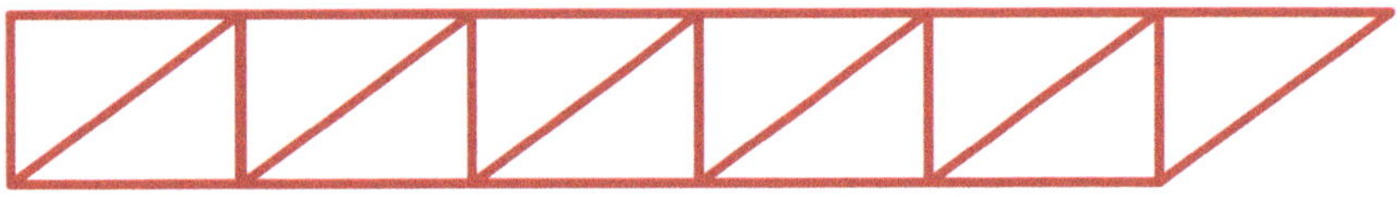

20. 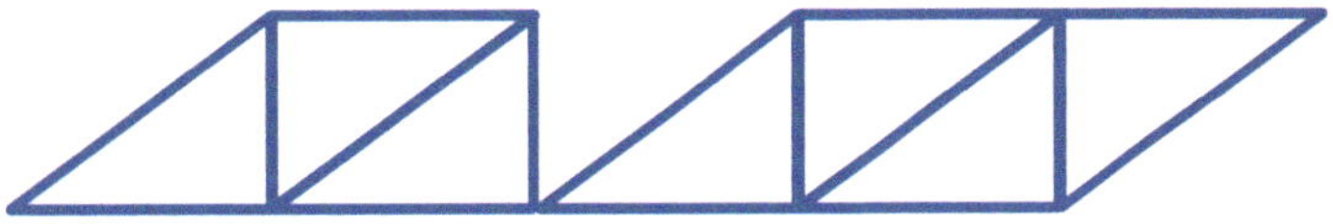

21. 

22. 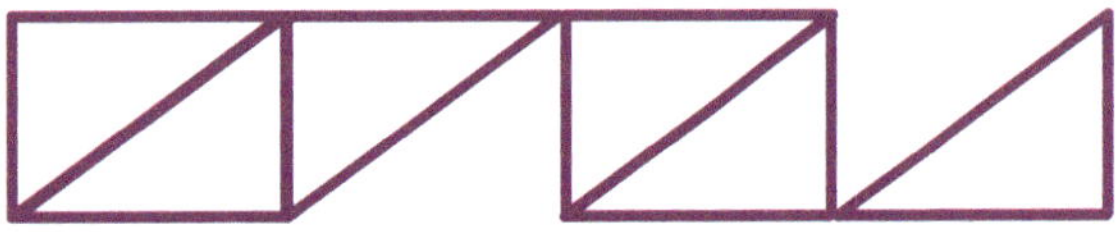

Practice rounding these numbers. Rounding helps us understand the size of numbers.

23. Round 4,248,922 to the nearest 100. ______________________

24. Round 4,248,922 to the nearest 10,000. ______________________

25. Round 4,248,922 to the nearest 100,000. ______________________

26. Round 7,931,377 to the nearest 1,000 ______________________

27. Round 7,931,377 to the nearest 100,000 ______________________

28. Round 7,931,377 to the nearest 1,000,000 ______________________

Fractions Create Numbers DAY 30

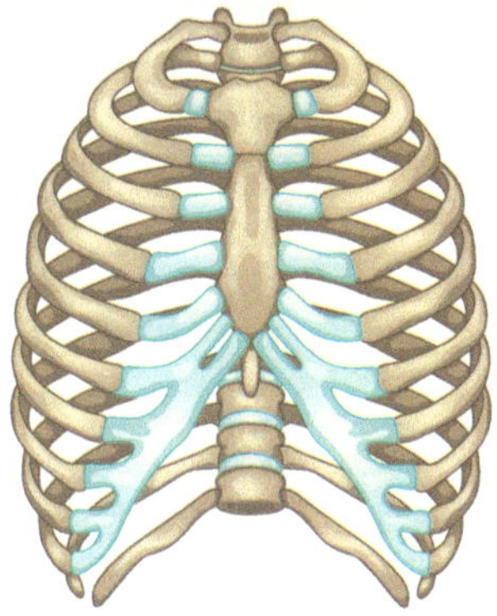

And the Lord God caused a deep sleep to fall on Adam, and he slept; and He took one of his ribs, and closed up the flesh in its place. Then the rib which the Lord God had taken from man He made into a woman, and He brought her to the man. (Genesis 2:21-22)

God makes whole things, and He makes the whole things out of parts. When God made the woman, He made her out of one part of the man. He took one of Adam's 24 ribs to make Eve! Today, most men and women, boys and girls have 24 ribs—12 on each side. God made Eve out of $\frac{1}{24}$ of Adam's ribs. Can you say one twenty-fourth?

When you use fractions, you are also using division. In a fraction, the numerator is being divided by the denominator. To help us understand this, let's think about sandwiches. If we have 8 half sandwiches, that equals 4 whole sandwiches. Eight half sandwiches can be written with a 2 in the denominator because each sandwich is divided into 2 pieces. The numerator will be written as an 8 because we are talking about 8 pieces of sandwich.

So you see that 8 over 2 equals 4. You already know that 8 divided by 2 also equals 4.

$$\frac{8}{2} = 4 \qquad 8 \div 2 = 4 \qquad 8 \div 2 = \frac{8}{2}$$

That means 8 divided by 2 is the exact same thing as 8 over 2. It's just a different way to write it. This is true for every fraction. Every fraction can also be written as division.

Can you see that the division symbol looks like a little fraction? The two dots stand for the numbers in a fraction.

$$8 \div 2 = \frac{8}{2}$$

With that in mind, can you figure out which whole number each of these fractions is equal to? The shapes may help you.

1. $\frac{5}{5} =$

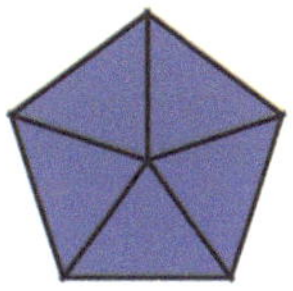

2. $\frac{24}{8} =$

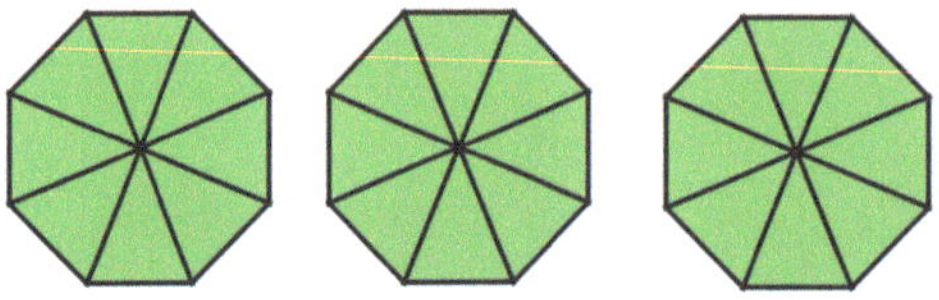

3. $\frac{40}{10} =$

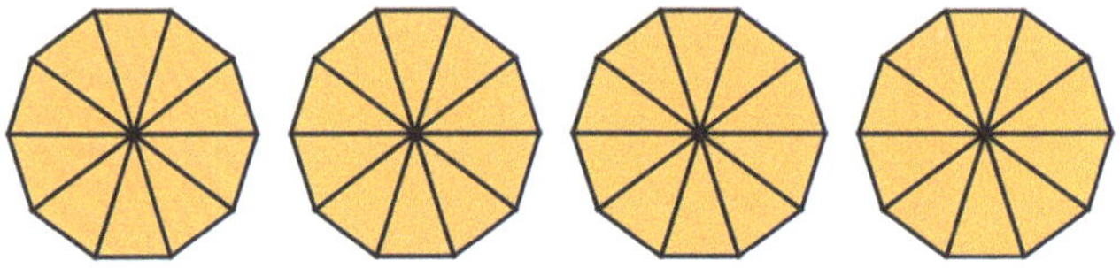

Sometimes fractions are a little more complicated. These circles show 5 halves shaded.

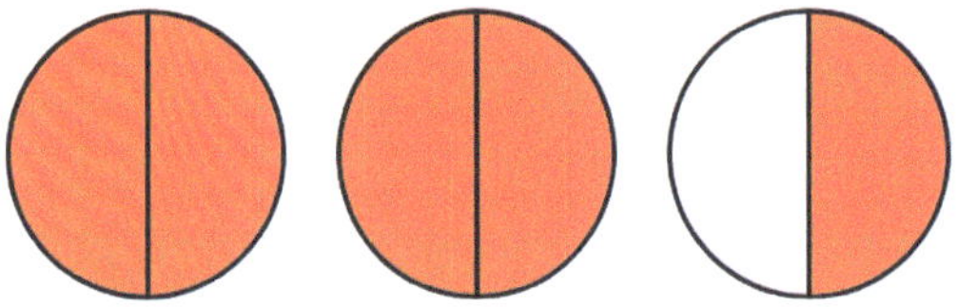

This is how we write the 5 halves as a fraction:

$$\frac{5}{2}$$

Now, if you change this into a division exercise, this is what it would look like: **5 ÷ 2**.

The answer to 5 ÷ 2 is not a whole number. You will not find the answer on your multiplication and division tables. The answer will be between 2 and 3. The answer is $2\frac{1}{2}$. Look at the circles again. There are 2 whole circles shaded and $\frac{1}{2}$ circle shaded.

Here is another way to think about it. On a ruler, count 5 half centimeters or 5 half inches. When you stop, you should be at $2\frac{1}{2}$. Can you find something in your house that is $\frac{5}{2}$ inches or $\frac{5}{2}$ centimeters long?

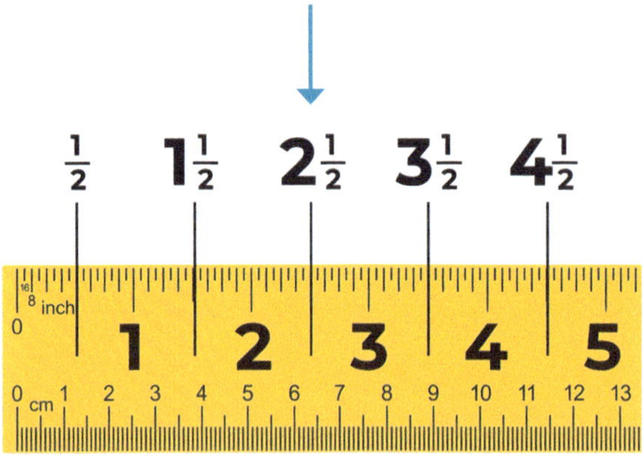

Student Exercises

We often use small numbers in God's world, but sometimes we need to use large numbers. God gave us addition and subtraction that works with all numbers. Find the answers to these math exercises. Some already have borrows and carries done for you.

1.

	3	0	5
	2	1	0
+	1	3	4

2.

	1	1	1
	4	6	0
+	3	2	8

3.

	1		
	4	3	0
	1	5	6
+	1	9	3

4.

	4	4	4
	3	3	3
	2	2	2
+	1	1	1

5.

	1	5	9	0
−		3	9	0

6.

	1	4	5	9
−		2	5	8

7.

			2	12
	4	1	~~3~~	~~2~~
−		1	2	6

8.

	2	9	9	9
−	1	9	9	8

9.

		1		
	7	1	3	1
+		3	8	6

10.

		1	1	
	3	0	9	5
+		4	7	7

11.

			5	14
	5	9	~~6~~	~~4~~
−		9	5	6

12.

	8	6	6	3
−		4	2	7

How many fruits and vegetables do you see here? Write what each picture shows in three different ways. The first example is provided for you.

13.		$2\frac{1}{2}$	$\frac{5}{2}$	$5 \div 2$
14.				
15.				
16.	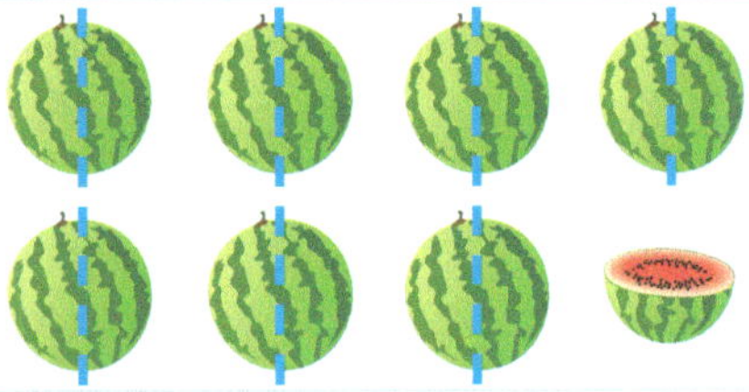			
17.	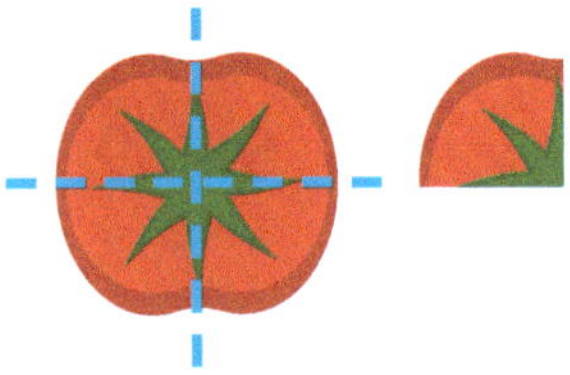			
18.				
19.				

Adding & Subtracting Fractions DAY 31

Memory Drill

These memory drills may be completed either in writing or orally with a parent/teacher. Your goal is to complete this drill in 60 seconds.

1. 4 + 4 =	5. 9 − 2 =	9. 9 + 6 =
2. 3 + 12 =	6. 8 − 5 =	10. 12 + 6 =
3. 7 x 4 =	7. 12 ÷ 3 =	11. 70 ÷ 7 =
4. 5 x 3 =	8. 81 ÷ 9 =	12. 45 ÷ 9 =

> The Lord has made everything for its purpose.
> (Proverbs 16:4a, ESV)

You've already learned about the four ways in which numbers dance together. You can add them. You can subtract them. You can multiply them. You can divide them. Today we are going to add and subtract pies! Or rather, let's add and subtract fractions!

Here is how to add fractions. Let's say that Maria has $\frac{1}{2}$ of a pie, and Juan has $\frac{1}{2}$ of a pie too. How many pies do they have all together? Remember, we combine the pies when we add. Put the two halves together, and what is the answer?

MARIA'S PIE		JUAN'S PIE		TOGETHER
	+		=	
$\frac{1}{2}$	+	$\frac{1}{2}$	=	$\frac{2}{2} = 1$

Now let's add quarter (¼) pizza pies. One child has 1 quarter of a pizza pie. Another child has 2 quarters of a pizza pie. Now, let's add them together!

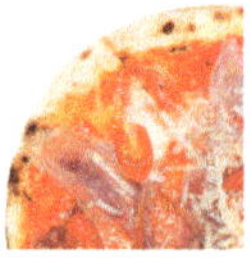	+	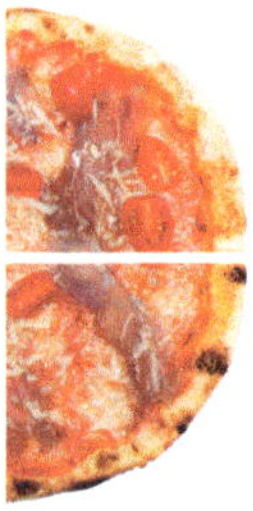	=	
$\frac{1}{4}$	+	$\frac{2}{4}$	=	$\frac{3}{4}$

When adding fractions, you can see that the denominator remains the same. But, we add the top numbers: 1 plus 2 equals 3. Think about it like this. Adding one-fourth of a pizza with two-fourths of a pizza will give you three-fourths of a pizza, because each of the pieces is still a fourth of the pizza.

Below is some leftover pizza pie. There used to be 8 pieces. Now there are only 6 of the 8 pieces left in the first picture (left). This is less than 1 pizza. This is $\frac{6}{8}$ of a pizza. Suppose then, the children ate 4 more of the pieces for lunch. How much of the pizza is left?

	–	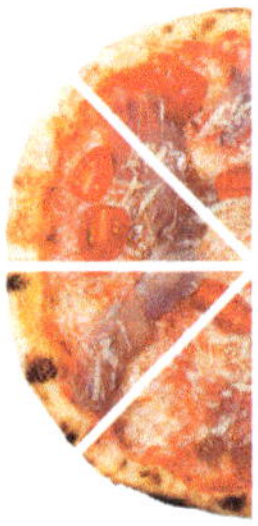	=	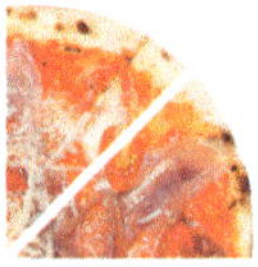
$\frac{6}{8}$	–	$\frac{4}{8}$	=	$\frac{2}{8}$

Let's say that you have 6 chairs around your table. See if you can explain the subtraction below to your parent/teacher using numbers of seats that have someone sitting in them. Maybe you can draw some pictures to help your explanation.

$$\frac{5}{6} - \frac{2}{6} =$$

Answer the fraction exercises below. Remember, the numerators are added and subtracted normally. The denominator stays the same!

1. $\frac{3}{5} + \frac{6}{5} = \frac{\square}{\square}$
2. $\frac{5}{10} + \frac{2}{10} = \frac{\square}{\square}$
3. $\frac{2}{3} + \frac{1}{3} = \frac{\square}{\square}$
4. $\frac{5}{4} - \frac{1}{4} = \frac{\square}{\square}$
5. $\frac{6}{2} - \frac{5}{2} = \frac{\square}{\square}$
6. $\frac{8}{7} - \frac{6}{7} = \frac{\square}{\square}$

Look at this fraction: $\frac{0}{2}$

Remember, a denominator tells us how many pieces something is divided into. The numerator tells us how many of those pieces we are talking about. Notice that this fraction has a zero in the numerator. In this faction, something has been cut into 2 pieces, but we are talking about none of those pieces. It's as though your mom cut a chocolate bar in half and said that you could come and have a piece. But when you got there, both pieces had already been eaten. So zero of the 2 pieces are left. That's 0 out of 2, or $\frac{0}{2}$.

Do you remember that a fraction is another way to show division? We saw that:

$$\frac{8}{2} = 8 \div 2$$

They both equal 4. If you understand this, you will understand that anytime there is a zero in the numerator, the fraction equals zero. This is because, if you divide zero into pieces, you still have the number zero. For example, if your mom asked you to share your zero amount of chocolate with her, you both would get zero chocolate.

Because of this, it doesn't matter what number is in the denominator. If zero is in the numerator, the fraction is still zero. If your mom asked you to divide your zero chocolate with her and your dad, you would each still get zero chocolate.

All of these equations are true:

$$\frac{0}{2} = 0 \qquad \frac{0}{11} = 0 \qquad \frac{0}{149} = 0 \qquad \frac{0}{3{,}672{,}409} = 0$$

Answer these exercises with a fraction and then either a zero or a 1.

7. $\frac{1}{3} + \frac{2}{3} = \frac{3}{3} = 1$

8. $\frac{2}{2} - \frac{2}{2} = \frac{0}{2} = 0$

9. $\frac{10}{10} - \frac{10}{10} = \frac{}{} = \square$

10. $\frac{4}{9} + \frac{5}{9} = \frac{}{} = \square$

11. $\frac{5}{5} - \frac{5}{5} = \frac{}{} = \square$

12. $\frac{9}{10} + \frac{1}{10} = \frac{}{} = \square$

Now, think about if you were doing math and you came up with the answer $\frac{2}{0}$. That's a fraction with a zero in the denominator. If that happened, you did something wrong. There is no such thing as $\frac{2}{0}$ because you can't divide by zero. We can't divide something into zero parts. We can't ask, "How many zeros are in the number 2?" When we talk about dividing by zero or about having a zero denominator, we say we can't have that because it's **undefined**. Circle the fractions below that are undefined.

13. $\frac{6}{3}$ 14. $\frac{3}{5}$ 15. $\frac{7}{0}$ 16. $\frac{1}{10}$ 17. $\frac{0}{10}$ 18. $\frac{10}{0}$

Student Exercises

1. Mom cut a pizza into 12 pieces. The children ate 4 pieces. What fraction of that pizza is left?

$$\frac{12}{12} - \frac{4}{12} = \frac{\square}{12}$$

2. Joe and Sam each had a pizza. Joe ate 3 out of 4 pieces of his pizza. Sam also ate 3 out of 4 pieces of his pizza. How much pizza did these boys eat? Hint: Together they ate more than 1 pizza.

JOE ATE THIS:		SAM ATE THIS:		
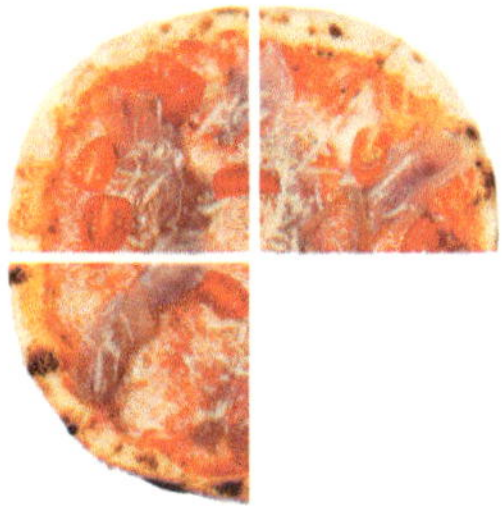	+		=	
$\frac{3}{4}$	+	$\frac{3}{4}$	=	$\frac{\quad}{4}$

3. Dad bought Mom a dozen (12) roses for their anniversary. Six of the roses are yellow and three of the roses are pink. **What fraction of the 12 roses are the yellow and pink roses together?**

$$\frac{\square}{\square} + \frac{\square}{\square} = \frac{\square}{\square}$$

4. There is $\frac{3}{4}$ of a gallon of milk in the refrigerator. Your sister wants to bake a cake that will use $\frac{1}{4}$ of a gallon of milk. How much milk will be left after your sister bakes the cake?

$\frac{\square}{\square} - \frac{\square}{\square} = \frac{\square}{\square}$

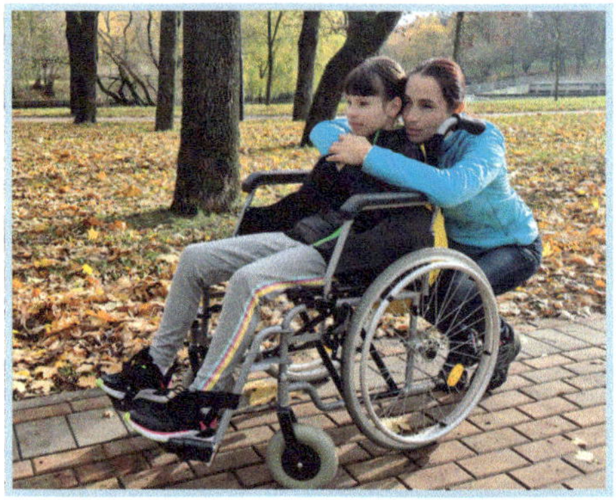

5. A park's trail is a loop. Once around the loop is $\frac{2}{3}$ of a mile. If you travel twice around the loop, how far would you go?

$\frac{\square}{\square} + \frac{\square}{\square} = \frac{\square}{\square}$

6. Karen asked her mom to sew her a new dress. She picked out $3\frac{1}{2}$ yards of one fabric and $2\frac{1}{2}$ yards of another fabric to be used for the dress. How much fabric did Mom use in all? Hint: think about how many half yards are in $3\frac{1}{2}$ and in $2\frac{1}{2}$. Add those fractions together in the boxes below.

$\frac{\square}{\square} + \frac{\square}{\square} = \frac{\square}{\square}$

Practice counting by chunks. This time you will be counting backwards.

7. Count by 2s: 20, 18, ______, ______, ______, ______

8. Count by 3s: 30, 27, ______, ______, ______, ______

9. Count by 5s: 50, 45, ______, ______, ______, ______

10. Count by 8s: 80, 72, ______, ______, ______, ______

11. Count by 10s: 100, 90, ______, ______, ______, ______

Just like these logs, numbers can be split apart in different ways—into two numbers that are added or into two numbers that are multiplied. Can you find different ways to split these numbers apart? Watch for whether you need to use addition or multiplication and fill in the blanks.

12. 12 = ___ + ___

13. 12 = ___ + ___

14. 12 = ___ x ___

15. 12 = ___ x ___

16. 18 = ___ + ___

17. 18 = ___ + ___

18. 18 = ___ x ___

19. 18 = ___ x ___

Add or subtract. Some of the borrows and carries have already been provided for you as hints.

20.

	1	1	
	1	5	5
+	1	5	5

21.

		6	13
	1	~~7~~	~~3~~
–	1	3	4

22.

		2	5
+	1	6	6

23.

	8	7
–	5	9

24.

	8	9
+	5	4

25.

	1	6	5
–	1	5	6

DAY 32 Practice

Find the area of these rectangles by multiplying. Write the number of rows of blocks and the number of columns of blocks in the blanks. Multiply and fill in the equations.

1.

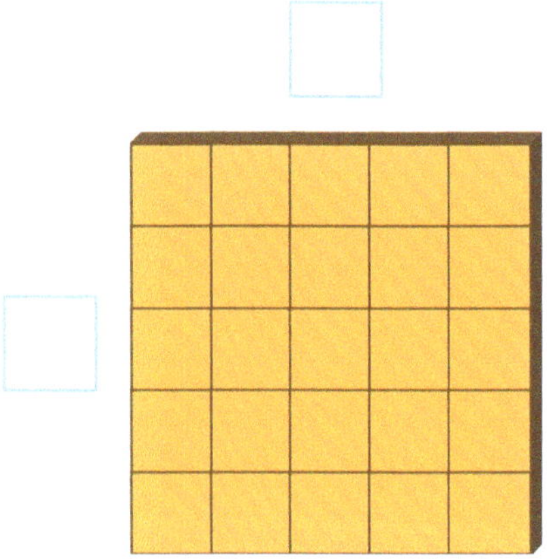

___ x ___ = ___

2.

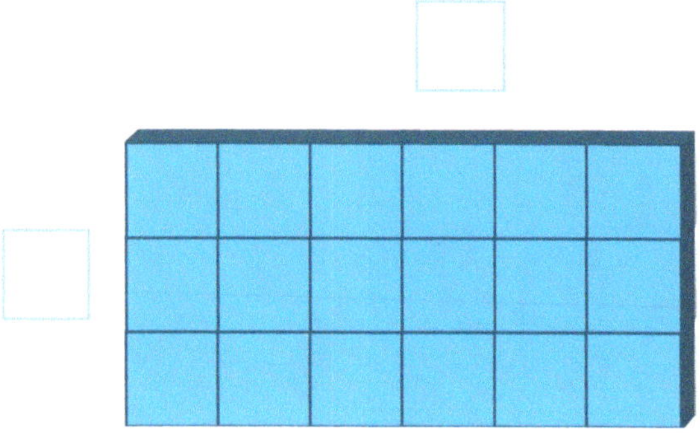

___ x ___ = ___

3.

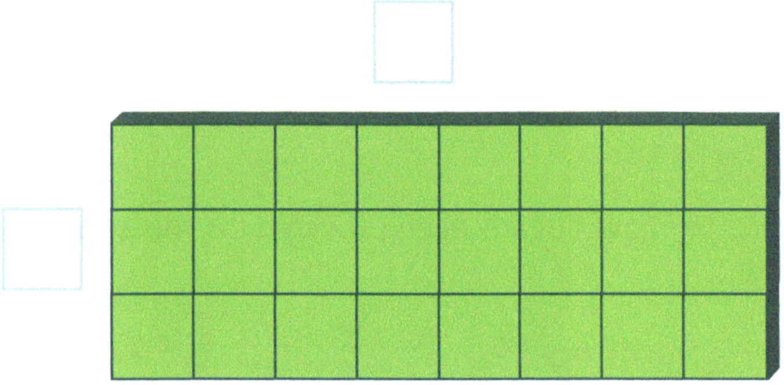

___ x ___ = ___

4.

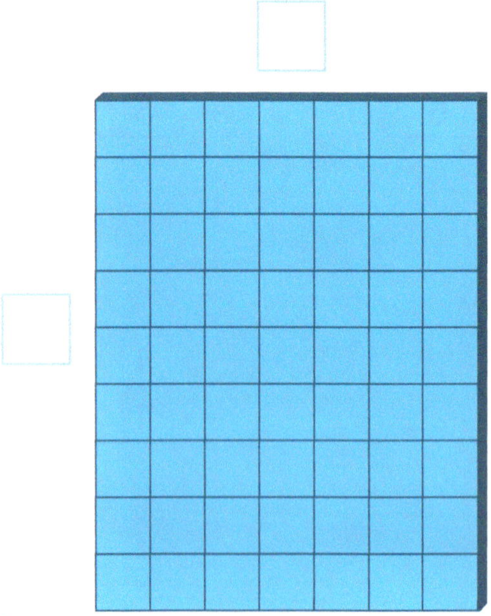

___ x ___ = ___

5.

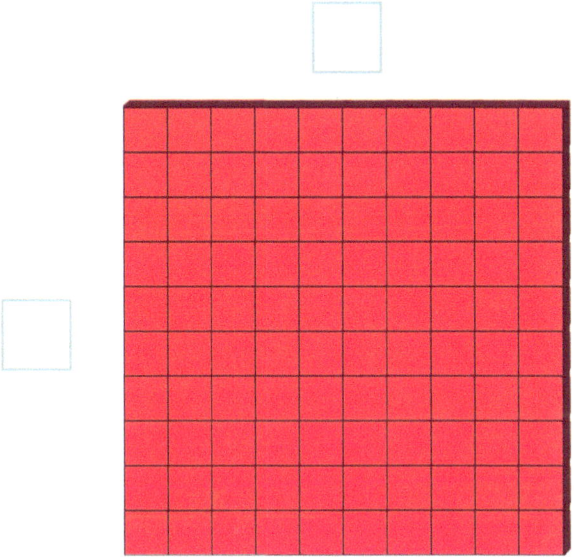

___ x ___ = ___

6.

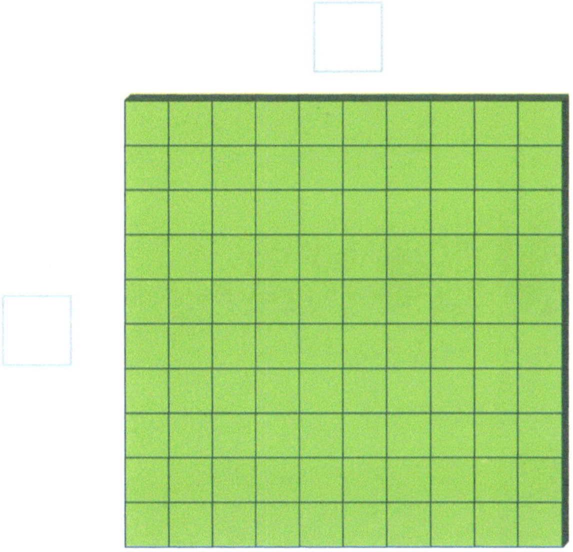

___ x ___ = ___

7.

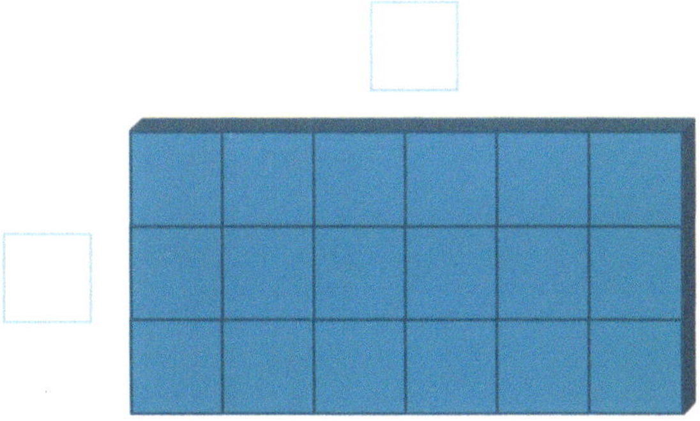

___ x ___ = ___

8.

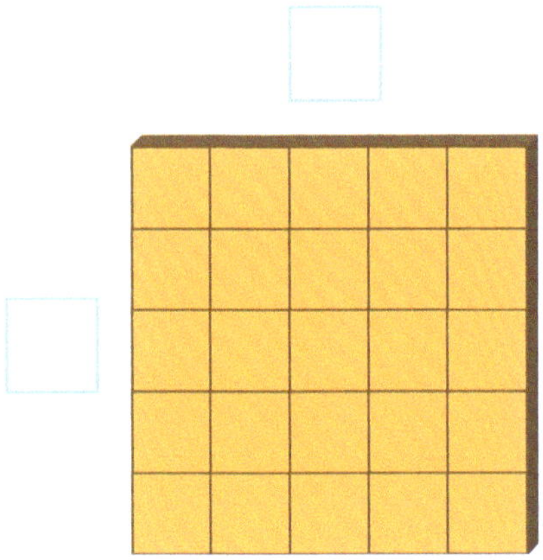

___ x ___ = ___

9.

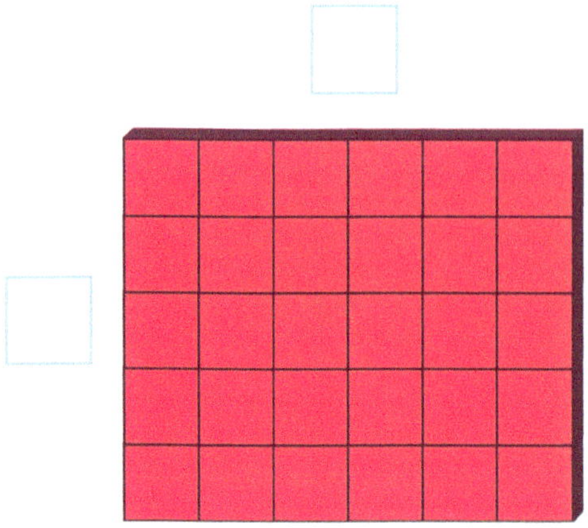

___ x ___ = ___

10.

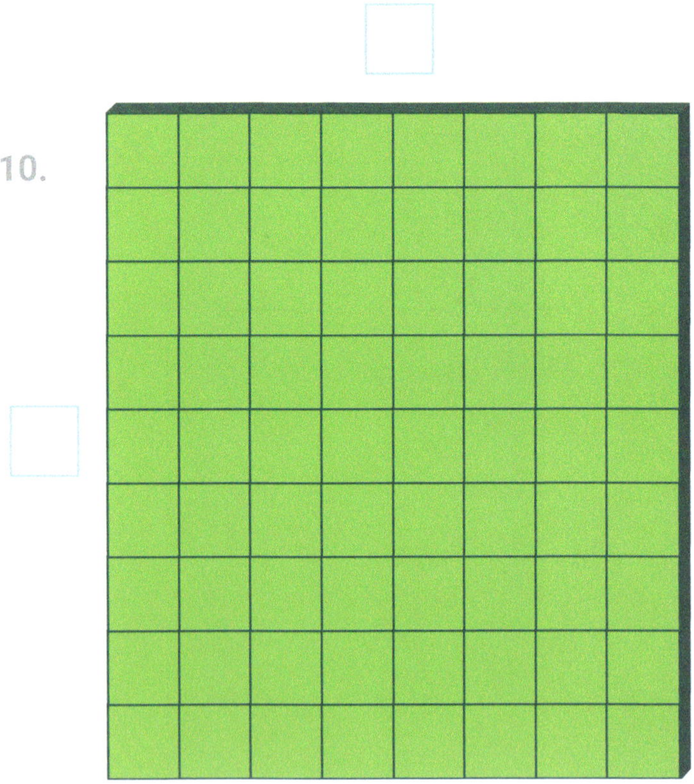

___ x ___ = ___

Let's round these numbers! Rounding helps us think about the large numbers in God's world.

11. Round 6,420,875 to the nearest 100. ______________________

12. Round 6,420,875 to the nearest 10,000. ______________________

13. Round 6,420,875 to the nearest 100,000. ______________________

14. Round 1,854,488 to the nearest 1,000. ______________________

15. Round 1,854,488 to the nearest 10,000. ______________________

16. Round 1,854,488 to the nearest 100,000. ______________________

Add and subtract some fractions!

17. $\frac{3}{5} + \frac{1}{5} = \square$

18. $\frac{2}{7} + \frac{3}{7} = \square$

19. $\frac{3}{6} + \frac{7}{6} = \square$

20. $\frac{4}{8} - \frac{2}{8} = \square$

21. $\frac{4}{2} - \frac{2}{2} = \square$

22. $\frac{3}{5} - \frac{3}{5} = \square$

Draw a line to match each fraction with the correct shaded shape.

23. $\frac{4}{6}$

a. 

24. $\frac{1}{3}$

b. 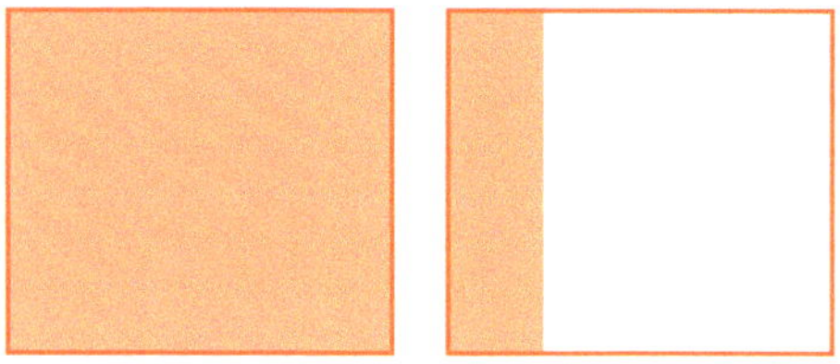

25. $1\frac{1}{4}$

c. 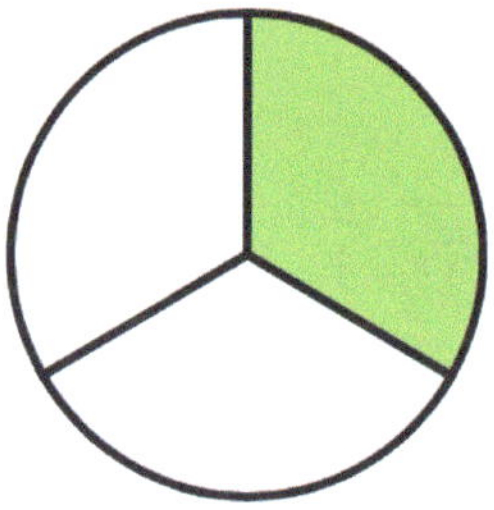

26. $2\frac{1}{3}$

d. 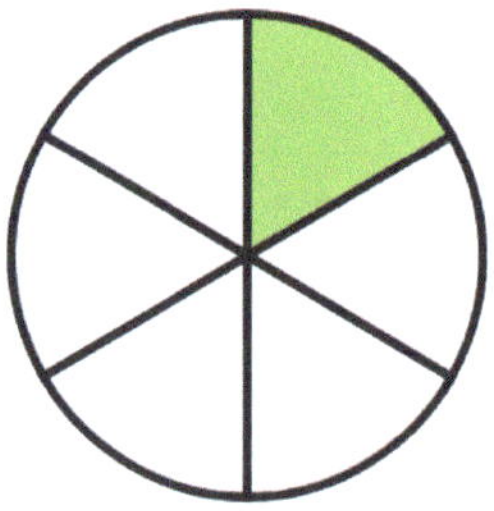

27. $\frac{1}{6}$

e. 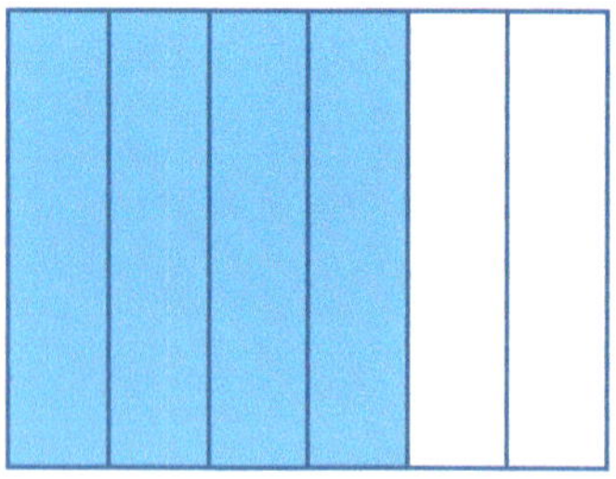

Fractions in the Bible

DAY 33

Memory Drill

These memory drills may be completed either in writing or orally with a parent/teacher. Your goal is to complete this drill of twelve exercises in 60 seconds.

1. 5 + 3 =	5. 6 – 3 =	9. 2 + 15 =
2. 7 + 4 =	6. 14 – 5 =	10. 5 + 6 =
3. 9 x 4 =	7. 24 ÷ 3 =	11. 81 ÷ 9 =
4. 5 x 11 =	8. 12 ÷ 2 =	12. 42 ÷ 7 =

"Remember the Sabbath day, to keep it holy. Six days you shall labor and do all your work, but the seventh day is the Sabbath of the Lord your God. In it you shall do no work: you, nor your son, nor your daughter, nor your male servant, nor your female servant, nor your cattle, nor your stranger who is within your gates. For in six days the Lord made the heavens and the earth, the sea, and all that is in them, and rested the seventh day. Therefore the Lord blessed the Sabbath day and hallowed it." (Exodus 20:8-11)

Fractions are important in the Bible. They are important to our daily life as children of our heavenly Father. In fact, God has called out one day in seven, or 1/7 of our week, as a rest day. We call this day the *Lord's Day*.

How many days do we give the Lord every week? How many days do we give to the Lord every month (or 28 days)? Fill in the fractions.

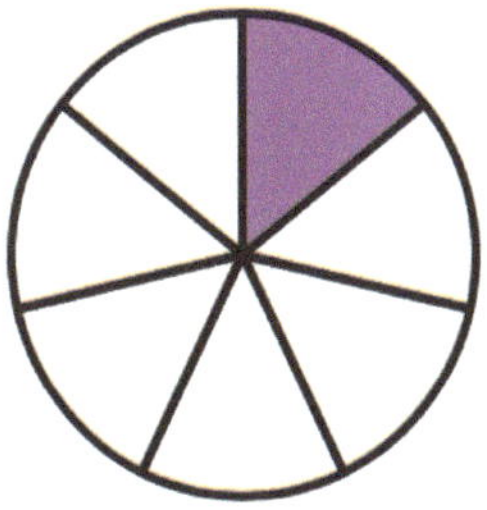

1. $\frac{\quad}{7}$

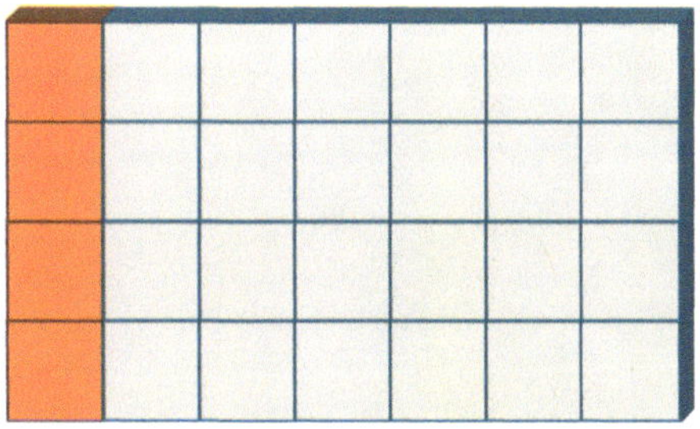

2. $\frac{\quad}{28}$

> "Bring the full [tenth] into the storehouse, that there may be food in my house. And thereby put me to the test, says the Lord of hosts, if I will not open the windows of heaven for you and pour down for you a blessing until there is no more need." (Malachi 3:10, ESV)

God has given us everything we have. But He tells us to give 1/10 of our money as a gift for His house (church). Sometimes we call this the *tithe*. The Old Testament word "tithe" means a "tenth." This green slice of the circle is how much we give God:

$\frac{1}{10}$

3. Suppose you earned 10 apples for working an hour in an orchard. Circle the number of apples you would give to God.

4. Suppose you earned 20 colored pencils for working an hour in a pencil factory. Circle the number of pencils you would give to God.

Student Exercises

Can you find more fractions in the Bible verses below? Circle the fraction and write the fraction as a number next to each verse. Try to understand the parts and wholes and how those parts and wholes are involved in what God is telling us.

1. The Lord said to [Abraham], "Bring me a heifer three years old, a female goat three years old, a ram three years old, a turtledove, and a young pigeon." And he brought him all these, cut them in half, and laid each half over against the other. But he did not cut the birds in half. (Genesis 15:9-10, ESV) $\frac{\quad}{\quad}$

2. Let Pharaoh proceed to appoint overseers over the land and take one-fifth of the produce of the land of Egypt during the seven plentiful years. (Genesis 41:34, ESV) $\frac{\quad}{\quad}$

3. Then the people of Israel were divided into two parts. Half of the people followed Tibni the son of Ginath, to make him king, and half followed Omri. (1 Kings 16:21, ESV) $\frac{\quad}{\quad}$

4. The first angel blew his trumpet, and there followed hail and fire, mixed with blood, and these were thrown upon the earth. And a third of the earth was burned up, and a third of the trees were burned up, and all green grass was burned up. (Revelation 8:7, ESV) $\frac{\quad}{\quad}$

Some numbers are small, but these numbers are large. Do the addition and subtraction exercises below.

5.

	4	6	3
+	6	1	6

6.

	5	4	8
−	1	7	0

7.

	5	6	1
+		7	2

8.

	4	1	5
+	5	4	7

9.

	8	9	5
−	1	8	3

10.

	1	2	3
−		3	1

DAY 34 Practice

Pretend that the shapes below are pieces of paper that have been folded and opened up again. The light blue lines are their creases. There is a smaller orange shape inside each larger shape. Figure out how many similar orange shapes fit in the big shape, and write the number next to the shape. A nice way to do this is to color in copies of the small shape (counting as you go) until you fill the whole space. For the first exercise, how many sets of two orange blocks can fit into this large shape? That's right: eight!

1. 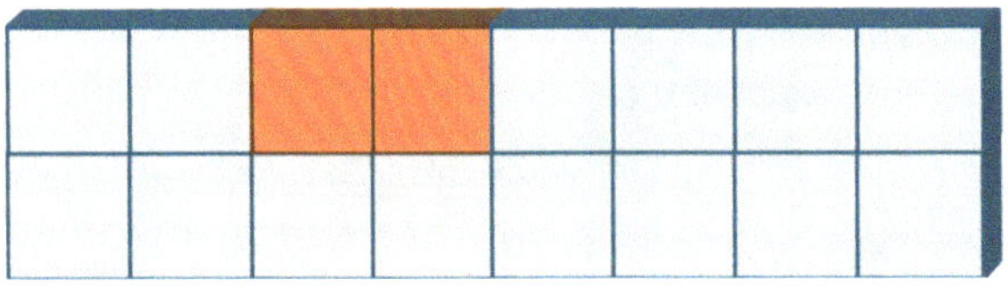 8

2. 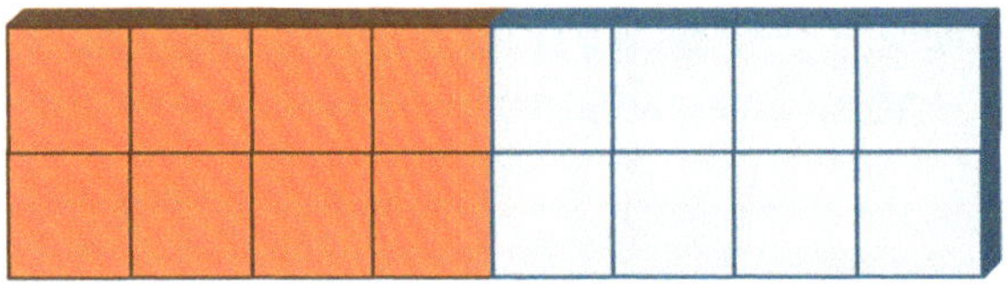

3. 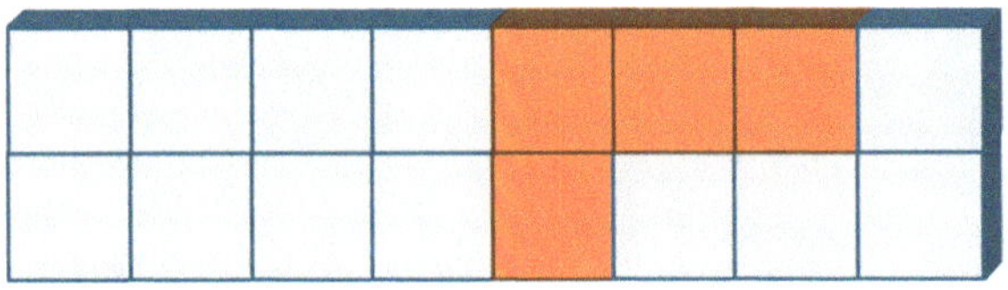

4. 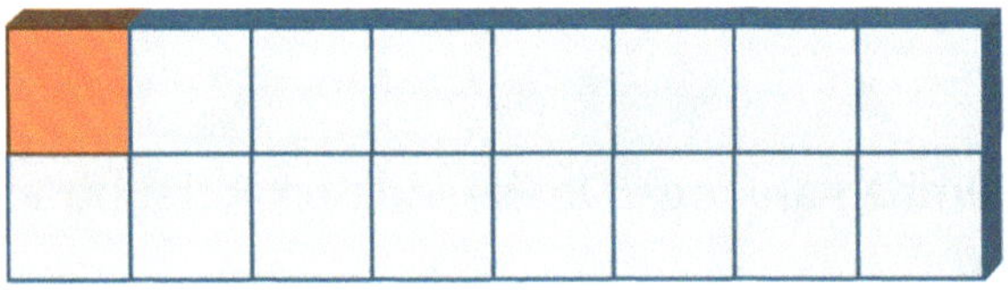

5. 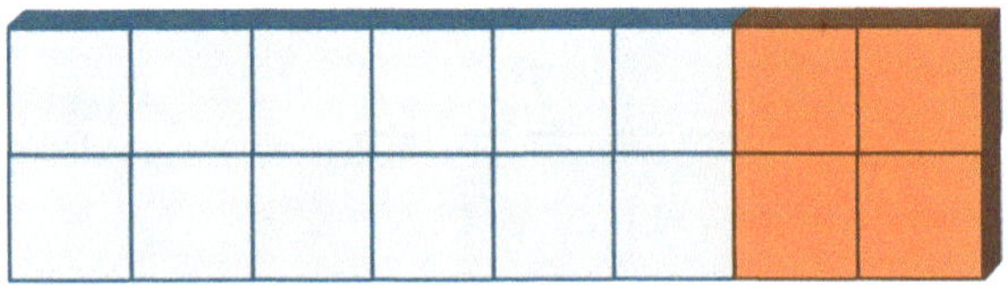

6. 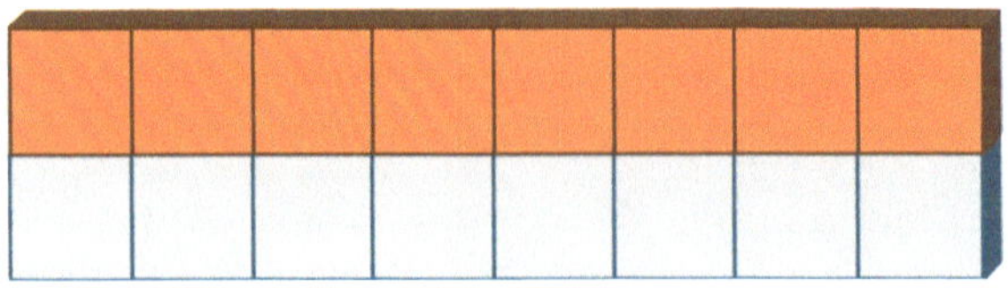

Here is some practice with division. Try answering the first set of exercises from memory. Then, you may use the chart to figure out the second set.

7. 6 ÷ 2 = ☐

8. 10 ÷ 2 = ☐

9. 6 ÷ 3 = ☐

10. 12 ÷ 3 = ☐

11. 8 ÷ 2 = ☐

12. 12 ÷ 2 = ☐

13. 9 ÷ 3 = ☐

14. 15 ÷ 3 = ☐

15. 42 ÷ 6= ☐

16. 50 ÷ 10 = ☐

17. 24 ÷ 6 = ☐

18. 24 ÷ 4 = ☐

19. 42 ÷ 7 = ☐

20. 50 ÷ 5 = ☐

21. 24 ÷ 8 = ☐

22. 24 ÷ 3 = ☐

X	1	2	3	4	5	6	7	8	9	10
1	1	2	3	4	5	6	7	8	9	10
2	2	4	6	8	10	12	14	16	18	20
3	3	6	9	12	15	18	21	24	27	30
4	4	8	12	16	20	24	28	32	36	40
5	5	10	15	20	25	30	35	40	45	50
6	6	12	18	24	30	36	42	48	54	60
7	7	14	21	28	35	42	49	56	63	70
8	8	16	24	32	40	48	56	64	72	80
9	9	18	27	36	45	54	63	72	81	90
10	10	20	30	40	50	60	70	80	90	100

DAY 35 Larger Fractions & Smaller Fractions

Memory Drill

These memory drills may be completed either in writing or orally with a parent/teacher. Your goal is to complete this drill of twelve exercises in 60 seconds.

1. 2 + 16=	5. 13 – 3 =	9. 6 + 11 =
2. 17 + 3 =	6. 20 – 5 =	10. 15 + 4 =
3. 4 x 8 =	7. 36 ÷ 6 =	11. 100 ÷ 10 =
4. 3 x 11 =	8. 16 ÷ 4 =	12. 18 ÷ 2 =

Then Jacob said to Joseph: "God Almighty appeared to me at Luz in the land of Canaan and blessed me. . . And now your two sons, Ephraim and Manasseh, who were born to you in the land of Egypt before I came to you in Egypt, are mine; as Reuben and Simeon, they shall be mine. Your offspring whom you beget after them shall be yours; they will be called by the name of their brothers in their inheritance. (Genesis 48:3-6)

In Genesis we read that Jacob had 12 sons. Normally, Jacob would have planned for all the things he owned to be divided equally between his 12 grown-up sons after he died. But Jacob decided to treat his son, Joseph, specially. Joseph himself had 2 sons. So, Jacob gave each of Joseph's sons a share of his inheritance as though they were his sons instead of his grandsons. Joseph did not get a share, but, in a way, he got a double share through his sons. Jacob blessed Joseph by including Joseph's sons in the inheritance.

Since the inheritance was divided between 13 people, each person received $\frac{1}{13}$ of the inheritance. But through his sons, Joseph received $\frac{2}{13}$ of the inheritance.

Can you read this math comparison? Two-thirteenths is larger than one thirteenth.

$$\frac{2}{13} > \frac{1}{13}$$

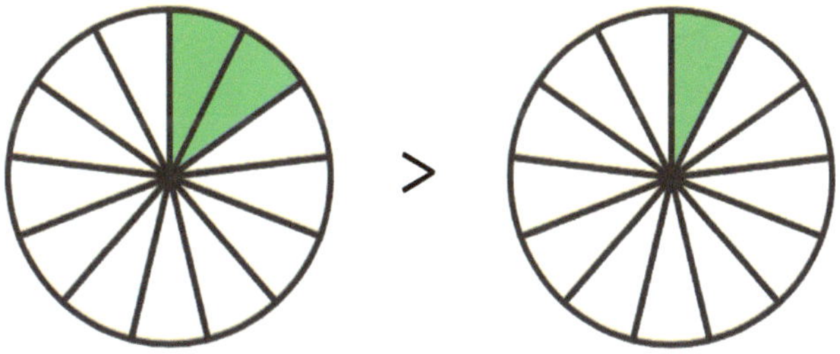

It's easy to compare fractions that have the same denominator.

$$\frac{4}{5} > \frac{2}{5}$$

Let's say you and a friend had a pizza cut into 8 pieces. If you ate 3 of the 8 pieces you would have eaten quite a bit more than your friend who only ate 1 of the 8.

$$\frac{3}{8} > \frac{1}{8}$$

When the denominators of two fractions are the same, the fraction with the larger numerator is the larger fraction. Can you say that in your own words?

Complete these fractions by placing $<$ or $>$ in the circle.

1. $\frac{2}{3} \bigcirc \frac{1}{3}$

2. $\frac{4}{5} \bigcirc \frac{3}{5}$

3. $\frac{3}{8} \bigcirc \frac{7}{8}$

Comparing those fractions was not very hard. But now let's learn how to compare fractions with different denominators and the same numerators. The best way to do this is to compare pies. Imagine that the fractions below are two parts of a blueberry pie. Which fraction of pie would be the larger piece? Circle the larger fraction. Write < or > in the box.

4. $\frac{1}{2}$ 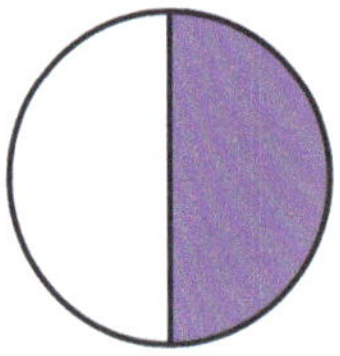 ☐ 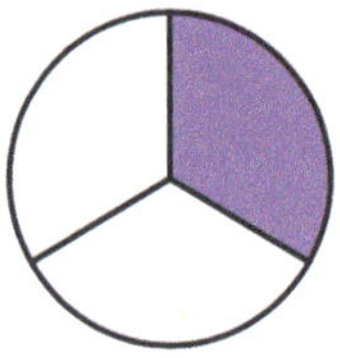 $\frac{1}{3}$

Let's see why $\frac{1}{2}$ is larger than $\frac{1}{3}$. You already know that a denominator of 2 means that something is divided into 2 pieces. A denominator of 3 would mean that it's divided into 3 pieces. Do you see that, if a pie is divided into less pieces, each piece is larger than if the pie was divided into more pieces? So even though 2 is a smaller number than 3, as a denominator, it makes larger pieces. That's why one-half of a blueberry pie is larger than one-third of it.

Now, let's compare a few more fractions with the same numerator but different denominators. You can imagine that each exercise is comparing two pizzas that are cut into a different number of pieces. For example the first two pizzas are cut into 3 pieces and 4 pieces. We'll say that one child is eating from the pizza that's cut into 3 pieces, and another child is eating from the pizza that's cut into 4 pieces. Both children eat the same number of pieces of their own pizza. Circle the larger fraction to show which child will eat the most pizza.

5. $\frac{2}{3}$ ☐ $\frac{2}{4}$

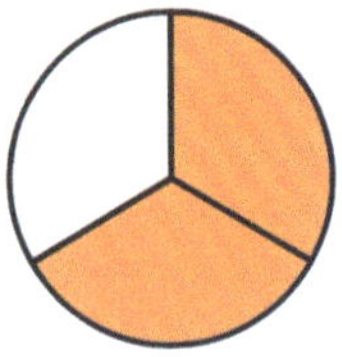 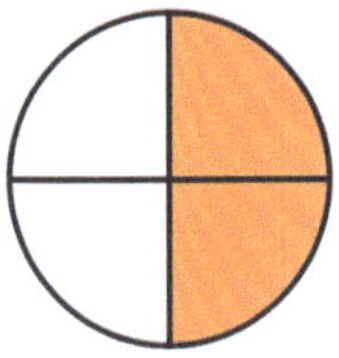

6. $\frac{3}{4}$ ☐ $\frac{3}{5}$

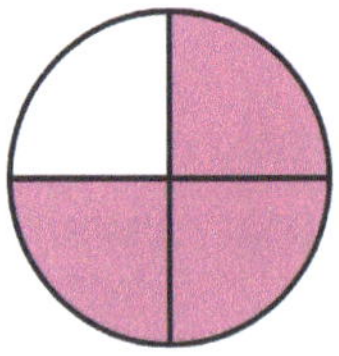 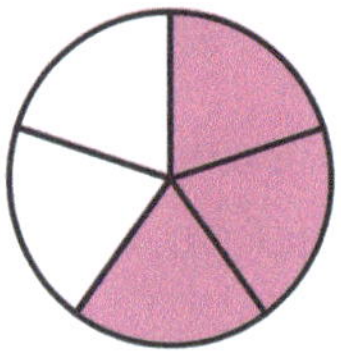

Let's say that your mom has invited 5 children to your birthday party. This means you must cut the cake into 6 pieces for you and your 5 friends. But what if your mom invited 19 children to your birthday party? Then you must cut the cake into 20 pieces. Will these be smaller pieces than the pieces cut for 6 children? What if your mom invited 100 friends to your birthday party? Everybody would only get a tiny sliver of the cake! Compare the size of these cake slices using the pictures below.

1/6 OF THE CAKE

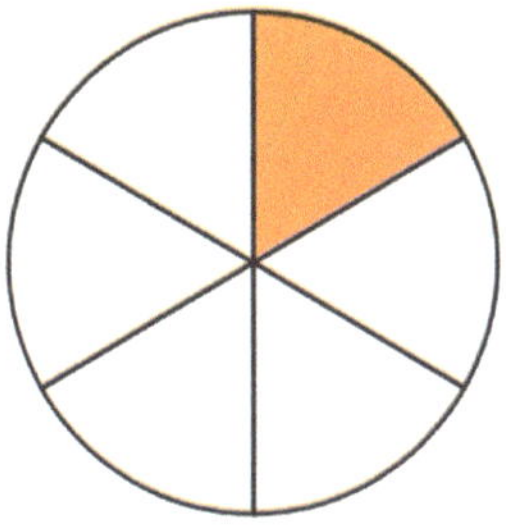

1/20 OF THE CAKE

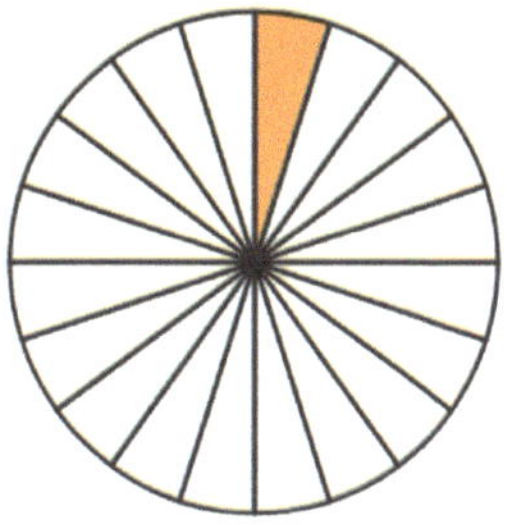

1/100 OF THE CAKE

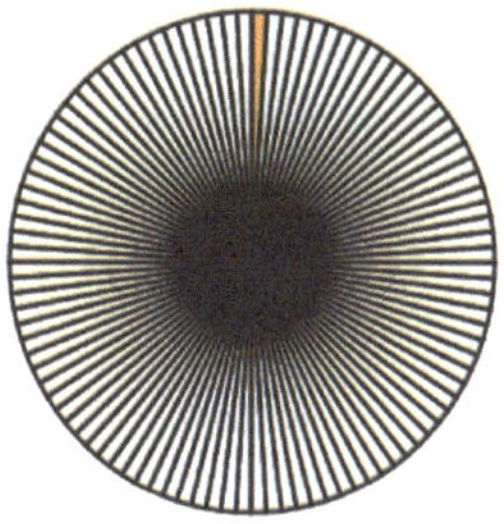

Remember, if the denominator of a fraction is really small, one piece of the whole thing is really big. If the denominator of a fraction is really big, one piece is going to be really small. One-sixth is way bigger than one one-hundredth!

$$\frac{1}{6} > \frac{1}{100}$$

Let's figure out which is the bigger fraction. If Jack eats $\frac{12}{3}$ pizzas, and John eats eats $\frac{12}{6}$ pizzas, who eats more pizza?

7. $\frac{12}{3}$ ☐ $\frac{12}{6}$ 8. $12 \div 3 =$ ☐ 9. $12 \div 6 =$ ☐

Student Exercises

Practice counting by chunks.

1. Count by 3s 15, ______, ______, ______, ______, ______, ______, ______

2. Count by 6s: 12, ______, ______, ______, ______, ______, ______, ______

3. Count by 9s: 0, ______, ______, ______, ______, ______, ______, ______

Let's explore God's patterns with sets. Think about how these sets come together to create new sets. For intersections, (∩) if there are no common members of the sets, write the word "empty" in the brackets.

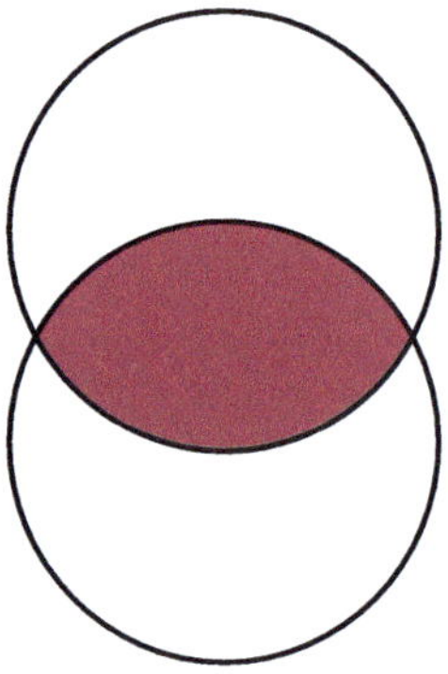

4. {1, 3, 5} U {2, 4, 6} = { }

5. {2, 3, 4} U {3, 4, 5} = { }

6. {3, 5, 7, 11} U {3, 5, 7, 11} = { }

7. {7, 8, 9} ∩ {9, 10, 11} = { }

8. {2, 4, 6} ∩ {8, 10, 12} = { }

9. {4, 5} ∩ {4, 5} = { }

Extra Challenge

Pactice a few more of these fraction exercises. Write < or > in the boxes. Try to imagine a pizza or a birthday cake. Which will be the larger amount?

10. $\frac{1}{100}$ ☐ $\frac{1}{10}$

11. $\frac{5}{20}$ ☐ $\frac{5}{6}$

12. $\frac{12}{4}$ ☐ $\frac{12}{3}$

Practice DAY 36

Complete these exercises.

1. 13 + 4 = ☐
2. 4 + 12 = ☐
3. 3 x 8 = ☐
4. 7 x 10 = ☐
5. 18 – 3 = ☐
6. 14 – 5 = ☐
7. 49 ÷ 7 = ☐
8. 21 ÷ 3 = ☐
9. 5 + 8 = ☐
10. 7 + 6 = ☐
11. 50 ÷ 0 = ☐
12. 24 ÷ 6 = ☐

Which fraction is bigger? Write a greater than (>) or less than (<) sign.

13. $\frac{1}{2}$ ◯ $\frac{1}{3}$

14. $\frac{2}{3}$ ◯ $\frac{4}{3}$

15. $\frac{5}{8}$ ◯ $\frac{7}{8}$

16. $\frac{10}{6}$ ◯ $\frac{1}{3}$

17. $\frac{6}{5}$ ◯ $\frac{3}{5}$

18. $\frac{5}{9}$ ◯ $\frac{7}{9}$

Finish each of these lists of shapes. Find each pattern, and then draw the next picture that keeps that pattern going.

19. 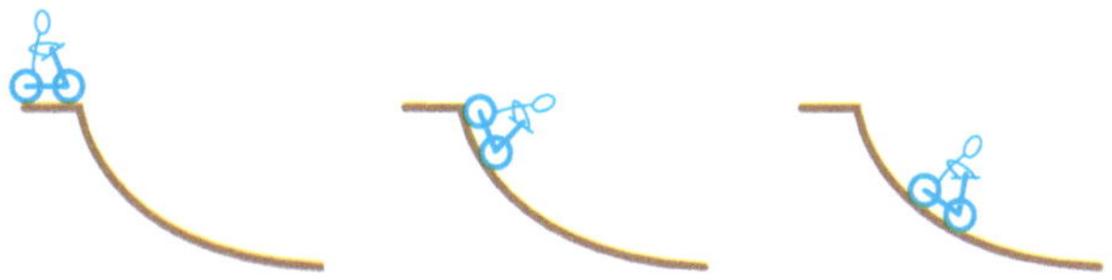

20. 

21.

22. 

23. 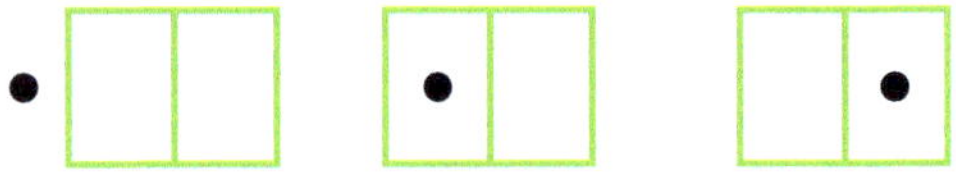

24. 

God gives us patterns in the Bible to discover. Read Psalm 136 with your parent.

25. What pattern do you notice?

26. How many times does it appear in that chapter?

27. What can you learn from seeing something repeated in the Bible?

Fractions in Music DAY 37

If you are learning to play an instrument, or if you like to sing, you've probably noticed patterns in music.

Songs can be broken up into many small pieces called bars. These bars give a pattern to music. Below, you will see how music looks written down. Each bar is the section of music inside two vertical lines.

When music is written, we use something that looks like a fraction to tell us a couple of things about the song. Can you find the numbers that look like a fraction in this music? These numbers are called the **time signature**.

The 4/4 tells us several things about the music. Although these numbers might look like a fraction, they do not act like the numbers in a fraction. The top number does not act like a numerator. Instead, it tells us how many beats are in each of the song's bars. The top number tells us there are 4 beats in each bar.

To understand what beats are, try clapping your hands together 4 times in a regular pattern. Repeat this several times, clapping a little louder on the first clap of each bar.

1, 2, 3, 4, 1, 2, 3, 4, 1, 2, 3, 4

These are the beats for a $\frac{4}{4}$ song. Did you hear the regular pattern of the beats? Each beat or clap was a fraction of a bar's total beats. In this song, each beat is $\frac{1}{4}$ of a bar. So even though the $\frac{4}{4}$ written on the song is not a real fraction, there are fractions in music.

When music is written, we use little symbols called **notes** to tell us different things about the music. If you look at the written music again, you will see the notes as round shapes, many with stems. The notes have several jobs. One of their jobs (along with the time signature's top number) is to help us know the beat of the song.

Music has whole notes, half notes, quarter notes, eighth notes, and sixteenth notes. Half notes are played twice as fast as whole notes. Quarter notes are played twice as fast as half notes. Eighth notes are played twice as fast as quarter notes, and of course, sixteenth notes are played twice as fast as eighth notes!

The bottom number of the music's time signature does not act like the denominator of a fraction. Instead, it tells us which kind of note gets one beat. If the bottom number is a 4, the quarter ($\frac{1}{4}$) note counts as one beat. (This means that an eighth note would get a half beat and would sound like a shorter, faster note.) If the bottom number is an 8, the eighth note ($\frac{1}{8}$) counts as one beat.

Activity

Here are some hymns with different numbers of beats in their bars. Your parent/teacher can help you find these hymns written in a hymnbook or online (i.e., hymnary.org or a similar site). For each hymn, fill in the number of beats per bar and the kind of note shown by the bottom number. If possible, listen to each hymn and see if you can pick out the speed, and the beats per bar.

- Amazing Grace
  Number of beats (top number): 4
  Kind of note (bottom number): quarter (1/4)
- Silent Night
  Number of beats (top number): ________
  Kind of note (bottom number): ________
- Come Thou Fount
  Number of beats (top number): ________
  Kind of note (bottom number): ________
- Christ the Lord is Risen Today
  Number of beats (top number): ________
  Kind of note (bottom number): ________
- Crown Him With Many Crowns
  Number of beats (top number): ________
  Kind of note (bottom number): ________
- Joy to the World
  Number of beats (top number): ________
  Kind of note (bottom number): ________
- Away in a Manger
  Number of beats (top number): ________
  Kind of note (bottom number): ________

Which of these hymns do you like best? Enjoy them!

> Sing to God, sing praises to His name;
> Extol Him who rides on the clouds,
> By His name Yah,
> And rejoice before Him. (Psalm 68:4)

Fractions and Prime Numbers DAY 38

Today we will introduce the idea of **prime numbers**.

But first, let's play a game. The rules are:

- You must divide these lollipops up evenly among your friends.
- You must give each friend more than one lollipop.

Think about how you can divide the lollipops between 2 people—yourself and your friend. How many candies will each person get? Each of you will get two.

Now, divide 5 lollipops up for 2 children. Remember, you must divide up evenly and each child must get more than 1 candy.

You can't divide these candies up evenly, can you? 5 cannot be divided by 2. It also can't be divided by 3 or 4. It can only be divided by 1 and by 5. That is because 5 is a **prime** number.

A **prime number** cannot be divided up into equal groups larger than 1. It can only be divided by 1 and by itself.

God made this lime with 10 sections. These sections can be divided up in different ways. You can give 2 children 5 sections each. Or you can give 5 children 2 sections each. If there were two children, they would each get 5/10 of the lime. With 5 children, they would each get 2/10 of the lime. We can see fractions in fruits, vegetables, flowers, and so many other parts of God's creation.

Check out this papaya. That's a 5 point star put into the papaya by the Creator Himself!

The papaya can be cut into 5 equal pieces. But you cannot slice it into groups of 2 as we did with the lime. This is what makes prime numbers special. They don't divide down into nice whole numbers.

Here are some more examples of numbers that are prime.

We can see that 3 is a prime number. This honeydew melon can only be split into 3 equal pieces—it can only be divided by 1 and by 3.

We know that 7 is also a prime number. It can only be divided by 1 and by 7.

Here are some examples of numbers that are not prime:

We find that 6 is not a prime number. It can be divided into more numbers than just 1 and 6. It can be divided by 2 and 3.

Look at this pepper which has 4 parts. Is 4 a prime number? No, 4 is not a prime number because it can also be split into equal chunks of 2 pieces for two people.

Activity

Check out these identical pairs. Write a fraction that shows the shaded portion of each picture. If the shaded portions are the same size in the identical pairs, see if you can write your second fraction with a different numerator and denominator than your first fraction has. The first one has been done for you. Parent/teacher: answers will vary.

1.		$\frac{2}{9}$		$\frac{1}{3}$
2.				
3.	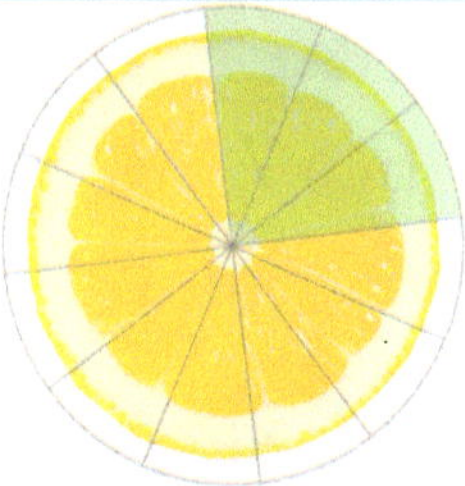		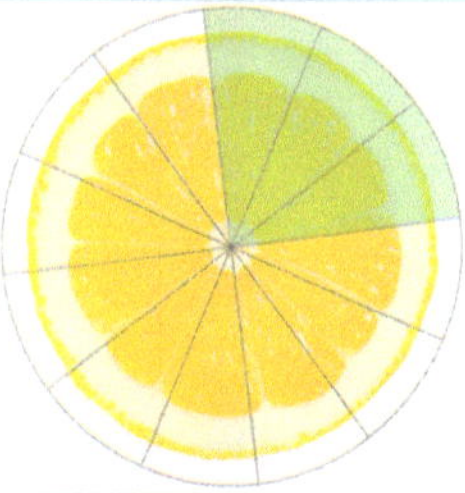	
4.				
5.			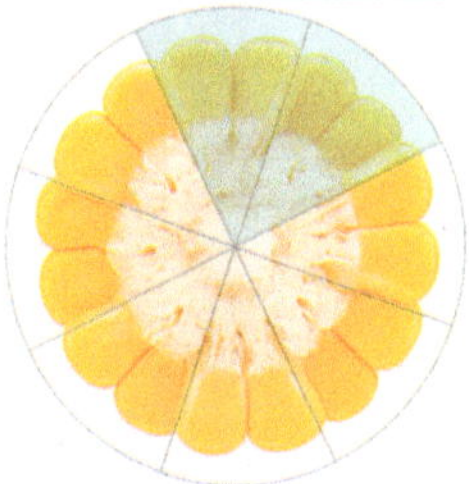	

Student Exercises

Here is some practice with division. Answer the first set of exercises from memory. Then, you may use the chart to figure out the second set.

1. 12 ÷ 3 = ☐
2. 15 ÷ 3 = ☐
3. 9 ÷ 3 = ☐
4. 16 ÷ 4 = ☐
5. 10 ÷ 2 = ☐
6. 20 ÷ 4 = ☐
7. 30 ÷ 5 = ☐
8. 18 ÷6 = ☐

9. 49 ÷7 = ☐
10. 48 ÷ 8 = ☐
11. 42 ÷ 7 = ☐
12. 64 ÷ 8 = ☐
13. 81 ÷ 9 = ☐
14. 36 ÷ 6 = ☐
15. 100 ÷ 10 = ☐
16. 25 ÷ 5 = ☐

X	1	2	3	4	5	6	7	8	9	10
1	1	2	3	4	5	6	7	8	9	10
2	2	4	6	8	10	12	14	16	18	20
3	3	6	9	12	15	18	21	24	27	30
4	4	8	12	16	20	24	28	32	36	40
5	5	10	15	20	25	30	35	40	45	50
6	6	12	18	24	30	36	42	48	54	60
7	7	14	21	28	35	42	49	56	63	70
8	8	16	24	32	40	48	56	64	72	80
9	9	18	27	36	45	54	63	72	81	90
10	10	20	30	40	50	60	70	80	90	100

Practice DAY 39

Try answering these multiplications from memory. You may use a chart to figure out the ones you don't know.

1. $7 \times 8 =$ ___
2. $8 \times 5 =$ ___
3. $5 \times 6 =$ ___
4. $8 \times 9 =$ ___
5. $7 \times 6 =$ ___
6. $3 \times 9 =$ ___
7. $10 \times 7 =$ ___
8. $6 \times 6 =$ ___

Practice rounding these numbers. Rounding helps us do mental math quickly out in God's world.

9. Round 525,739 to the nearest 100. ______________
10. Round 525,739 to the nearest 10,000. ______________
11. Round 525,739 to the nearest 100,000. ______________
12. Round 7,511,034 to the nearest 1,000 ______________
13. Round 7,511,034 to the nearest 100,000 ______________
14. Round 7,511,034 to the nearest 1,000,000 ______________

Here is some practice with addition.

15. $111 + 222 + 333 =$ ___
16. $444 + 333 + 222 =$ ___
17. $1234 + 1111 =$ ___
18. $1000 + 2000 + 3000 + 4000 =$ ___
19. $1111 + 2222 + 3333 =$ ___
20. $987 + 111 =$ ___
21. $11 + 22 + 33 =$ ___
22. $321 + 432 + 543 =$ ___

DAY 40 Go Find Fractions!

This lesson integrates math into everyday life. With this activity, the student will apply God's patterns and wisdom in meaningful ways to home and community. The following are suggestions or examples. Other ideas may be added to the list.

We use fractions every day around the house and in other places. Let's think about ways we can use fractions in God's world. The more you know fractions, the more you can serve and bless others in the kitchen or workplace.

1. **Cooking and baking in the kitchen.** Get familiar with every measuring cup and measuring spoon in the kitchen. Be careful when you cook. You will only use a little bit of salt, baking powder, and spices in recipes. You will usually use quite a bit more of flour, sugar, and liquids.

Usually, we use teaspoons (smaller measuring spoons) to measure out salt and pepper.

Usually, we use tablespoons (bigger measuring spoons) to measure out butter or cooking oil.

Usually, we use cups to measure out flour, sugar, and liquids.

Play with the teaspoons and tablespoons.

a. How many teaspoons of water make up a whole tablespoon? ________

b. How many teaspoons of water make up ½ tablespoon? ________

c. How many tablespoons make up ¼ cup? ________

With your parent/teacher's supervision, follow a recipe and prepare food for the family. Bake or cook something today!

2. **Fixing things in the garage.** Drill bits, wrenches, and sockets are sometimes measured in inches, sometimes in millimeters. With your parent/teacher's supervision, look at the drill bits, wrenches, and sockets from the tool box. Use a ruler to check the size of each of these tools.

a. Make a list of some tool sizes you would need for some of the common bolts on your bicycle or car (such as the wheel bolts).

Now that you understand fractions, there are all kinds of projects you can help out with around the house!

3. **Tracking Progress.** You don't really pay attention to your height or weight in order to get taller or heavier. God does though! But sometimes, you want to work on growing in your abilities. Imagine that you want to learn to play basketball. For practice, you might shoot 20 shots and make 13 of them. That's 13 parts out of a whole 20. Knowing that fraction, you can make a goal for yourself: "By the end of the week I want to be able to make 15/20 shots." There's also a prayer in there: "God, would you help me grow to be able to make 15/20 shots?"

a. What kind of growth would you like to track?

b. How can you use fractions to do that?

When you're finished, spend some time in prayer. What thanks do you have to share with God? What would you like to ask him to help you with? Has this chapter helped you see any ways in which He is wonderful? Let Him know! He is always happy to hear from you.

Additional Activities

For extra activities, games, and helpful videos on these math concepts, refer to **generations.org/math3**.

This recipe for hot fudge sauce sounds delicious! Aren't you glad you are learning math so you can measure fractions in recipes?

CHAPTER 4

Fractions and Decimals

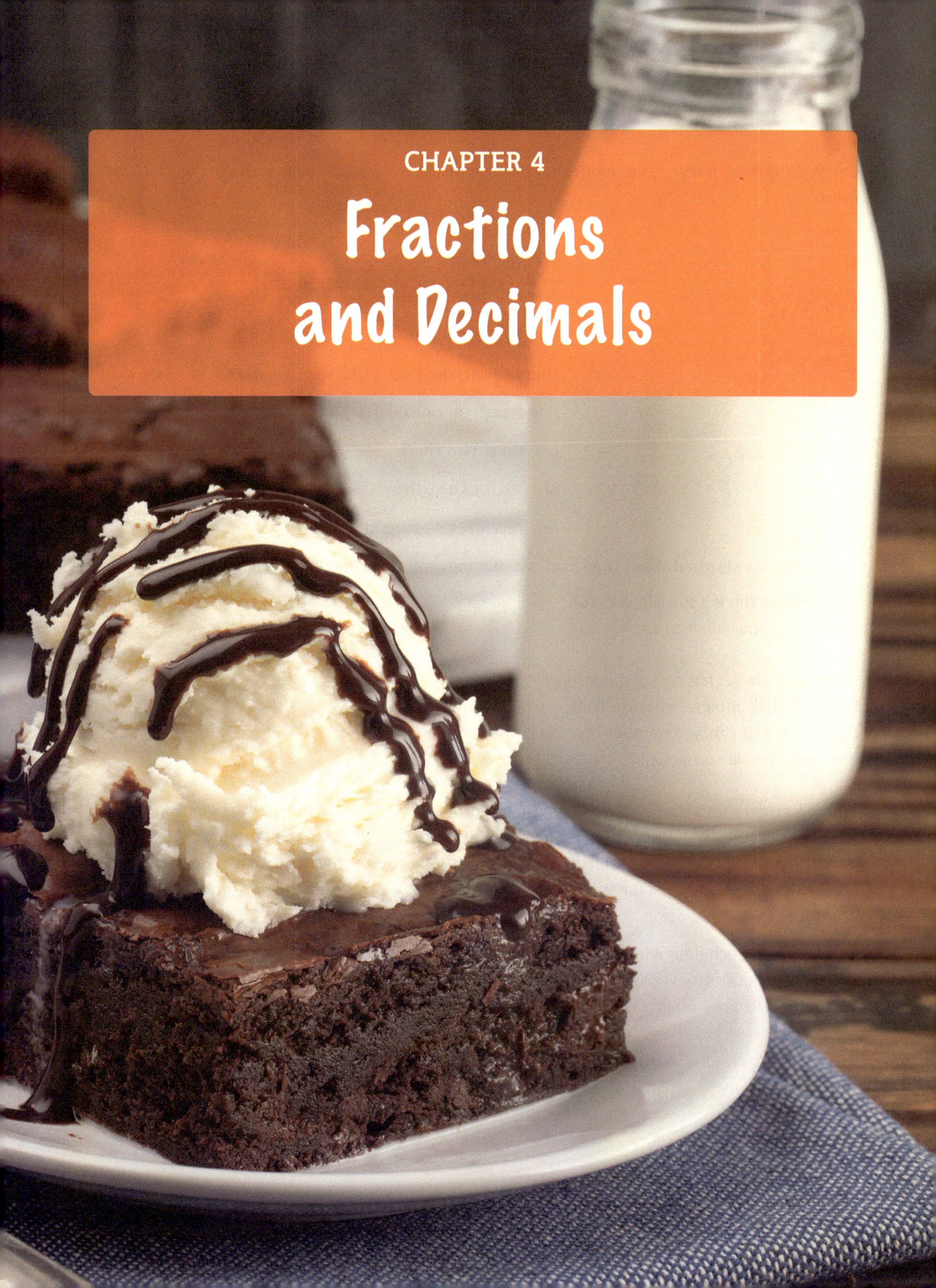

DAY 41 Cooking with Fractions

Can you say these fractions out loud? Draw lines to match the fraction pairs with their written-out form.

1. $\frac{1}{2}$, $\frac{2}{2}$	one-eighth, two-eighths
2. $\frac{1}{3}$, $\frac{2}{3}$	one-half, two-halves
3. $\frac{1}{4}$, $\frac{2}{4}$	one-fifth, two-fifths
4. $\frac{1}{5}$, $\frac{2}{5}$	thirty-seven over forty-eight ⇐ *(With some large denominators, it's easiest to say "over.")*
5. $\frac{1}{6}$, $\frac{2}{6}$	one-fourth (or one quarter), two-fourths
6. $\frac{1}{7}$, $\frac{2}{7}$	one-third, two-thirds
7. $\frac{1}{8}$, $\frac{2}{8}$	one-sixth, two-sixths
8. $\frac{37}{48}$	one-seventh, two-sevenths

So far, you have learned about large fractions of things and small fractions of things.

For example, this is a small fraction of something: one half of an apple.

$\frac{1}{2}$

And this is a larger fraction of something: two and a half apples. Two and a half ($2\frac{1}{2}$) is called a **mixed number** because it contains a whole number and a fraction.

$2\frac{1}{2}$

You've learned about fractions by thinking about fruits and vegetables. But, now let's look at fractions of liquids and dry goods used in baking. We measure things we can pour or scoop, like milk or flour, using teaspoons, tablespoons, and measuring cups. Sometimes we use amounts that are fractions of teaspoons, tablespoons, and cups.

Let's see how much flour is in this measuring cup. You can see that the flour comes up to the ½ cup line, so this is ½ of a cup of flour. Recipes would say "one-half cup flour" for short.

Liquids can be measured in several ways. In the United States, liquid measurements are made in teaspoons, tablespoons, cups, quarts, and gallons. Fluid ounces are also used in the U.S. Most other countries measure liquids in milliliters and liters.

Now let's look at two measuring cups which contain water. Let's see how much water is in both cups put together. We can write the amount this way:

$$1 + \frac{2}{3}$$

But we can also write the amount as a mixed number: 1 ⅔. We have just made this simple addition equation:

$$1 + \frac{2}{3} = 1\frac{2}{3}$$

Activity

Mixed numbers are part of God's world. We use mixed numbers often for cooking and doing science. Let's bake a simple recipe, that uses fractions. Ask your parent/teacher for help, and get ready to bake vanilla sugar cookies!

Another recipe can be substituted as the parent/teacher directs.

Step 1 — Preheat your oven to 375°F (190°C).

Step 2 — Measure the following 4 ingredients into a bowl and mix lightly.

2 ¾ cups all purpose flour 

1 teaspoon baking salt 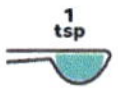

1 teaspoon baking soda 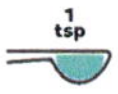

½ teaspoon baking powder 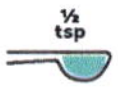

Step 3 — For the following ingredients:

1 cup butter, softened to room temperature 

1 ½ cups sugar 

1 teaspoon vanilla 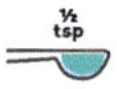

1 egg 

In a separate bowl, mix the butter and sugar until smooth. Add the egg and vanilla to the butter mixture and beat together.

Step 4 — Blend the dry mixture into the butter mixture.

Step 5 — Shape the dough into 1-inch balls and place them 2 inches apart on a greased cookie sheet. If the dough is too sticky to handle, you may use a spoon to scoop out similar amounts of dough. Scrape the dough off the spoon onto the cookie sheet with the back of another spoon.

Step 6 — Bake for 8-10 minutes until lightly browned.

Step 7 — After taking the cookies out of the oven, let them sit on the cookie sheet about two minutes. Then remove the cookies and place them on a rack to cool completely.

Let's do an outdoor activity! Find a large area in your yard or a park where you can imagine a big square. You can pick the 4 corners of the square using trees or bushes, or you can mark the corners yourself with sticks, or small rock piles. Besides having 4 corners, your square will have 4 equal sides.

Once you've marked your square, start at one corner and run to the right to the next corner. You will have run along 1 side which is $\frac{1}{4}$ of the way around your square.

Now run to the right again until you reach the next corner. You will have gone $\frac{2}{4}$ of the way around your square. Continue until you have run $\frac{3}{4}$ and then $\frac{4}{4}$ of the way. At $\frac{4}{4}$, you will have run the whole way around the square.

Thinking this way, travel around your square $\frac{3}{4}$ times. Then add to the $\frac{3}{4}$ by running until you reach $1\frac{3}{4}$ times. Then run until you reach $2\frac{3}{4}$ times. Do you notice a pattern? How does adding 1 more whole number to the mixed number change where you end up on the square?

DAY 42 Practice Using Small Numbers

Memory Drill

These memory drills may be completed either in writing or orally with a parent/teacher. Your goal is to complete this drill in 60 seconds.

1. 3 + 8 =	6. 9 + 9 =	11. 15 – 7 =
2. 2 + 9 =	7. 0 + 9 =	12. 14 – 3 =
3. 8 + 3 =	8. 7 + 7 =	13. 14 – 4 =
4. 5 + 5 =	9. 16 – 7 =	14. 17 – 8 =
5. 7 + 8 =	10. 17 – 9 =	15. 14 – 6 =

Can you find different ways to split these numbers apart? Watch for whether you need to use addition or multiplication.

1. 6 = ☐ + ☐	5. 12 = ☐ + ☐
2. 10 = ☐ + ☐	6. 15 = ☐ + ☐
3. 6 = ☐ x ☐	7. 12 = ☐ x ☐
4. 10 = ☐ x ☐	8. 15 = ☐ x ☐

Count the portions of the shapes that are shaded. Write the fractions. Some hints are given.

9.  $\frac{\quad}{\quad}$

10. 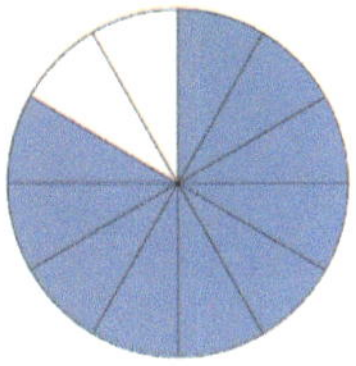 $\frac{\quad}{\quad}$

11. 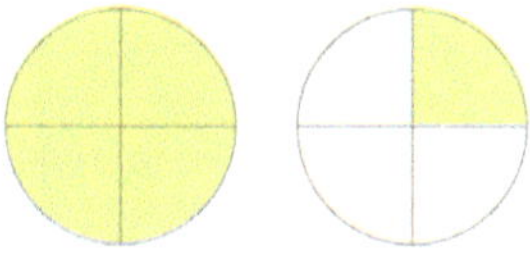 $1\frac{\quad}{\quad}$

12. 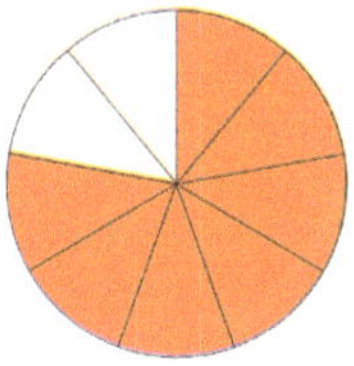

13. 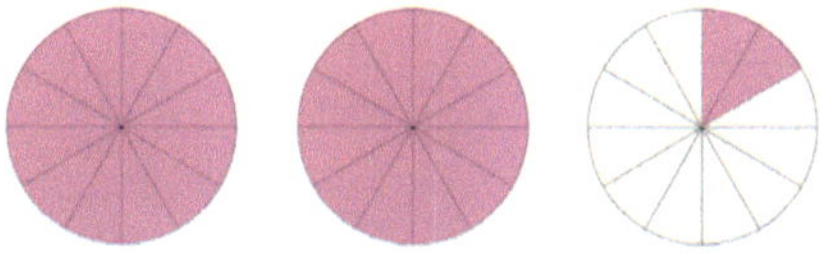 $2\frac{\quad}{12}$

14. 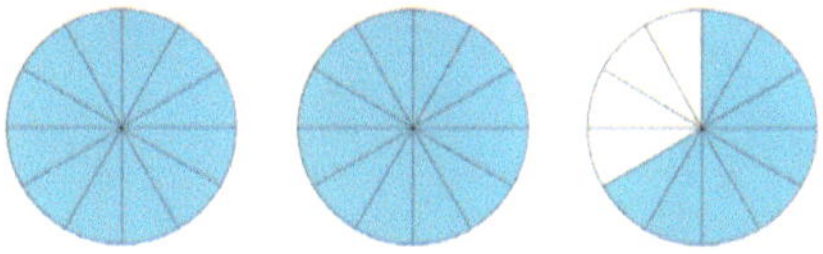 $\frac{\quad}{12}$

Let's practice measuring angles, lengths, and area. First, guess a measurement for each of the shapes. Then measure to see how close your guess was.

15. Your guess: _____ Your measurement: _____

16. Your guess: _____ Your measurement: _____

17. Your guess: ______ Your measurement: ______

18. Your guess: ______

Your measurement: ______

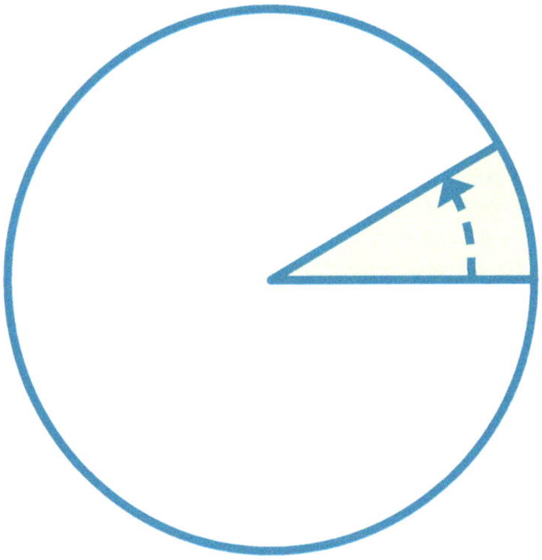

19. Your guess: ______

Your measurement: ______

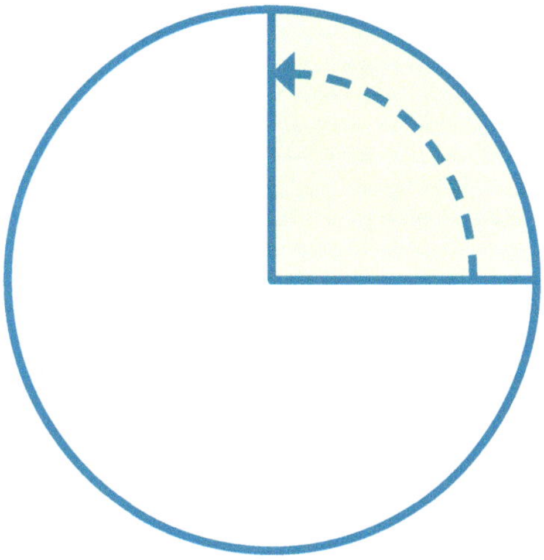

20. Your guess: ______

Your measurement: ______

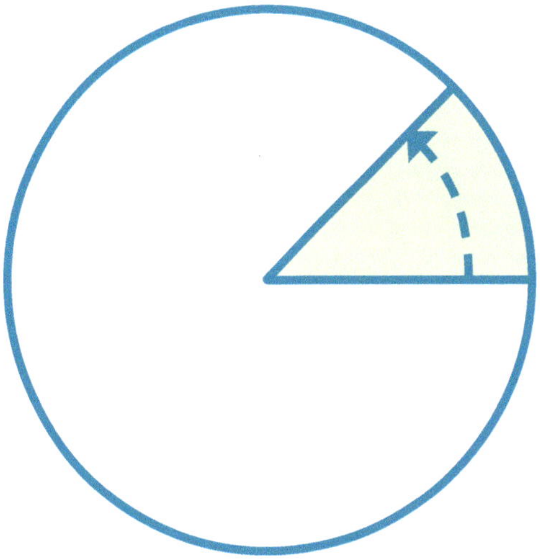

21. Guess the area of the rectangle: ________ sq. in.

Your measurement: ________ sq. in.

Practice Multiplication & Order

DAY 43

Here is some practice with multiplication. Try answering the first set of exercises from memory. You may use the multiplication chart in the back of the book to figure out the second set if you need to.

1. 8 × 2 = ☐
2. 3 × 3 = ☐
3. 3 × 4 = ☐
4. 5 × 4 = ☐
5. 2 × 6 = ☐
6. 1 × 12 = ☐
7. 7 × 3 = ☐
8. 10 × 1 = ☐

9. 4 × 8 = ☐
10. 8 × 8 = ☐
11. 5 × 4 = ☐
12. 7 × 9 = ☐
13. 7 × 6 = ☐
14. 3 × 6 = ☐
15. 9 × 7 = ☐
16. 10 × 6 = ☐

Put each of the following sets of numbers in order from least to greatest.

17. 1,000 750 1,500 ________, ________, ________

18. 1,250 1,340 1,520 ________, ________, ________

19. 17,005 16,020 15,900 ________, ________, ________

20. 451 541 145 ________, ________, ________

Do these addition and subtraction exercises. Some will require borrowing and carrying.

21.
$$\begin{array}{r} 38 \\ +\ 57 \\ \hline \end{array}$$

22.
$$\begin{array}{r} 47 \\ -\ 21 \\ \hline \end{array}$$

23.
$$\begin{array}{r} 235 \\ -\ 168 \\ \hline \end{array}$$

24.
$$\begin{array}{r} 57 \\ +\ 18 \\ \hline \end{array}$$

25.
$$\begin{array}{r} 58 \\ -\ 29 \\ \hline \end{array}$$

26.
$$\begin{array}{r} 147 \\ +\ 122 \\ \hline \end{array}$$

27.
$$\begin{array}{r} 159 \\ -\ 127 \\ \hline \end{array}$$

Building & Breaking Mixed Numbers DAY 44

The wise writer of Ecclesiastes says,

> To everything there is a season,
> A time for every purpose under heaven:
> A time to break down,
> And a time to build up. (Ecclesiastes 3:1,3)

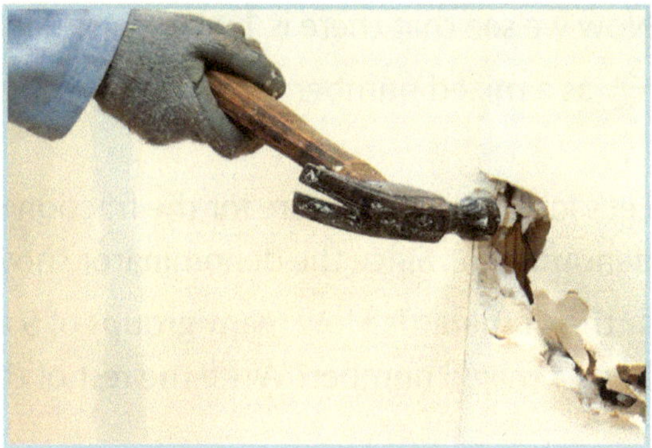

Today you will learn to build up mixed numbers, and then you will learn how to break them down. You will see that every fraction can be written as a mixed number, and every mixed number can be written as a fraction.

Check out the fraction, $\frac{3}{7}$. We can treat $\frac{3}{7}$ like a mixed number if we write it like this:

$$0\frac{3}{7}$$

Check out a fraction that's a little more interesting: $\frac{5}{3}$. Do you notice that the numerator of $\frac{5}{3}$ is larger than the denominator? When a fraction's numerator is larger than its denominator, it means there is at least 1 whole number hidden in the fraction. Think about it this way: The numerator in $\frac{5}{3}$ shows that there are 5 parts. But the denominator shows that you only need 3 parts to make a whole. That means that $\frac{5}{3}$ is equal to 1 whole with a little left over.

To help us understand this, let's turn $\frac{5}{3}$ into a neat little picture made up of dots.

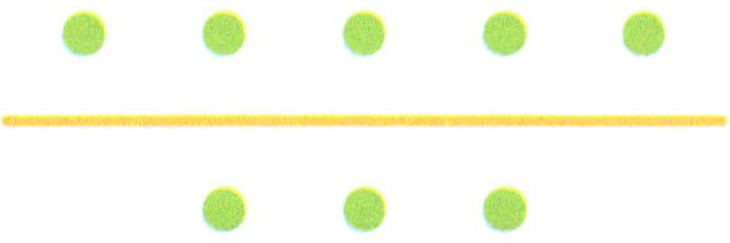

The 5 dots in the numerator are 5 parts. Since the denominator has 3 dots, we know that 3 parts make 1 whole. So we can circle 3 of the numerator's dots. This shows 1 whole because $\frac{3}{3}$ = 1.

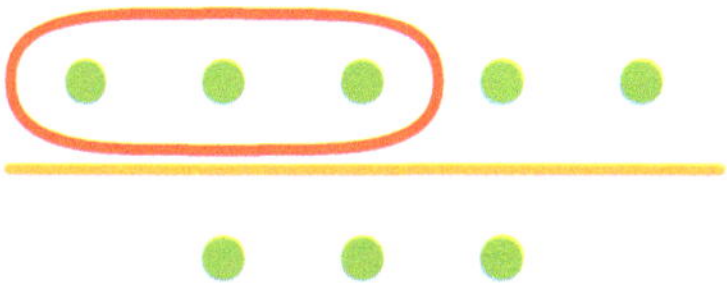

Now we see that there is 1 whole and 2 left-over dots. This helps us write $\frac{5}{3}$ as a mixed number:

$$\frac{5}{3} = 1\frac{2}{3}$$

Let's look at a dot picture for the fraction $\frac{12}{5}$. We'll draw 12 dots in the numerator and 5 dots in the denominator. Since the denominator shows that 5 parts make a whole, we'll circle groups of 5 dots in the numerator. How many groups of 5 are circled in the numerator? Can you tell what $\frac{12}{5}$ would be as a mixed number? Write the rest of the mixed number in the box.

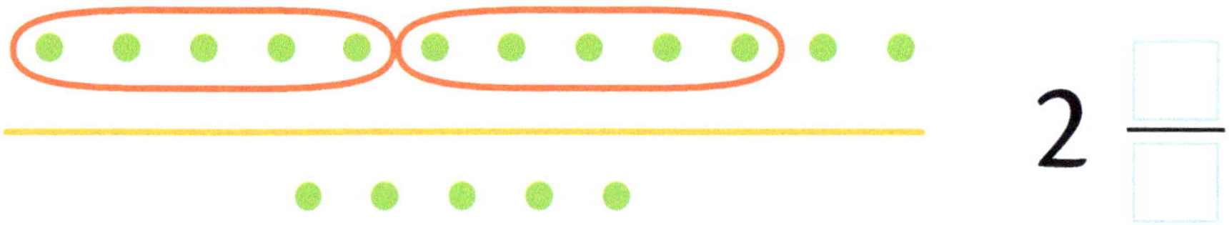

Now you know how to build mixed numbers from fractions whose numerators are larger than their denominators. Next let's learn how to break down mixed numbers.

Let's break down the mixed number $2\frac{1}{4}$ into a fraction. Again, we'll use dot pictures to help.

The first question to ask is, "How many parts make a whole?" In $2\frac{1}{4}$, the denominator of $\frac{1}{4}$ is 4. This means you will draw 4 dots in the denominator of your dot picture (below). Then, in your dot picture, you will draw sets of dots in the numerator until you have the same number of sets as your whole number. So, for $2\frac{1}{4}$ you will draw 2 sets (each with 4 dots) in the numerator. The last thing to draw in the numerator is the same number of dots that you see in the numerator of the small fraction from your mixed number. So, for $2\frac{1}{4}$, you see the 1 in the numerator of the $\frac{1}{4}$. This means you will draw one more dot in the numerator of your dot picture.

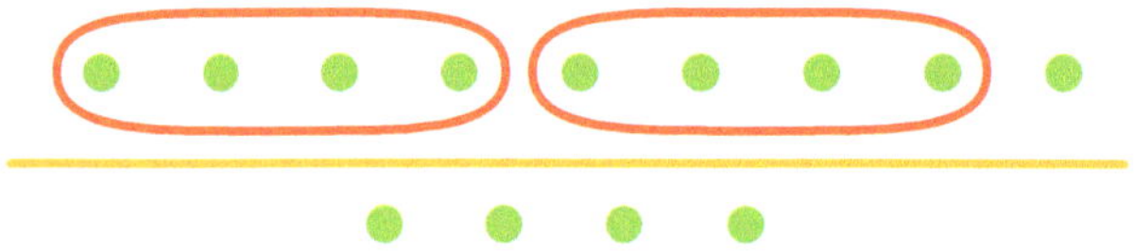

Now you can count the dots in the numerator and write your fraction: $\frac{9}{4}$.

$$2\frac{1}{4} = \frac{9}{4}$$

Now let's try breaking down some mixed numbers on your own, remembering that, with God, you are never alone. He can help you.

Student Exercises

Draw dot pictures to change the fractions below into mixed numbers numbers (1-6) and mixed numbers into fractions (7-12). Think about how many dots you will draw in each denominator.

		DOT PICTURES	MIXED NUMBERS/FRACTIONS
1.	$\frac{7}{2}$		
2.	$\frac{11}{2}$		
3.	$\frac{17}{3}$		
4.	$\frac{4}{3}$		
5.	$\frac{6}{4}$		
6.	$\frac{14}{10}$		
7.	$1\frac{1}{2}$		
8.	$5\frac{1}{2}$		
9.	$2\frac{2}{3}$		
10.	$1\frac{2}{3}$		
11.	$3\frac{1}{4}$		
12.	$1\frac{3}{4}$		

Practice counting by chunks. For each list of numbers, figure out if the list is counting up or counting down. Then, fill in the blanks to complete the pattern.

13. Count by 2s — 30, 32, ______, ______, ______, ______

14. Count by 3s: — 24, 21, ______, ______, ______, ______

15. Count by 5s — 50, 55, ______, ______, ______, ______

16. Count by 8s: — 16, 24, ______, ______, ______, ______

17. Count by 10s: — 150, 140, ______, ______, ______, ______

Draw hands on these clocks to show the time.

Start eating lunch at 12:00:

30 minutes after starting lunch:

18.

Start eating dinner at 6:00:

19.

45 minutes after starting dinner:

20.

21. What time do you begin your bedtime prayers? Draw the hands on the clock.

Rejoice always, pray without ceasing, in everything give thanks; for this is the will of God in Christ Jesus for you. (1 Thessalonians 5:16-18)

Finish the lists of shapes. Find each pattern, and then draw the next picture that keeps that pattern going.

22. 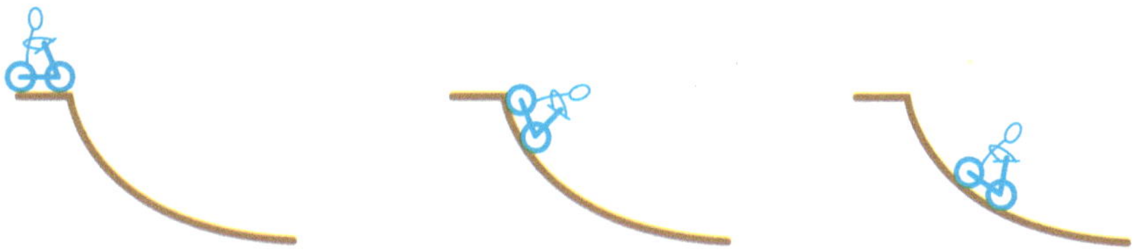

23. 

24. 

25. 

26. 

27. 

DAY 45 Practice

Numbers can be split apart with addition and multiplication. Can you find different ways to split these numbers apart? Watch for whether you need to use addition or multiplication.

1. 4 = ☐ + ☐
2. 4 = ☐ x ☐
3. 9 = ☐ + ☐
4. 9 = ☐ x ☐
5. 12 = ☐ + ☐
6. 12 = ☐ + ☐
7. 12 = ☐ x ☐
8. 12 = ☐ x ☐

Let's add and subtract more fractions. Remember, when we add and subtract fractions, the denominator stays the same. First add and write the answer in the first box. Then, draw a dot picture of the fraction in the second box. Circle sets of dots in the numerators and write the mixed numbers in the far right box.

		FRACTION	DOTS HERE	MIXED NUMBERS
9.	$\frac{10}{4} + \frac{3}{4} =$			
10.	$\frac{2}{7} + \frac{9}{7} =$			
11.	$\frac{3}{10} + \frac{9}{10} =$			

Put each set of numbers in order from least to greatest.

12.	555	5555	55	_____, _____, _____
13.	123	132	321	_____, _____, _____
14.	1,000	100	10,000	_____, _____, _____
15.	1,240	1,140	1,040	_____, _____, _____

Everyday Math

Let's say you have decided to go into the cookie business. So far you have baked cookies and sold them door-to-door on four different streets for two Saturdays. You have kept track of the number of cookies you sold. Here is the data you collected.

Cookies Sold

	Saturday 1	Saturday 2
Street 1	22	10
Street 2	10	16
Street 3	5	12
Street 4	28	18

16. Which street produced the highest sales of cookies? Add them up!

Street 1

22 + 10 = _______ cookies

Street 2

___ + ___ = _____ cookies

Street 3

___ + ___ = _____ cookies

Street 4

___ + ___ = _____ cookies

17. Which Saturday produced the highest sale of cookies? Add them up!

Saturday 1

___ + ___ + ___ + ___ = _______ cookies

Saturday 2

___ + ___ + ___ + ___ = _______ cookies

18. How many more cookies were sold on the Saturday with the most sales compared to the other Saturday? Subtract!

___ – ___ = _______ cookies

DAY 46 Sharing Equally

> Whoever has a bountiful eye will be blessed, for he shares his bread with the poor. (Proverbs 22:9 ESV)

God wants us to share our things. If you have 5 families in need in your church, and God has blessed you with a little extra, you may wish to share what you have. Let's think about how you might share some apples equally with the 5 families.

First, let's learn a little more about fractions. You have already learned something you can do with fractions whose numerators are larger than their denominators. You learned how to change those fractions into mixed numbers. But sometimes fractions with large numerators and smaller denominators become whole numbers instead of mixed numbers. Let's look at the fraction below:

$$\frac{12}{4} =$$

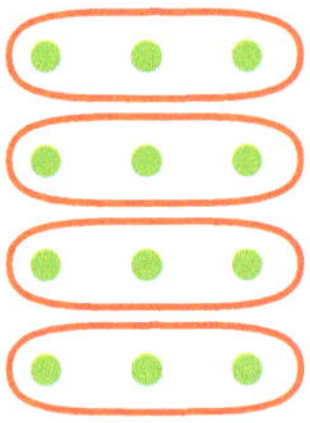

$$= 3\frac{0}{4} = 3$$

Now suppose that you wanted to share 20 kilograms of apples equally with the 5 families? Let's break these 20 kilograms into 5 groups using dots.

$$\frac{20}{5} = 20 \div 5$$

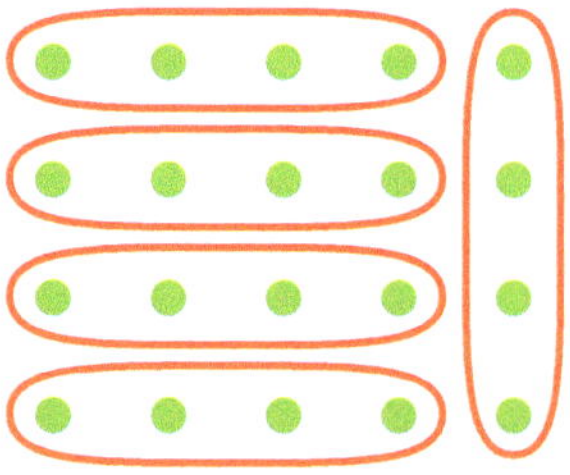

That's easy. Each of the 5 families will receive 4 kilograms of apples.

Now, let's make this a little more challenging. Suppose you have a 22 kilogram sack of apples to share with the 5 families.

$$\frac{22}{5} = 22 \div 5$$

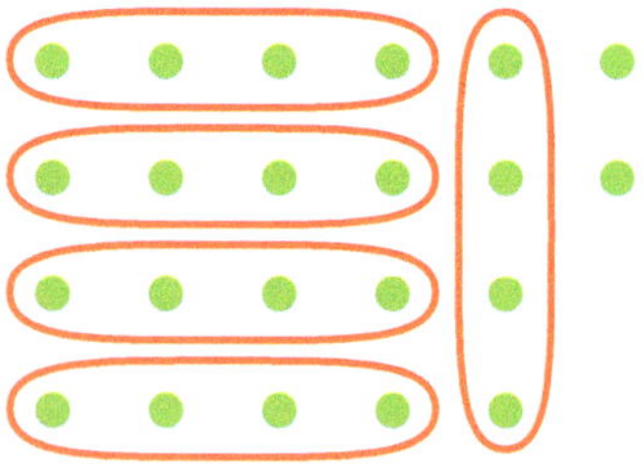

Our dots show that each family will receive 4 kilograms plus a little more. To figure out exactly how much each family receives, let's turn $\frac{22}{5}$ into a mixed number.

$$\frac{22}{5} = 4\frac{2}{5}$$

You can tell each family they will receive $4\frac{2}{5}$ kilograms of apples. $4\frac{2}{5}$ sounds less confusing than $\frac{22}{5}$. Mixed numbers help us understand amounts better than if they were left as fractions. Fractions like $\frac{22}{5}$, $\frac{35}{6}$, and $\frac{28}{3}$ are hard to understand. Changing these fractions into mixed numbers gives us a better idea of how God made them to fit into His ordered world. This helps you share your extra amounts with others and love your neighbor.

Fill in the following table with your own dots to break up the fraction into a whole or mixed number.

	FRACTION	DOTS HERE	WHOLE OR MIXED NUMBER
1.	$\frac{10}{3}$	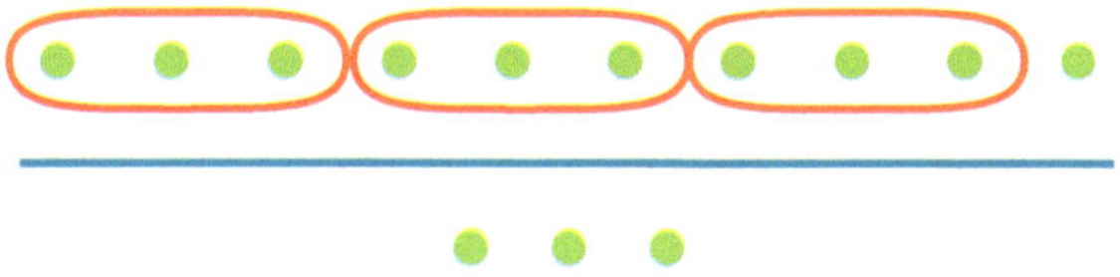	3 1/3
2.	$\frac{10}{2}$		
3.	$\frac{10}{4}$		
4.	$\frac{22}{8}$		
5.	$\frac{24}{6}$		
6.	$\frac{7}{1}$		
7.	$\frac{7}{5}$		

Memory Drill

These memory drills may be completed either in writing or orally with a parent/teacher. Your goal is to complete this drill in 60 seconds.

1. 5 + 6 =	6. 8 + 5 =	11. 18 – 6 =
2. 7 + 8 =	7. 9 + 6 =	12. 14 – 5 =
3. 7 + 5 =	8. 9 + 8 =	13. 16 – 4 =
4. 6 + 7 =	9. 18 – 8 =	14. 17 – 6 =
5. 6 + 8 =	10. 14 – 6 =	15. 16 – 9 =

Optional Exercise

Now that you've learned something about fractions and mixed numbers, it's time to learn more about measuring.

Fractions and mixed numbers play a big part in the way you've already learned to measure in God's world.

You already know how to measure amounts of things you can scoop or pour. For example, a recipe might call for $1\frac{1}{4}$ cups of sugar. No problem! You know exactly what that means.

When you measure lengths of things, fractions and mixed numbers are just as important, but they aren't always so easy.

Today you'll need a ruler and a tape measure.

How many inches are in a foot? If you don't know, check your ruler or tape measure to find out.

On your tape measure, find 3 feet and 4 inches. Write 3 feet, 4 inches as a mixed number. (Hint: You will use the number you wrote above as a denominator).

Now let's zoom in. Imagine the width of your eye's pupil, a ladybug, or a blade of grass. You would need to use something smaller than an inch. Let's break down an inch.

Look at a single inch on your ruler. Do you see the biggest line that is right in the middle of that inch? That's the halfway mark. It marks $\frac{1}{2}$ inch. This means that two of those $\frac{1}{2}$-inch spaces make 1 whole inch. Can you see those two $\frac{1}{2}$-inch spaces within 1 inch?

Now look even closer, in the middle of each $\frac{1}{2}$ inch. You'll see two more lines that are a little shorter than the $\frac{1}{2}$-inch lines, but still longer than the rest of the little lines within each inch. These are the $\frac{1}{4}$-inch lines. Four of these $\frac{1}{4}$-inch lengths make up 1 whole inch. Did you notice that the $\frac{1}{2}$ -inch mark is the same as the $\frac{2}{4}$ -inch mark? That's because $\frac{2}{4}$ is the same as $\frac{1}{2}$. This may not make sense now, but you will learn about it another day.

Now, the next smallest lines are $\frac{1}{8}$-inch marks. $\frac{1}{8}$-inch marks are used often in woodworking and other construction arts. Can you tell your parent/teacher how many $\frac{1}{8}$-inch lengths would be in 1 inch? Can you find $\frac{3}{8}$ of an inch? How about $\frac{7}{8}$? If you understood the $\frac{1}{2}$ inches and the $\frac{1}{4}$ inches, you should be able to figure out the $\frac{1}{8}$ inches!

You are learning much about working in God's world. Numbers are everywhere, and the more you understand them, the more you will be able to use numbers to help others.

Now let's learn about centimeters. When we broke down an inch, we broke it into fractions like $\frac{1}{2}$, $\frac{1}{4}$, and $\frac{1}{8}$. We can break down centimeters into fractions too, but centimeters are always broken into 10 pieces, not 2, 4, and 8 pieces. Each piece is $\frac{1}{10}$ of a centimeter. $\frac{1}{10}$ of a centimeter is called a **millimeter**. Look at your centimeter ruler. Do you see the millimeter marks between the centimeter marks? Count to make sure there are 10 spaces between each centimeter mark. See if you can find a 4-millimeter length and a 7-millimeter length.

Try finding a 35-millimeter length. You know that the number 35 is 3 tens and 5 ones, but now let's see this in a new way. 35 millimeters is 3 centimeters and 5 millimeters. We can write it as the mixed number, $3\frac{5}{10}$ centimeters. Now you have a lot of fun ways to talk about the lengths in God's world.

Activity

Go measure some lengths and practice your new understanding of fractions and mixed numbers.

- Find and measure a few things whose lengths are less than an inch.
- Find and measure a few things whose lengths are between 1 foot and 5 feet. Use mixed numbers to talk about how many feet long they are.
- Find and measure a few things whose lengths are between 1 and 20 centimeters. Record the lengths with mixed numbers.

DAY 47 Practice

Here is some practice with division. Try answering these exercises from memory. You may use the multiplication chart in the back of this book if you need to.

1. 10 ÷ 2 = ☐
2. 14 ÷ 2 = ☐
3. 6 ÷ 3 = ☐
4. 15 ÷ 3 = ☐
5. 12 ÷ 4 = ☐
6. 4 ÷ 4 = ☐
7. 15 ÷ 5 = ☐
8. 20 ÷ 5 = ☐
9. 30 ÷ 10 = ☐
10. 45 ÷ 5 = ☐
11. 32 ÷ 8 = ☐
12. 49 ÷ 7 = ☐
13. 36 ÷ 6 = ☐
14. 56 ÷ 8 = ☐
15. 36 ÷ 4 = ☐
16. 60 ÷ 6 = ☐

Practice counting by chunks. For each list of numbers, figure out if the list is counting up or counting down. Then, fill in the blanks to complete the pattern.

17. Count by 2s — 50, 48, ______, ______, ______, ______
18. Count by 4s: — 0, 4, ______, ______, ______, ______
19. Count by 4s — 20, 24, ______, ______, ______, ______
20. Count by 6s: — 6, 12, ______, ______, ______, ______
21. Count by 6s: — 36, 30, ______, ______, ______, ______

Fill in the following table with your own dots to break up the fraction into a whole or mixed number.

	FRACTION	DOTS HERE	WHOLE OR MIXED NUMBER
22.	$\frac{7}{4}$		
23.	$\frac{10}{3}$		
24.	$\frac{18}{7}$		
25.	$\frac{21}{3}$		
26.	$\frac{5}{2}$		
27.	$\frac{12}{5}$		
28.	$\frac{8}{1}$		
29.	$\frac{9}{3}$		

DAY 48 Money and Fractions

> Command those who are rich in this present age not to be haughty, nor to trust in uncertain riches but in the living God, who gives us richly all things to enjoy. (1 Timothy 6:17)

Money can be a blessing. Money can solve a few of our problems. We use money to buy food. We use money to pay for medicine. God gives us money to help us through life. But we must be careful not to trust in money. Instead, we must trust in God, who is the source of every good gift for us. These are the lessons from these verses.

Today, we will learn more about money.

Many countries use fractions when counting money. This includes America (the dollar), England (the pound), Europe (the Euro), and Canada (the Canadian dollar).

Let's look at fractions of the American dollar. This picture shows 1 dollar as a paper "dollar bill." It also shows a coin we call a half dollar. A half-dollar coin is worth 50 cents because 50 is one half of 100.

$1\frac{1}{2}$ dollars = \$$1\frac{1}{2}$ = \$1.50

Notice that we can write amounts of money as mixed numbers. We can also write amounts of money as decimals. In \$1.50, the dot that looks like a period is called a decimal point. \$1.50 is called a decimal because it has a decimal point. The decimal point is important when we write money because it separates the dollars and the fractions of dollars which we call cents.

The next picture shows $1\frac{1}{4}$ dollars. The coin you see is worth $\frac{1}{4}$ of a dollar. It's worth 25 cents because a dollar is 100 cents. We call this coin a quarter because quarter is another way to say one-fourth. ⇨

$1\frac{1}{4}$ dollars = \$$1\frac{1}{4}$ = \$1.25

Do you see the mixed number and the decimal in the different ways to write $1\frac{1}{4}$ dollars?

There are other coins that are worth smaller fractions of the dollar. For these coins, we don't usually say what fraction they are of a dollar like we did with the half dollar and the quarter. Instead, they have different sorts of names and we think of them according to how many cents they are worth. Look at each of these coins and see how we write their value when they are with 1 dollar.

= $1.10 ⇦ **A dollar and a dime**
= 1 dollar and 10 cents = $1.10

= $1.05 ⇦ **A dollar and a nickle**
= 1 dollar and 5 cents = $1.05

= $1.01 ⇦ **A dollar and a penny**
= 1 dollar and 1 cent = $1.01

Now let's add up some American money, including quarters and dimes. Remember, a dime is worth 10 cents and a quarter is worth 25 cents. The decimal point should always line up when you add money.

$1.00
.25
.25
.25
+ .10
$1.85

You can count money in your head if you can count by 10s and 25s. The best way to count money in your head is by starting with the big dollar bills. To count the money pictured here, we say: One dollar, one dollar twenty-five, one dollar fifty, one dollar seventy-five, and one dollar eighty-five.

$1.00
.25
.25
.10
.10
+ .10

Try this one on your own:

Now, try adding money together on paper instead of by counting.

$2.11
+ $1.50

Go ahead and add this up yourself!

Student Exercises

Practice adding money with the following exercises:

1. $3.75 + $1.21

2. $4.82 + $3.17

3. $4.85 + $1.11

4. $7.76 + $2.22

5. $3.29 + $1.13

6. $8.28 + $1.62

Let's practice measuring angles, lengths, and areas. First guess a measurement for each of the shapes. Then measure to see how close your guess was.

7. Guess the length of this line in inches: ______ Measure the length of the line: ______

8. Guess the length of this line in inches: ______ Measure the length of the line: ______

9. Guess the size of this angle: _____

Measure the size of the angle: _____

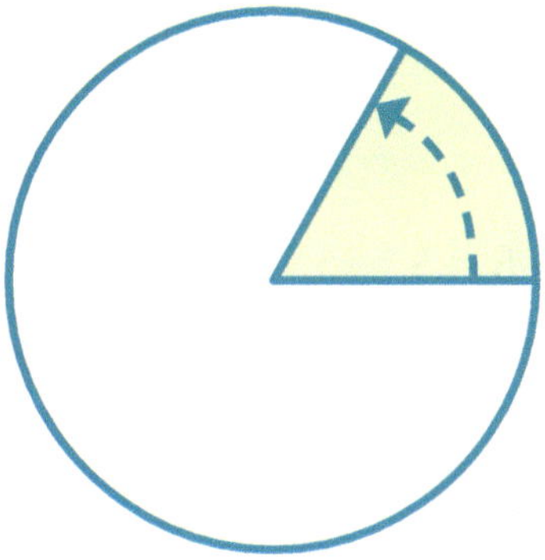

10. Guess the size of this angle: _____

Measure the size of the angle: _____

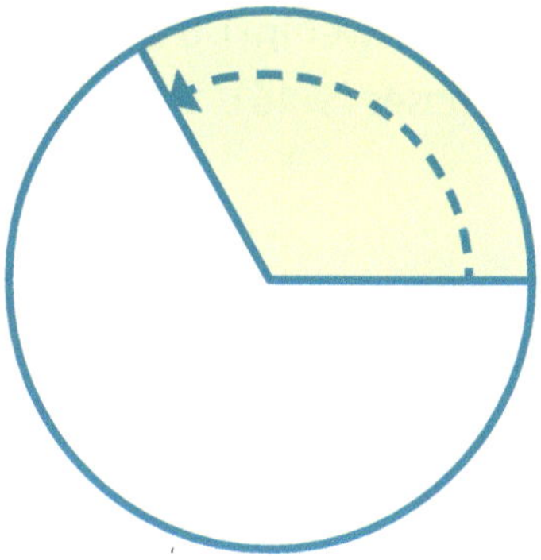

11. Guess the area of this rectangle in square inches: _____

Measure the rectangle's area: _____

12. Guess the area of this rectangle in square inches: _____

Measure the rectangle's area: _____

DAY 49 Practice

If an answer in 1-8 is wrong, circle it and write the correct answer in the blank. Then do the division exercises (9-16).

1. 1 x 8 = 1 ☐
2. 3 x 9 = 12 ☐
3. 9 x 5 = 45 ☐
4. 10 x 4 = 14 ☐
5. 6 x 0 = 6 ☐
6. 16 ÷ 4 = 12 ☐
7. 25 ÷ 5 = 5 ☐
8. 10 ÷ 5 = 50 ☐
9. 30 ÷ 10 = ☐
10. 45 ÷ 5 = ☐
11. 32 ÷ 8 = ☐
12. 49 ÷ 7 = ☐
13. 36 ÷ 6 = ☐
14. 56 ÷ 8 = ☐
15. 36 ÷ 4 = ☐
16. 48 ÷ 4 = ☐

Find the right number to complete these math equations.

17. What plus 4 equals 13? ________ + 4 = 13
18. What plus 10 equals 13? ________ + 10 = 13
19. What minus 8 equals 8? ________ − 8 = 8
20. What minus 7 equals 8? ________ − 7 = 8
21. What times 2 equals 10? ________ × 2 = 10
22. What times 4 equals 8? ________ × 4 = 8
23. What divided by 2 equals 10? ________ ÷ 2 = 10
24. What divided by 5 equals 4? ________ ÷ 5 = 4

Adding Larger Amounts of Money DAY 50

> But those who desire to be rich fall into temptation and a snare, and into many foolish and harmful lusts which drown men in destruction and perdition. For the love of money is a root of all kinds of evil, for which some have strayed from the faith in their greediness, and pierced themselves through with many sorrows. (1 Timothy 6:9–10)

Money can become too big of a deal. We may come to love money too much, and desire to be rich. We must be careful that money does not turn into an idol for us. God gives us good gifts, but we must always worship God and not the gifts. We should thank God for His good gifts. The more gifts and money we receive, the more we should be willing to share with others.

Every country uses its own money. Here are several examples of what money looks like from different nations around the world.

Money From Around the World

Canadian Money

Brazilian Money

South African Money

British Money

Australian Money

Mexican Money

Today, we're going to look at larger amounts of American money and practice adding it up.

We often need to use more than a single 1-dollar bill at a time. But instead of using a lot of 1-dollar bills to buy groceries, we can use fewer bills that are each worth more than 1 dollar. You can see how many dollars each bill is worth by looking at the number in its corners. Find the corner numbers in the pictures of the bills below.

Let's add some larger American bills together now. If we add up \$20, \$10, \$5, and \$1, and how much money do we have all together? We have \$36.

$$\begin{array}{r} \$20 \\ \$10 \\ \$\ 5 \\ +\ \$\ 1 \\ \hline \$36 \end{array}$$

How much money is pictured in each of the sets below? You can add them up by long addition or you can count them up in your head. Start with the large bills first. Add in the smaller bills and then the coins. The first example is filled in for you.

1.	\$62.50	
2.		
3.		
4.		

Let's add this money up. The dollar sign stands for American money. Be sure to keep the decimal point in the right place.

5. \$1.50 + \$1.25	7. \$4.50 + \$2.00	9. \$3.25 + \$1.54	11. \$30.75 +\$60.00
6. \$3.75 + \$1.21	8. \$6.85 + \$3.14	10. \$10.50 +\$20.25	12. \$20.85 +\$70.14

Let's explore God's patterns with sets. How do these sets come together to create new sets?

13. {1, 3, 5} U {2, 4, 6} =
{ }

14. {2, 3, 4} U {3, 4, 5} =
{ }

15. {3, 5, 7, 11} U {3, 5, 7, 11} =
{ }

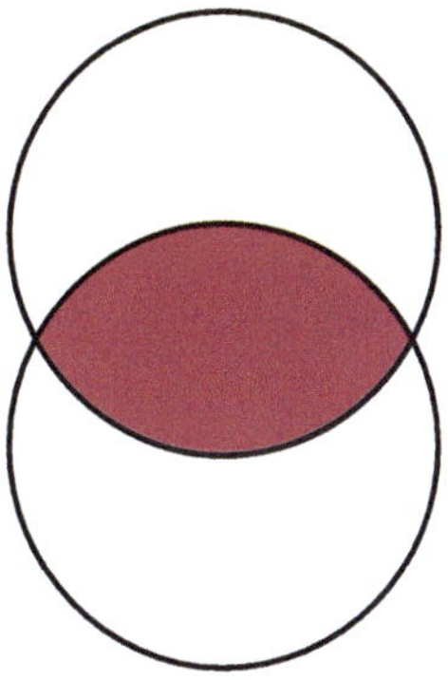

16. {7, 8, 9} ∩ {9, 10, 11} =
{ }

17. {2, 4, 6} ∩ {8, 10, 12} =
{ }

18. {4, 5} ∩ {4, 5} =
{ }

DAY 51 Practice

1. 15 + 8 =	5. 13 − 6 =	9. 6 x 10 =	13. 48 ÷ 6 =
2. 4 + 12 =	6. 14 − 9 =	10. 5 x 12 =	14. 56 ÷ 7 =
3. 14 + 3 =	7. 20 − 9 =	11. 3 x 8 =	15. 36 ÷ 3 =
4. 5 + 7 =	8. 11 − 2 =	12. 4 x 7 =	16. 6 ÷ 6 =

17. $\begin{array}{r} \$5.50 \\ +\ \$4.50 \\ \hline \end{array}$

18. $\begin{array}{r} \$9.35 \\ -\ \$2.75 \\ \hline \end{array}$

19. $\begin{array}{r} \$8.20 \\ +\ \$1.65 \\ \hline \end{array}$

20. $\begin{array}{r} \$10.25 \\ -\ \$7.05 \\ \hline \end{array}$

Everyday Math

Solve these exercises.

21. a. The camp chair you would like to buy your dad for his birthday costs $20.00. You have $8.50 already. Your grandfather will pay you $15.00 to wash his car. After the car is washed, will you be able to buy the camp chair?

b. How much money is either left over or still needed for the camp chair?

22. Mary received $25.00 as a gift. She gave a $2.50 tithe to God at church. How much money does she have left to save or spend?

23. Mom is excited to be having a new baby. She buys a new bottle for the baby for $4.95 and a onesie for $9.99. How much does Mom spend all together for the baby items?

24. You are earning money to send to a missionary that your church supports. You raked the yard for $4.50, folded the laundry for $1.25, and ran a lemonade stand that made $7.25. How much money will you be able to give the missionary?

DAY 52

Go Earn, Buy, Save, and Serve with Money

This lesson integrates math into everyday life. With this activity, the student will apply God's patterns and wisdom in meaningful ways to home and community. Take a break from memory work and academic exercises to make math part of everyday life. The following are suggestions or examples. Other ideas may be added to the list.

Money has a lot of uses! We can earn money by working. We can save money. We can buy things with our money. And we can share our money with others. Here are some practical things you can do with money today!

1. **Plan ahead for a future purchase.** Set aside some of your money every month to save for something you need or something you want to buy for someone else. Plan out how much money you can save each month and add it up here.

2. **Make a shopping list and plan your next shopping trip.** Plan a trip to the store to buy a few things you need. First write a list of those things and mark down how much you think each item will cost. Add up the costs and compare with the amount of money you can spend today. Can you make the trip, or must you wait until you have saved more money?

3. **Buy sale items and save money.** Buy a few items that are on sale. Compare the amount you spend to the amount you would have spent if you had paid regular price. How much did you save?

Additional Activities

For extra activities, games, and helpful videos on these math concepts, refer to generations.org/math3.

God's fruits are made to easily divide into equal portions for our delight!

CHAPTER 5

More Decimals

DAY 53 Equal Fractions

When you learned about money, you found out that 1½ dollars should be written like this: \$1.50. But when we talk about other fractions, we can use the mixed number 1½ as well as the number 1.5. We can ignore the zero. The zero at the end of a decimal does not change its value. Let's write these numbers in math equations:

$$1\tfrac{1}{2} = 1.50 = 1.5$$

$$2\tfrac{1}{2} = 2.50 = 2.5$$

$$\tfrac{1}{2} = .50 = .5$$

You remember that a fraction is part of a whole. A fraction might be 1 half of an apple or 1 quarter (1 fourth) of an orange. Let's learn more about how fractions can be written as decimals.

Suppose you eat 1 piece of this pizza pie. What is the fraction you have eaten? You've eaten 1 out of 2 pieces or one-half of the pizza.

1. $\frac{1}{2}$

Now suppose you eat 3 pieces of this pizza. What is the fraction you have eaten? You've eaten 3 out of the 6 pieces. Finish writing the fraction.

2. $\frac{3}{\ \ }$

And now suppose you have eaten four pieces of this pizza. What is the fraction you have eaten? You've eaten 4 out of the 8 pieces. Finish writing the fraction.

3. $\frac{4}{\ \ }$

In each case, you have eaten $\frac{1}{2}$ of the pizza pie. So now you can see that one-half ($\frac{1}{2}$) is the same as three-sixths ($\frac{3}{6}$) which is the same as four-eighths ($\frac{4}{8}$).

$$\frac{1}{2} = \frac{3}{6} = \frac{4}{8}$$

Suppose the pizza was cut into 10 pieces, and you ate 5 pieces. Once again, you have eaten half the pizza. What if the pizza was cut into 4 pieces, and you ate 2 of the pieces? Again, you have eaten half the pizza. That's a lot of pizza! We can write an equation for this.

$$\frac{1}{2} = \frac{2}{4} = \frac{3}{6} = \frac{4}{8} = \frac{5}{10}$$

And remember, the way to write ½ as a decimal is:

.5 or 0.5 (said as "five-tenths")

We often write a zero in front of decimals that have no whole number. This helps draw our attention to the decimal point so we don't miss it.

Now let's learn more about decimals using these two circles. Each of the circles are broken into 10 pieces. Let's figure out the total amount of the circles that are colored purple.

One full circle is colored purple, and 2 of the 10 slices in the other circle are colored purple. We can write this amount two different ways—as a fraction and as a decimal.

$$1\frac{2}{10} = 1.2$$

Student Exercises

What total amount of the circles below are colored? Fill in the fractions and the decimals.

	PICTURES	FRACTIONS	DECIMALS
1.	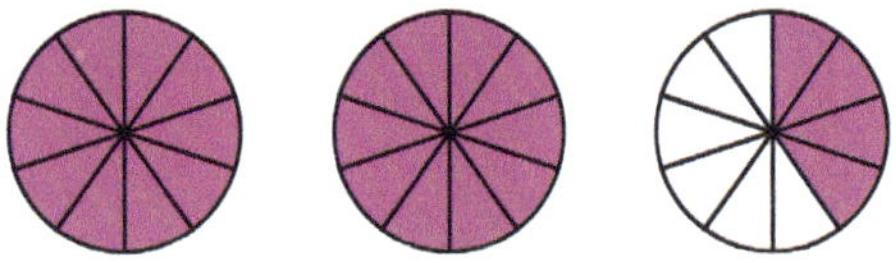	$2\frac{4}{10}$	2.4
2.	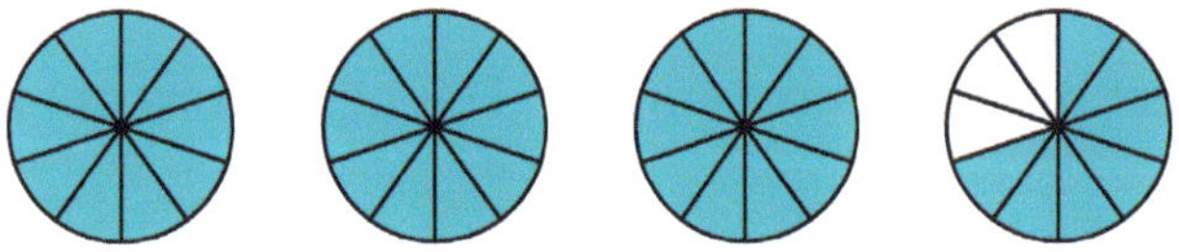		
3.	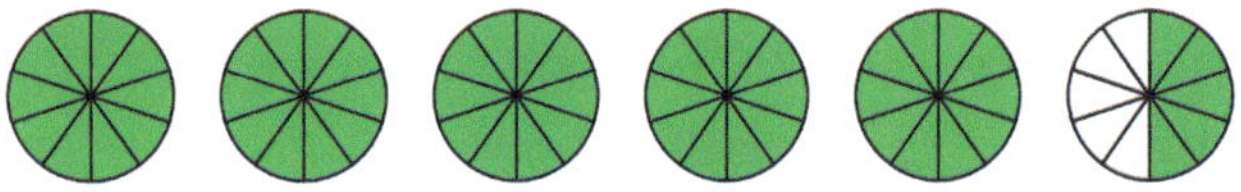		
4.	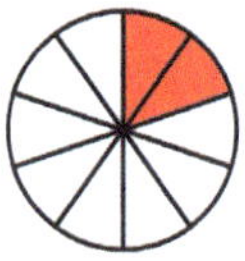		

Did you notice that the number after the decimal point always tells how many tenths we are talking about? There are two ways to say the decimal in number 1 above:

"2 and 4 tenths" or "2 point 4"

Write the decimal for the following.

1. 1 and 1 tenth ______
2. 9 point 2 ______
3. 4 point 7 ______
4. 5 and 5 tenths ______
5. 4 and 7 tenths ______
6. 10 point 3 ______

Why is the fraction $\frac{1}{2}$ written as 0.5 in decimal form?

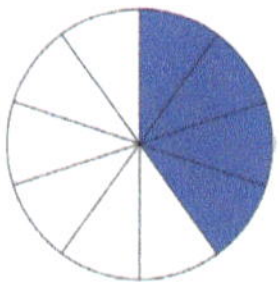

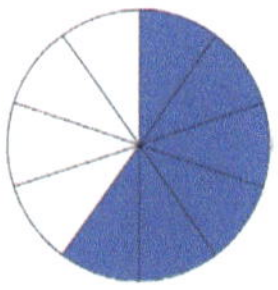

Remember that 0.5 is "5 tenths." That is because the first number after the decimal point tells us how many tenths of a whole number we have. Do you think "4 tenths" could equal $\frac{1}{2}$? No, $\frac{4}{10}$ is less than $\frac{1}{2}$. Could we say $\frac{6}{10}$ is $\frac{1}{2}$? No, $\frac{6}{10}$ is more than $\frac{1}{2}$. Because $\frac{5}{10} = \frac{1}{2}$, then we can write 0.5 for $\frac{1}{2}$.

Activity

Let's talk about fractions as a part of real life. You might say, "I ate 4 slices of an orange. Assuming there were 10 slices all together, I ate four-tenths of the orange. As a decimal, that's 0.4."

Now create three different stories yourself about dividing a group in half. For each story, make the group a different size, and create a fraction that represents half the group. For example, you could use $\frac{2}{4}$, $\frac{3}{6}$, and $\frac{5}{10}$.

1.

2.

3.

DAY 54 Practice

Let's explore God's patterns with sets. Let's see how these sets come together to create new sets.

1. $\{4, 5, 6\} \cup \{5, 6, 7\} =$
$\{ \quad \}$

2. $\{x, y\} \cup \{w, z\} =$
$\{ \quad \}$

3. $\{1, 2, 4\} \cup \{1, 3, 4\} =$
$\{ \quad \}$

4. $\{4, 8, 12\} \cap \{8, 12, 16\} =$
$\{ \quad \}$

5. $\{0, 4, 8\} \cap \{1, 5, 9\} =$
$\{ \quad \}$

6. $\{x, y, z\} \cap \{z, x, y\} =$
$\{ \quad \}$

Here is some practice with adding and subtracting. Try to complete this page by memory. Getting these down now will help you in big ways in the future.

7. 4 + 9 = ☐
8. 17 – 9 = ☐
9. 15 – 7 = ☐
10. 7 + 11 = ☐
11. 9 + 9 = ☐
12. 18 – 4 = ☐
13. 7 + 8 = ☐
14. 13 – 11 = ☐
15. 15 – 7 = ☐
16. 10 + 4 = ☐
17. 6 + 9 = ☐
18. 11 – 9 = ☐

Check out each of these numbers and answer the following questions. Write in commas to help you read the number more easily.

19. 7 6 3 8 0 5 1

How many 10s? ______
How many 1,000s? ______
How many 10,000s? ______

20. 4 7 2 9

How many 1s? ______
How many 100s? ______
How many 1,000s? ______

21. 9 0 4 7 3 5 2 3 4

How many 100s? ______
How many 1,000s? ______
How many 1,000,000s? ______

Everyday Math

22. Your mother has asked you to pick up carrots and celery at the store for the soup she's making for dinner. She needs 2 pounds of carrots which cost $1.00 per pound. She also needs 1 pound of celery, which costs $1.49 per pound. How much money will you need to bring with you to the store?

+ ______

23. You built two bird houses, and sold them to the neighbors for $6.50 and $7.25. How much money did you earn from the bird houses?

+ ______

24. You plan to buy some gifts to fill up a Christmas box for a poor child somewhere in the world. You will buy some colored pencils for $3.67, and some socks for $5.32. How much money will you spend on these gifts?

+ ______

DAY 55 Equal Fractions

Then Jacob made a vow, saying, "If God will be with me, and keep me in this way that I am going, and give me bread to eat and clothing to put on, so that I come back to my father's house in peace, then the Lord shall be my God. . . . and of all that You give me I will surely give a tenth to You." (Genesis 28:20-22)

Jacob promised to give God 1/10 of his flocks and herds. This was his tithe. Today we will look at tenths in measurements.

When taking measurements, you will sometimes find that the length or weight of something falls between two numbers. When that happens, you must write the measurement as a fraction or a decimal. Today you will learn to make measurements that need to be written as a decimal with one number after the decimal point. This means you will be measuring tenths.

Let's begin by measuring this pencil using centimeters. You can see there are 10 little lines between each numbered centimeter mark on the ruler. These little lines mark tenths of a centimeter. Each mark between 5 and 6 is 0.1 centimeter. Remember, we also call a tenth of a centimeter a millimeter. Notice that the length of this pencil lies somewhere between 5 and 6 centimeters.

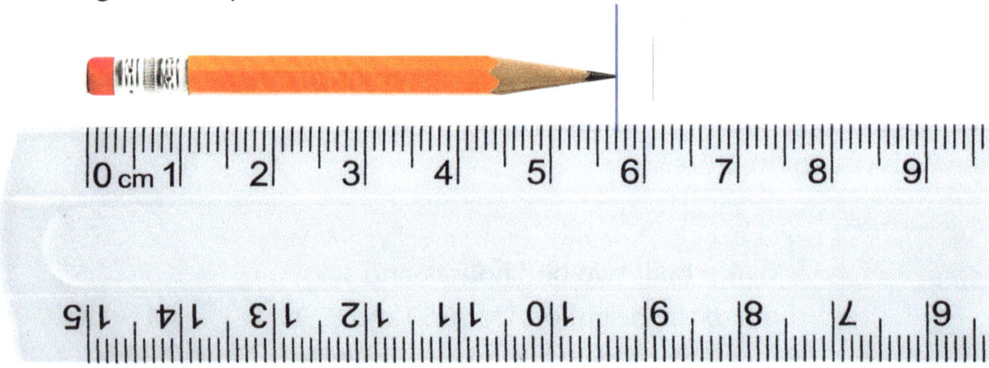

Count each mark between 5 and 6 this way, starting with 5.1. You could count by saying "5 and 1 tenth, 5 and 2 tenths . . . ," but it's faster to count this way:

"5 point 1, 5 point 2 . . . "

5.1, 5.2, 5.3, 5.4, 5.5, 5.6, 5.7, 5.8, 5.9, 6.0

Now count and write down how long the pencil is. ______ Did you measure it to be 5.7 centimeters? Let's write 5.7 as a fraction.

$$5\frac{7}{10}$$

We say that we have measured the pencil "to the tenths place" because we measured its length past the whole number to the number of additional tenths of a centimeter.

Now let's weigh something to the tenths place. Let's see how many kilograms of peppers we are going to buy today. Read the scale on the right. The red pointer is pointing between 0.5 and 1 kilogram. If you look carefully, you will see some thick lines that divide each kilogram into tenths. The weight of these peppers is very close to 0.6 kilograms.

Radio dials use decimals too! This FM radio dial is set to 107.6. Let's write this as a fraction.

$$107\frac{6}{10}$$

Activity

Let's find a few real life examples of numbers that use the tenths place. Can you find any tenths-place decimals in your house. What about a thermometer in the medicine chest? Maybe you can weigh some fruits or vegetables on a kilogram scale. Check out your own height using a centimeter tape measure. Write down your results using decimals to the tenths place.

Student Exercises

We are going to pretend that each chunk below is a block of 1. Each block is divided into 10 tenths. Color in the following decimals. The first one is completed for you.

1.6
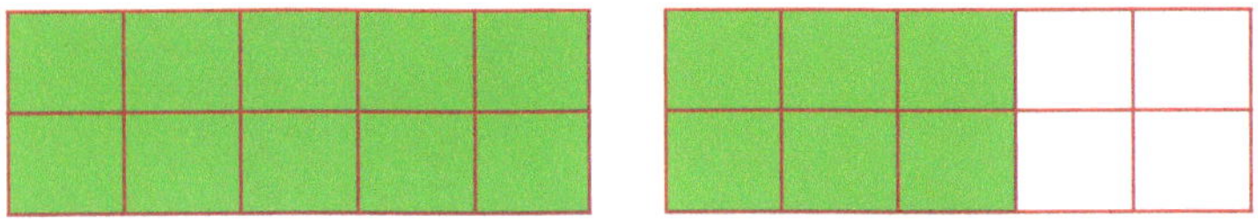

1. 1.3
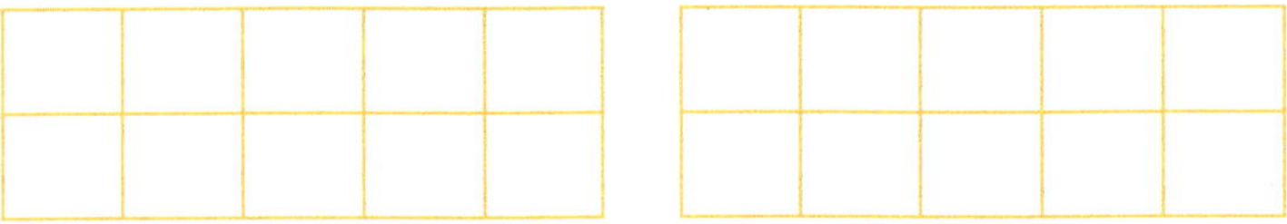

2. 2.5
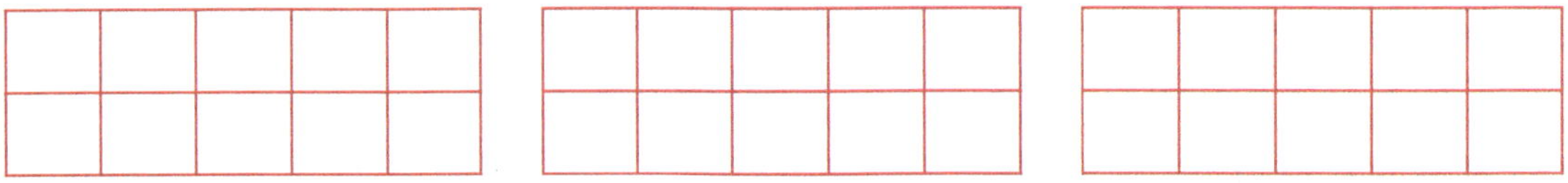

3. 4.9
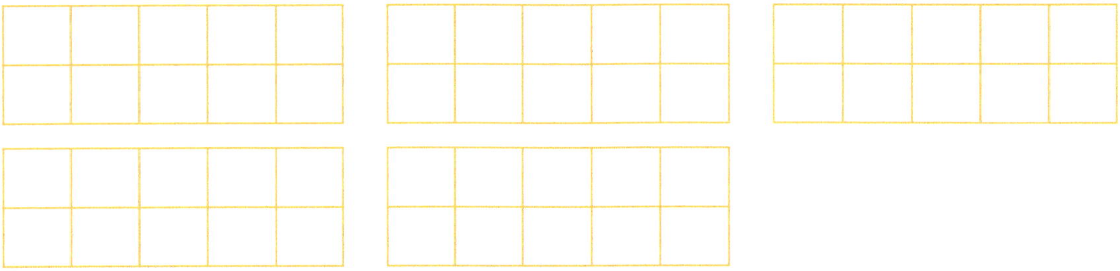

You already know about chunks of numbers. For example, 345 has 3 chunks of 100, 4 chunks of 10, and 5 single 1s. We can think of decimals in chunks too. For example 2.7 has 2 single 1s and 7 chunks of $\frac{1}{10}$. This will be helpful for you later when we do more math with decimals.

Practice counting by chunks. For each list of numbers, figure out if the list is counting up or counting down. Then, fill in the blanks to complete the pattern. Count by 2s.

4. Count by 2s 40, 38, ______, ______, ______, ______

5. Count by 3s: 30, 27, ______, ______, ______, ______

6. Count by 5s 100, 105, ______, ______, ______, ______

7. Count by 7s: 0, 7, ______, ______, ______, ______

8. Count by 9s: 0, 9, ______, ______, ______, ______

For the number 1,207,453.9:

9. What digit is in the tens place? ______

10. What digit is in the tenths place? ______

11. What digit is in the millions place? ______

12. What digit is in the ten thousands place? ______

DAY 56 Practice

Practice multiplying and dividing these numbers using your memory.

1. 5 × 2 = ☐
2. 15 ÷ 3 = ☐
3. 12 ÷ 3 = ☐
4. 8 × 3 = ☐
5. 2 × 4 = ☐
6. 20 ÷ 2 = ☐
7. 10/5 = ☐
8. 1 × 13 = ☐
9. 4 × 4 = ☐
10. 18/2 = ☐

> And let us not grow weary while doing good, for in due season we shall reap if we do not loose heart. (Galatians 6:9)

Numbers can be split apart with addition and multiplication. Fill in the blanks.

11. 8 = ____ + ____
12. 10 = ____ + ____
13. 8 = ____ x ____
14. 10 = ____ x ____
15. 16 = ____ + ____
16. 20 = ____ + ____
17. 16 = ____ x ____
18. 20 = ____ x ____

Everyday Math

19. You have invited friends over for a pizza party. The pizza will be cut into 8 pieces for you and your 3 friends.

How many pieces will each child get to eat? ☐

What fraction of the pizza will each child be eating? ☐

20. For the upcoming Christmas holiday, you plan to purchase a soccer ball for your brother and a teddy bear for your sister. The soccer ball costs $12.95. The bear costs $14.50. How much will you spend?

____ + ____ = ☐

21. You plan to buy a total of 3 kites: one for yourself and one for each of your two friends. Each kite costs $9.00. How much will you spend?

____ x ____ = ☐

22. If you were to take 3 weeks of vacation this winter, how many days of vacation would this be?

____ x ____ = ☐

DAY 57 Two Decimal Places

Pretend this circle is a pizza pie. If we were to cut this pie into 10 pieces to share with your 9 friends, everyone would get a small slice. Write the fraction.

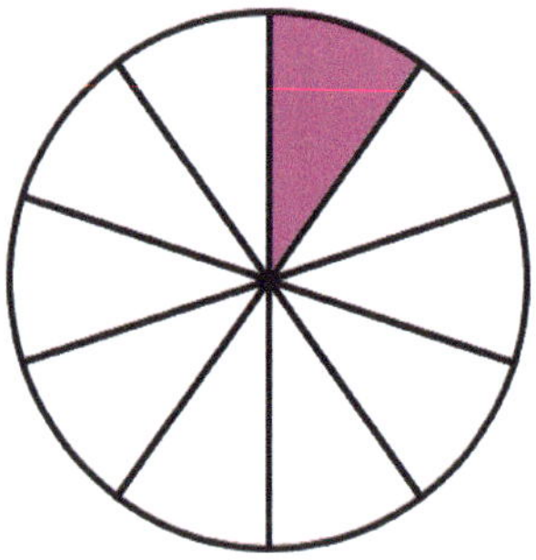

or 0.1 of this pie

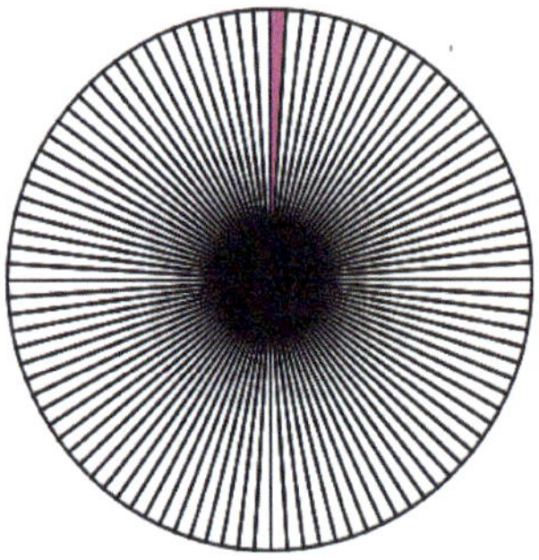

But, if we were to cut this same pizza pie into 100 pieces to share with 99 friends, you would get a very small slice of the pizza!

1/100 of this pizza
That's a tiny slice!

Here is another way to write 1/100th of a pie:

0.01

Do you see how 0.01 of the pie is much smaller than 0.1 of the pie? If you have 0.01 of a pie, you only have a very small piece of it—one little tiny bite!

Now let's think more about 1/100 as the decimal 0.01. In 0.01, the number 1 is in the hundredths place. "Hundredths" is a hard word to say because of the "dths" at the end.

Have you noticed that all the decimal places sound similar to whole number places? But decimal places always have a "ths" on the end. The tens place is the first place to the left of the decimal, and the tenths place is the first place to the right of the decimal. In the same way, the hundreds place is two places to the left of the decimal, and the hundredths place is two places to the right of the decimal.

Remember, we use decimals to write amounts of money. One American dollar equals 100 pennies (or 100 cents). Each penny is 1/100 of a dollar. $0.01 is the way to write one cent as a decimal. Do you see that the decimal for 1 cent has 2 places? The 1 cent is in the hundredths place because it is worth 1/100 of a dollar. We can picture it this way:

Let's use some chunks of blocks to help us understand decimal places. We'll pretend that each large box below is a single block of 1 whole. The whole blocks are divided into 100 small blocks arranged in 10 columns (of 10 small blocks each). Each column of 10 equals 1 tenth of the whole block. Each small block is 1 hundredth of a large block. Now let's color in the number 1.34. First, we'll color in the whole large block on the left. That's 1 whole. Then we will color in 3 of the tenths in the middle box. Finally, we will color 4 hundredths in the last box.

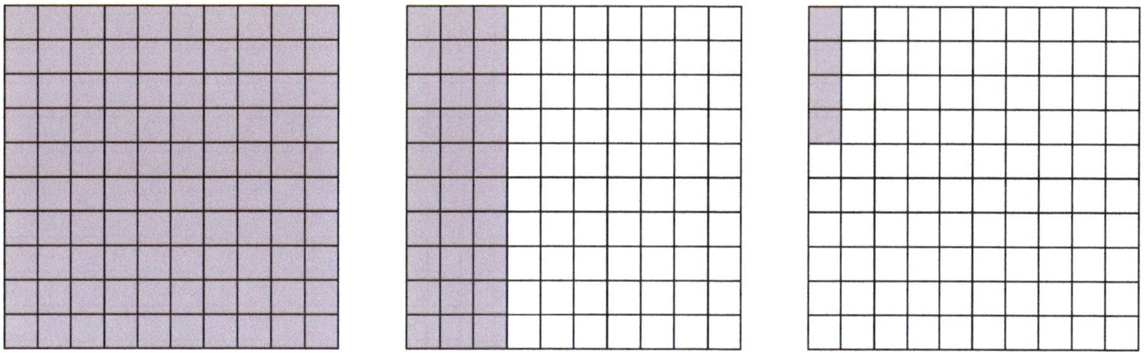

1.34

This illustrates the number 1 and 3 tenths and 4 hundredths. We can take a shortcut and color 34 blocks in the same large box. This displays the same number. But, we would call this number 1 and 34 hundredths. We can also read the number as one point three four.

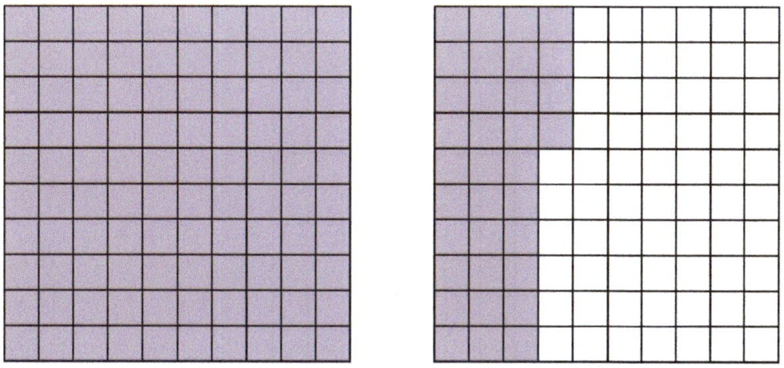

1.34

Try shading a few of these numbers yourself. Use the shortcut and color the tenths and hundredths together in the same box.

1.
1.76

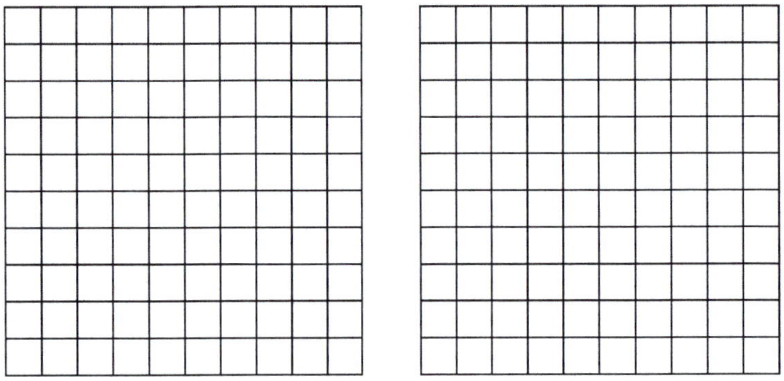

2.
2.38

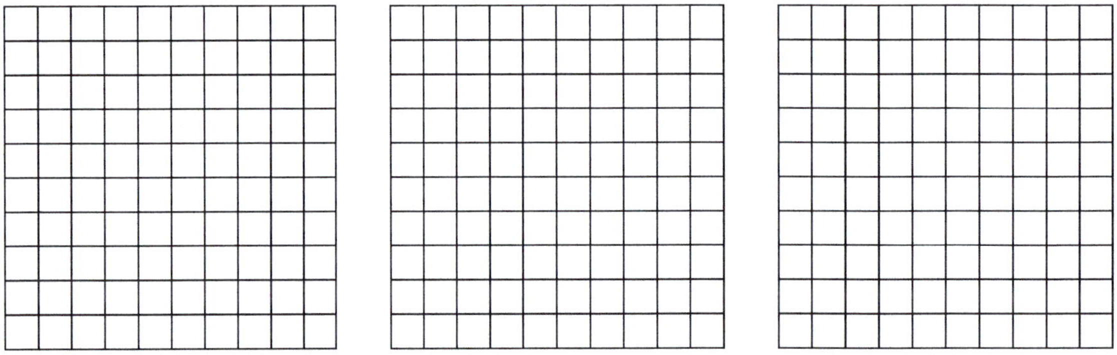

3.
3.49

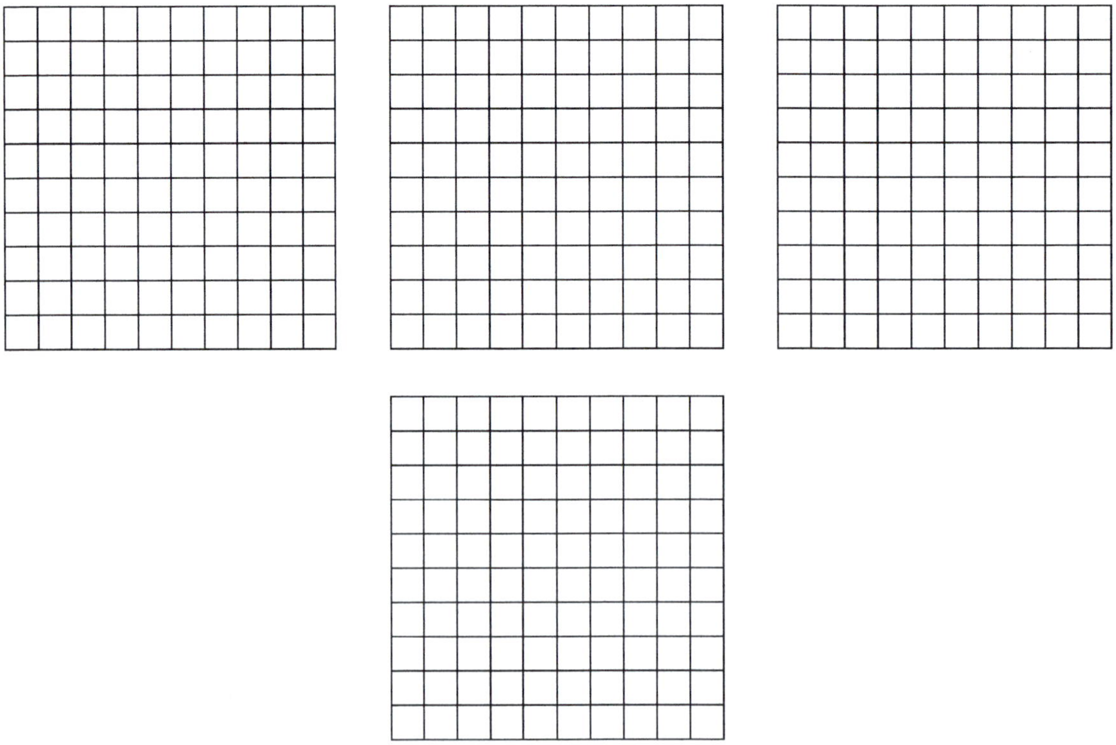

Activity

Let's find a few real-life examples of numbers that use two decimal places. Look around your house and write a list of the places you see decimals to the hundredths place. Maybe your family has a digital scale that reads out to two decimal places. Shopping receipts will contain double decimals because most money is written out to the hundredths place.

Student Exercises

Finish each of these lists of shapes. First find the pattern, and then draw a picture that keeps that pattern going.

1. 

2. 

3. 

4. 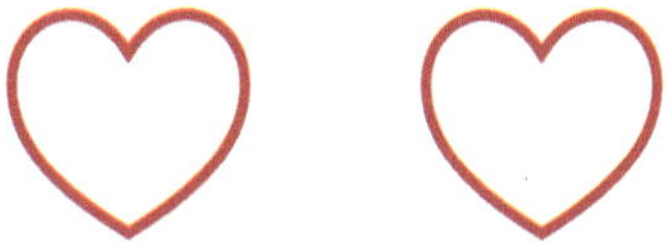

5. 

6. 

Read the following sentences and decide how the numbers in the story are combining–addition, subtraction, multiplication, or division. Circle the correct one.

7. Every day of the week I brush my teeth 3 times. (+ − × ÷)

8. Two new kids started coming to our Sunday school class this week. (+ − × ÷)

9. My backpack had a hole in it, and I lost three pencils. (+ − × ÷)

10. I split my candy bar into 3 equal pieces to share with my friends. (+ − × ÷)

11. I'll be ready to go play soccer a few hours from now. (+ − × ÷)

12. My 100-page book has 4 sections that are all the same size. (+ − × ÷)

13. I gave 2 Valentine candies to each of my friends. (+ − × ÷)

14. I spent 10 dollars at the store using money from my birthday. (+ − × ÷)

Solve this everyday math problem.

15. The donut shop has a discount for purchasing a dozen donuts. If you were to buy the 12 donuts individually, they would add up to $15.00. But a box of a dozen donuts costs $13.80. How much do you save by buying a dozen donuts instead of buying them individually?

___ − ___ = ☐

Check out each of these numbers and answer the questions. Write in commas to help you read the number more easily.

1. 763805.1

How many tenths? ________ How many 1,000s? ________ How many 10,000s? ________

2. 47.29

How many hundredths? ________ How many 1s? ________ How many tenths? ________

3. 904735234

How many 100s? ________ How many 1,000s? ________ How many 1,000,000s? ________

Let's practice multiplying and dividing. You may use the multiplication chart in the back of the book. The more you use it, the sooner you will have these facts memorized. Remember that the fraction line is the same as the ÷ symbol. For example $\frac{50}{10}$ means 50 divided up into groups of 10, and $\frac{24}{6}$ means 24 divided into groups of 6.

4. 3 × 10 = ☐

5. 18 ÷ 3 = ☐

6. 24 ÷ 3 = ☐

7. 7 × 5 = ☐

8. 42 ÷ 7 = ☐

9. 63/7 = ☐

10. 8 × 7 = ☐

11. 6 × 6 = ☐

12. 54/9 = ☐

Everyday Math

13. How many fingers are on your hands? Now you can figure out how many fingers there are in your family.

10 x ____ = ☐

14. How many ears do you have? How many ears are in your family?

2 x ____ = ☐

15. If spiders have eight legs, and there are nine spiders in your house, how many spider legs are in your house?

____ x ____ = ☐

16. Doctors say that children should drink 40 ounces of water every day. If your cup holds 8 ounces, how many cups should you drink each day?

____ x ____ = ☐

Adding and Subtracting Decimals DAY 59

Now that you have learned something about decimals, let's try working with them. First take a look at this number:

234.56

The 2 in the hundreds place is worth 10 times more than if it were in the tens place.

The 3 in the tens place is worth 10 times more than if it were in the ones place.

The 4 in the ones place is worth 10 times more than if it were in the tenths place.

The 5 in the tenths place is worth 10 times more than if it were in the hundredths place. Each decimal place is 10 times larger than the one on its right.

Let's show 234.56 by thinking of it as blocks.

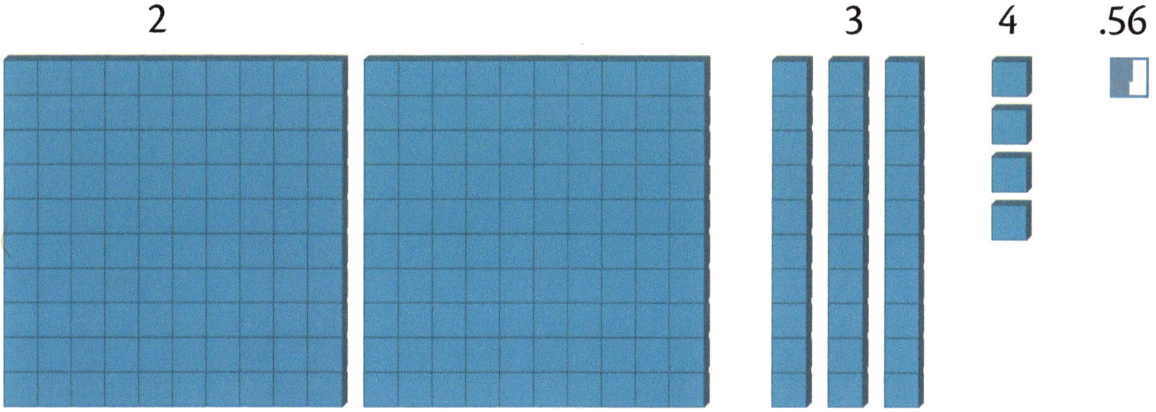

The 0.56 at the end is so small because it's just a tiny piece of a single block. If we zoomed in, this is what it would look like:

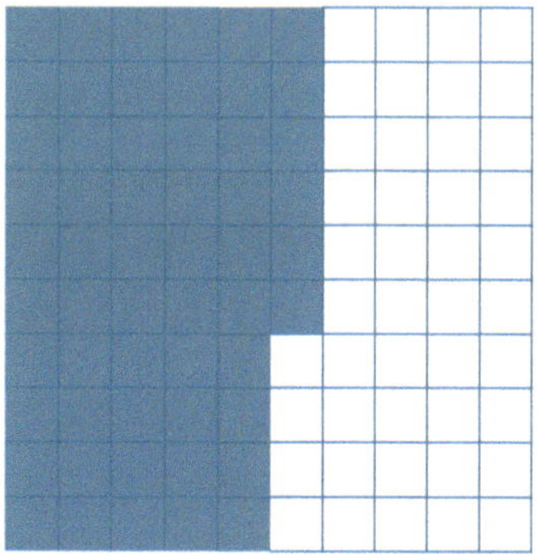

Adding and subtracting numbers with decimals is a lot like adding and subtracting whole numbers. But with decimals, we have to be sure to keep the decimal point in the right place. This time, let's subtract money. We subtract money when making change. If we were to give \$4.50 to the cashier when purchasing apples for \$3.25, the cashier would give us change. Let's subtract \$3.25 from \$4.50 to find out how much change we would get.

$$\begin{array}{r} \$4.50 \\ -\ \$3.25 \\ \hline \$1.25 \end{array}$$

Notice that the decimal in the answer must also line up with the decimals in the numbers that are being added or subtracted above it.

Student Exercises

Now add and subtract a few numbers with decimals on your own.

1. $$\begin{array}{r} 3.56 \\ +\ 1.24 \\ \hline \end{array}$$

2. $$\begin{array}{r} 8.53 \\ -\ 1.62 \\ \hline \end{array}$$

3. $$\begin{array}{r} 4.59 \\ +\ 0.25 \\ \hline \end{array}$$

4. $$\begin{array}{r} 3.27 \\ +\ 2.33 \\ \hline \end{array}$$

5. $$\begin{array}{r} 8.06 \\ +\ 1.73 \\ \hline \end{array}$$

6. $$\begin{array}{r} 9.55 \\ +\ 4.65 \\ \hline \end{array}$$

7. $$\begin{array}{r} 2.44 \\ +\ 8.26 \\ \hline \end{array}$$

8. $$\begin{array}{r} 7.32 \\ +\ 0.89 \\ \hline \end{array}$$

Now may He who supplies seed to the sower, and bread for food, supply and multiply the seed you have sown and increase the fruits of your righteousness. (2 Corinthians 9:10)

Practice multiplying and dividing these numbers using your memory. If you get stuck, your multiplication and division table will help you.

9. $4 \times 3 =$ ☐
10. $16 \div 4 =$ ☐
11. $8 \div 4 =$ ☐
12. $4 \times 5 =$ ☐
13. $5 \times 2 =$ ☐
14. $10 \div 2 =$ ☐
15. $\frac{18}{3} =$ ☐
16. $40 \times 1 =$ ☐
17. $9 \times 3 =$ ☐
18. $\frac{40}{1} =$ ☐
19. $\frac{16}{4} =$ ☐
20. $6 \times 7 =$ ☐

Practice DAY 60

Practice counting by chunks. For each list of numbers, figure out if the list is counting up or counting down. Then, fill in the blanks to complete the pattern.

1. Count by 2s: 34, 32, _____, _____, _____, _____

2. Count by 5s: 65, 70, _____, _____, _____, _____

3. Count by 6s 0, 6, _____, _____, _____, _____

4. Count by 7s: 35, 42, _____, _____, _____, _____

5. Count by 10s: 530, 520, _____, _____, _____, _____

Check out each of these numbers and answer the questions. Write in commas to help you read the number more easily.

6. 7 4 0 8 3 5

How many 10s? ______ How many 1,000s? ______ How many 10,000s? ______

7. 7 0 3 4 5 1 . 4 5

How many 1s? ______ How many hundredths? ______ How many 100,000s?

8. 4 8 9 0 0 5

How many 100s? ______ How many 1,000s? ______ How many 1,000,000s? ______

Everyday Math

9. A neighbor asks you to work in his yard and says he will pay you $10.00 per hour. After you work for 8 hours, how much will you have earned?

____ x ____ = ☐

10. Karen wants to buy a book for $14.69. She hands the cashier two ten-dollar bills and three quarters. How much money will she have left over? (Hint: Remember, when you write money as decimals, the decimal point goes between the dollars and the cents.)

____ – ____ = ☐

Math Game

11. Computers and maps use **coordinates** to tell where things are found. The chart below looks kind of like a multiplication table with numbers down the side and letters (instead of numbers) along the top. In this game, the numbers and the letters are called coordinates. If we want to say where the giraffe is, we look up at the letter above him and see the coordinate E. Then we look sideways and see that he is even with the coordinate 3. Now we can say that the giraffe is at E3. Fill in the coordinates to tell where the animals are. Be sure to write the letter first and the number second.

ANIMAL	HOME
	E3

	A	B	C	D	E	F	G	H	I	J
1										
2										
3										
4										
5										
6										
7										
8										
9										
10										

Decimals in Real Life

Today let's study some scientific data that uses decimals. Our data will show how much water is needed by certain food crops. We will also show rainfall data for certain areas.

Vegetable crops need water during their growing season. Some plants need more water than others. We thank God for rain, which helps the plants grow! But sometimes crops need extra watering during dry spells. Farmers can measure rainfall with rain gauges. Then they can do some math to figure out how much more water to give the crop.

This rain gauge has collected 1.8 inches of rain.

First, let us take a look at how much water these vegetables need during the summer months.

VEGETABLE	TOTAL WATER FOR SUMMER
Asparagus	4.5 inches
Beans	18 inches
Cabbage	9 inches
Carrots	4.5 inches
Onions	12 inches
Potatoes	13.5 inches

Now let's look at one summer's rainfall data for two cities:

MONTHS	DENVER, COLORADO RAINFALL	SPRINGFIELD, ILLINOIS RAINFALL
June	0.4 inches	2.6 inches
July	1.1 inches	6.5 inches
August	1.0 inches	2.5 inches
Total Summer	**2.5 inches**	

Add up the summer rainfall for Springfield above. Denver has been added for you. You will use these total summer rainfall amounts later.

Now let's see how much additional water farmers near these two cities might have to give each vegetable crop. Look at the chart for Denver below. For each vegetable crop, we have subtracted the total summer rainfall from the total summer water needed for that crop.

Growing Vegetables Near Denver, CO

VEGETABLE	WATER NEEDED IN SUMMER	NORMAL SUMMER RAINFALL RECEIVED	ADDITIONAL WATER NEEDED
Asparagus	4.5 inches	2.5	2
Beans	18 inches	2.5	15.5
Cabbage	9 inches	2.5	6.5
Carrots	4.5 inches	2.5	2
Onions	12 inches	2.5	9.5
Potatoes	13.5 inches	2.5	11

Notice that, for some of these vegetable crops, we had to subtract a decimal from a whole number. To see how this is done, look below to see how we get ready to subtract 18 – 2.5 for beans:

$$\begin{array}{r} 18.0 \\ -\ \ 2.5 \\ \hline \end{array}$$

Notice that we have put a decimal point after the whole number 18, and we have written a zero in the tenths place. We wrote a zero here so we could borrow from it when we subtract. Putting a zero at the end of a decimal does not change how much the decimal is worth. 18 and zero tenths is still just 18. Also notice that, when we set up the numbers, we made sure our decimal points were lined up.

Now we can borrow as usual. It's a new thing for you to borrow 10 tenths from the 1s place, but you can do it the same way as you do whole-number borrowing.

$$\begin{array}{r} {\scriptstyle 7\ \ 10} \\ 1\not{8}.\not{0} \\ -\ \ 2.5 \\ \hline 15.5 \end{array}$$

Sometimes a crop gets more rain than it needs. You will discover this as you try to set up some of these subtractions. If you find that there is more than enough rainwater for a certain crop, the farmer won't have to add any extra water. Write a "0" in the box.

Growing Vegetables Near Springfield, IL

VEGETABLE	WATER NEEDED IN SUMMER	NORMAL SUMMER RAINFALL RECEIVED	ADDITIONAL WATER NEEDED
Asparagus	4.5 inches		
Beans	18 inches		
Cabbage	9 inches		
Carrots	4.5 inches		
Lettuce	13.5 inches		
Onions	12 inches		

1. If you were a farmer who would rather not water your crops, which vegetables do you think would be best to grow in the Denver, Colorado area?

2. Which would be the best crops to grow in the Springfield, Illinois area if you would rather not water?

Student Exercises

Numbers can be split apart with addition and multiplication. Split these numbers apart and fill in the blanks.

1. 7 = ____ + ____
2. 11 = ____ + ____
3. 7 = ____ x ____
4. 11 = ____ x ____
5. 20 = ____ + ____
6. 12 = ____ + ____
7. 20 = ____ x ____
8. 12 = ____ x ____
9. 9 = ____ x ____

Add or subtract. Some of the borrows and carries have been filled in for you.

10.

		1	
	3	3	4
+	4	3	7

11.

		4	13
	1	~~5~~	~~3~~
−		4	4

12.

		1	
	1	2	3
+	4	3	7

13.

	4	12	17
	~~5~~	~~3~~	~~7~~
−	1	8	8

14.

	6	3	1
+	1	4	9

15.

	3	4	5
−	1	3	9

Here is some practice with adding and subtracting. Try to complete this page by memory. Getting these down now will help you in big ways in the future.

1. $5 + 7 =$

2. $15 - 8 =$

3. $13 - 6 =$

4. $6 + 8 =$

5. $5 + 5 =$

6. $20 - 10 =$

7. $4 + 9 =$

8. $16 - 9 =$

9. $\frac{4}{5} - \frac{1}{5} =$

10. $\frac{3}{7} + \frac{8}{7} =$

11. $\frac{5}{2} + \frac{9}{2} =$

12. $\frac{10}{3} - \frac{3}{3} =$

Each square below has a different purple shape inside. The white area of each square can be filled with more copies of its own purple shape. For each square, let's pretend that we have a lot of different-colored tiles that are the same shape as its purple shape. Gather some crayons or colored pencils and make tile designs below. In each square, be sure to make its colored tiles the same shape as its purple tile. Make the tiles different colors within each square so you can tell the tiles apart.

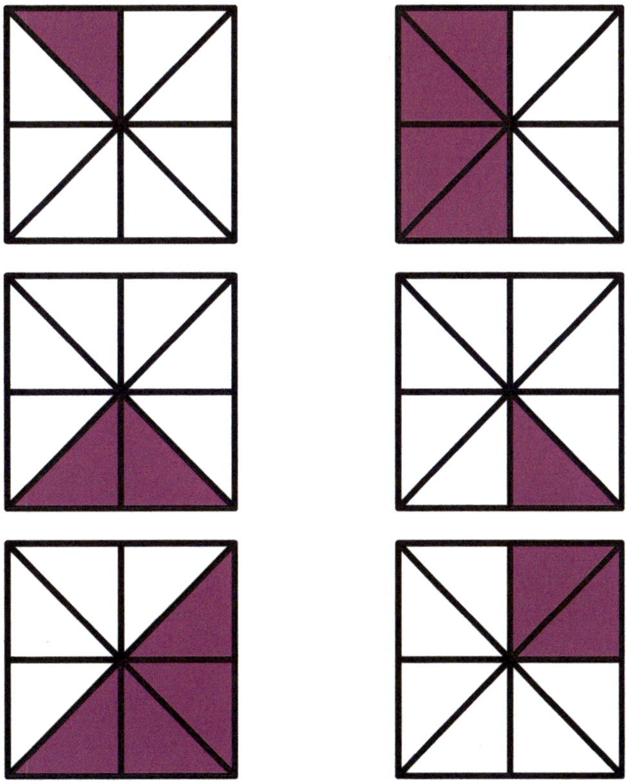

Everyday Math

13. Let's say you made 9 jars of strawberry jam and sold them for $8 each. To make the jam, you had to spend $30 for the strawberries, $10 for the jars, and $3.00 on sugar and other ingredients. How much did you earn?

____ x ____ = ☐

____ + ____ + ____ = ☐

____ – ____ = ☐

14. Imagine that you earned $140 last month by helping neighbors with yard work. You gave God one tenth of what you earned by giving $14 to some missionaries. Then you bought a new lawnmower for $99. How much money would you have left from your earnings last month?

140 – ____ – ____ = ☐

15. Suppose your parents earned $948 this month in their side business. They must set aside $379 for paying taxes. How much will your parents have left to spend?

____ – ____ = ☐

Let's practice adding large numbers.

16.
$$\begin{array}{r} 34.5 \\ +\ 56.7 \\ \hline \end{array}$$

17.
$$\begin{array}{r} 334 \\ +\ 287 \\ \hline \end{array}$$

18.
$$\begin{array}{r} 1306.99 \\ +\ 117.52 \\ \hline \end{array}$$

Larger and Smaller Decimals DAY 63

Which do you think is larger: 0.10 or 0.01? How about 0.25 or 0.05? Which of these is larger: 356.67 or 356.52? Comparing numbers with decimals works the same way as comparing all other numbers. You start with the biggest number on the left. If they are equal, you move to the right, to the next number. Look at the number comparisons below and make sure you understand how to know which is the larger number for each.

1,200 > 1,100

1,230 > 1,220

1,234 > 1,233

1,234.5 > 1,234.4

1,234.56 > 1,234.55

Notice in the top comparison that 1,200 is 100 more than 1,100. Notice in the bottom comparison that 1,234.56 is only 0.01 more than 1,234.55.

In this lesson, we will learn to count money faster. We'll count from greatest value to least value as it's arranged from left to right below.

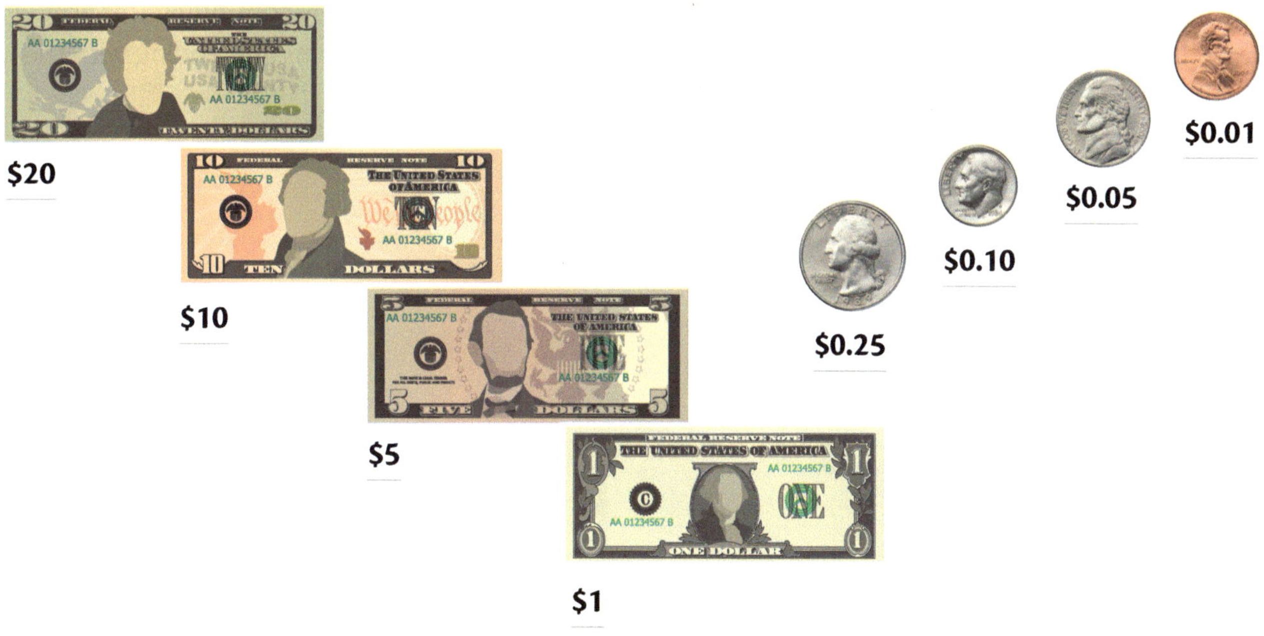

Let's pretend that you emptied your pockets and found the money below. Let's count it.

Count the bills first. Count from largest to smallest. Say:

20 + 10 = 30

30 + 1 = 31

31 + 1 = 32

Or you could say it faster without saying "plus" and "equals." You add silently in your head and say out loud: 20 30 31 32

Saying it this way is more like counting. You can keep counting the money by continuing this way with the coins. You will add the coins to the bills starting with the largest coins and ending with the smallest:

Add on the quarters and say: 32.25 32.50

Add on the dimes: 32.60 32.70 32.80 32.90 33

Add on the nickels: 33.05 33.10 33.15

Add on the pennies: 33.16 33.17 33.18 33.19 33.20 33.21 33.22

Now try counting the money below.

Student Exercises

Count the money from greatest to least and write the amount in the box.

1. 

2. 

3. 

4. 

Finish each of these lists of shapes. Find the pattern and then draw a picture that keeps that pattern going.

5. 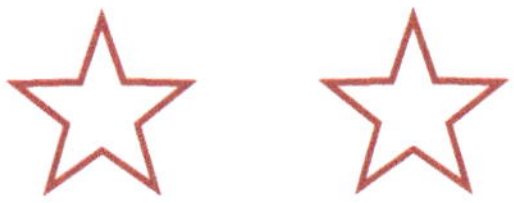

6. 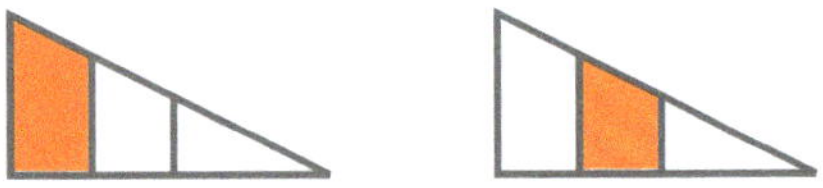

7. 

8. 

9. 

10. 

Add or subtract these fractions. Color in the correct number of pie slices for each answer. The first has been completed as an example.

11. $\frac{4}{4}+\frac{3}{4}=\frac{7}{4}=$ $1\frac{3}{4}$

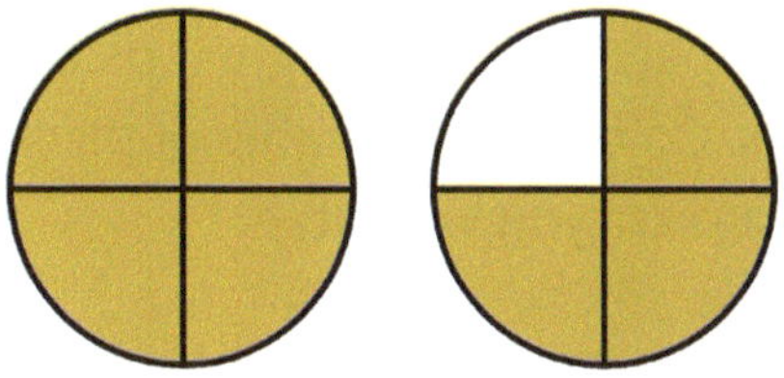

12. $\frac{4}{7}+\frac{4}{7}=\frac{}{7}=$ ☐

13. $\frac{3}{12}+\frac{11}{12}=\frac{}{12}=$ ☐

14. $\frac{7}{6}+\frac{6}{6}=\frac{}{6}=$ ☐

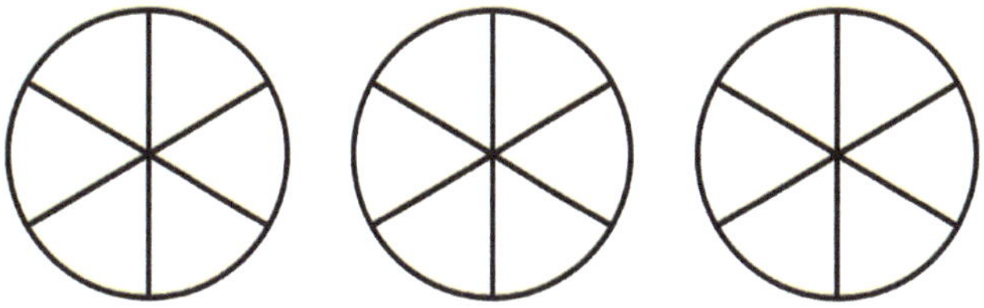

15. $\frac{5}{9}-\frac{2}{9}=\frac{}{9}=$ ☐

16. $\frac{11}{7}-\frac{4}{7}=\frac{}{7}=$ ☐

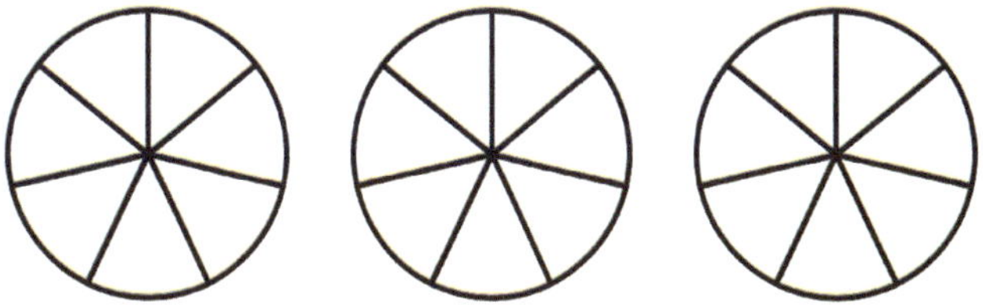

17. $\frac{17}{5}-\frac{7}{5}=\frac{}{5}=$ ☐

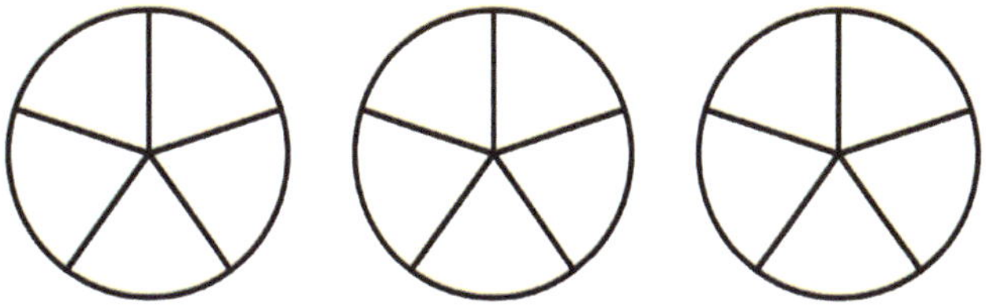

Everyday Math

18. Let's pretend your parents have bought 2 pizzas for you and your three friends. Each pizza is cut into eight pieces. Color in the number of pieces in each child's share using a different color. Write the fraction of a pizza you will eat.

= ———

19. Your mother has cut three apples into eight pieces each. How many pieces of apples are there all together? If she wants to divide them evenly among four children, how many pieces will each child get?

_____ ÷ 4 = ______

20. Suppose that you have 8 kilograms of apples to split up among 3 poor families in your church. Let's see how many kilograms of apples will each family receive. Hint: Think of this as the fraction 8/3. Pretend that each of these 8 circles represents one kilogram, and each color represents a family's share. Notice that, because we are dividing all the apples between 3 families, we have shown all the circles divided into 3 parts. You see that each family gets at least two kilograms. Now you can split up the remaining two kilograms of apples? Color the remaining two circles with equal amounts of green, orange, and blue. You see that each family gets $\frac{8}{3}$ of a kilogram. That's 8 divided by 3 or eight-thirds of a kilogram.

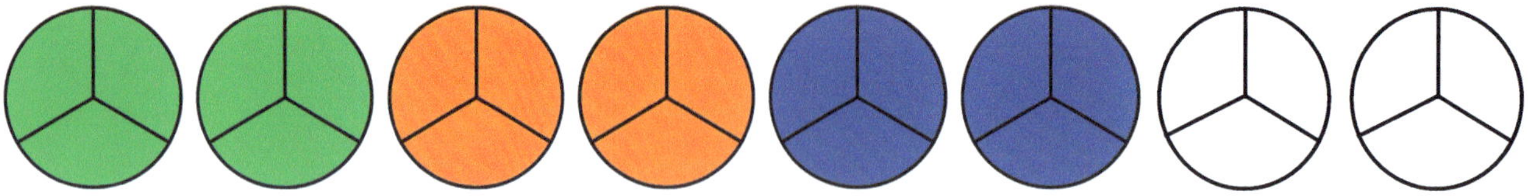

$$\frac{8}{3} = 2\frac{\square}{3}$$

Practice DAY 64

Read the following sentences and decide how the numbers in the story are combining—addition, subtraction, multiplication, or division. Circle the correct symbol.

1. I took the dog for a walk three hours ago. (+ − × ÷)

2. At the grocery store, Mom said I could pick out 3 boxes of cereal to add to the cart. (+ − × ÷)

3. Each of my 3 bookshelves holds 10 books. (+ − × ÷)

4. I have 10 apples and plan on eating 2 per day. I wonder on what day I will eat my last apples. (+ − × ÷)

5. Dividing a snack equally can make everyone happy. (+ − × ÷)

6. When I walk on the beach, I pick up every cool shell I see. (+ − × ÷)

Let's explore God's patterns with sets. Let's see how these sets come together to create new sets.

7. {a, b, c} U {c, a, d} = { }	10. {7, 8, 9} ∩ {9, 10, 11} = { }
8. {1, a, b} U {2, a, c} = { }	11. {2, 4, 6} ∩ {8, 10, 12} = { }
9. {0, 1, 2} U {0, 1, 2} = { }	12. {4, 5} ∩ {4, 5} = { }

Solve these exercises.

13. 7 x 5 = ☐

14. 14 ÷ 7 = ☐

15. 24 ÷ 3 = ☐

16. 9 x 3 = ☐

17. $\frac{8}{2} - \frac{6}{2} =$ ☐

18. $\frac{3}{5} + \frac{4}{5} =$ ☐

19. $\frac{10}{8} + \frac{4}{8} =$ ☐

20. $\frac{9}{3} - \frac{3}{3} =$ ☐

DAY 65 Go Find More Fractions and Decimals in God's World

This lesson integrates math into everyday life. With this activity, the student will apply God's patterns and wisdom in meaningful ways to home and community. Take a break from memory work and academic exercises to make math part of everyday life. The following are suggestions or examples. Other ideas may be added to the list.

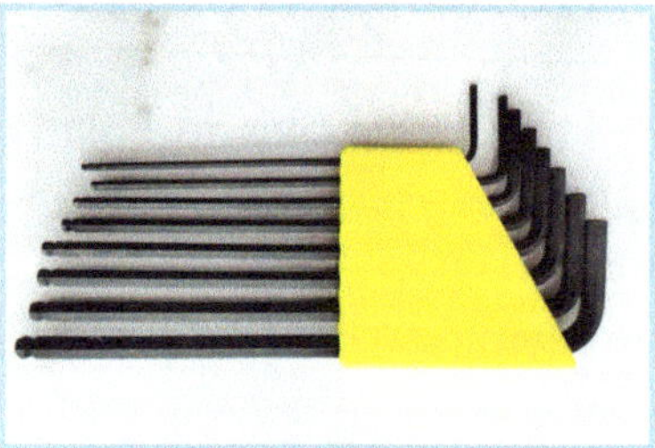

1. Identify common tools that your parents will need when they do jobs in the home and yard. Ask what tools they use when they do things like changing the oil in the car, replacing gaskets in faucets, replacing batteries in clocks, or any other common tasks. Write down the size of Allen wrenches or sockets needed and keep this information in a place where you can find it. These tools often have their sizes shown as decimals. This will make the process easier when your parents need to do this work.

2. Create a budget for your family with help from a parent. A budget is a plan for managing money, usually for a month or for a year. The amounts earned and spent in different categories are written in spaces on a budget chart. The chart also includes spaces to subtract and write how much more (or less) was earned (or spent) than what was planned.

Make a chart like the one shown but adapt it for your family's needs. Write down how much will be earned each month. This money coming in is called *income*. Then plan how much will be given to your church, how much will be spent, and how much will be set aside as savings each month. This money going out is called *expense*.

Parents may wish to provide estimates or make-believe numbers for this exercise.

Monthly Household Budget

MONEY COMING IN (INCOME)			
CATEGORY	BUDGETED AMOUNT	ACTUAL AMOUNT	DIFFERENCE
INCOME			
Job Earnings	$4,000		
Other Income	$500		
TOTAL INCOME	**$4,500**		

MONEY GOING OUT (EXPENSES)			
CATEGORY	BUDGETED AMOUNT	ACTUAL AMOUNT	DIFFERENCE
HOUSING			
Rent/Mortgage	$1,200		
Utilities (Electric, Gas, Water)	$200		
Internet/Phone	$100		
TOTAL HOUSING	**$1,500**		
CATEGORY	BUDGETED AMOUNT	ACTUAL AMOUNT	DIFFERENCE
TRANSPORTATION			
Car Fund	$300		
Fuel	$150		
Public Transportation	$50		
Maintenance/Insurance	$100		
TOTAL TRANSPORTATION	**$600**		

MONEY GOING OUT (EXPENSES)			
CATEGORY	**BUDGETED AMOUNT**	**ACTUAL AMOUNT**	**DIFFERENCE**
GROCERIES	**$400**		
DINING OUT	**$100**		
ENTERTAINMENT	**$100**		
HEALTHCARE			
Insurance	$150		
Medications/Doctor Visits	$50		
TOTAL HEALTHCARE	**$200**		
PERSONAL CARE	**$50**		
CLOTHING	**$75**		
EDUCATION	**$50**		
CATEGORY	**BUDGETED AMOUNT**	**ACTUAL AMOUNT**	**DIFFERENCE**
SAVINGS			
Emergency Fund	$200		
Future Plans	$150		
TOTAL SAVINGS	**$350**		
TITHE	**$450**		
MISCELLANEOUS	**$100**		
TOTAL EXPENSES	**$3,975**		

In this budget chart, if we subtract the total planned expenses from the total planned income, we find that there is $525 more income than expenses. But it's good to have extra because sometimes unexpected things like car repairs and doctor visits cost more than planned in the budget. This extra is called net income.

$$\$4500 - \$3975 = \$525$$

$$\text{Income} - \text{Expenses} = \text{Net Income}$$

Additional Activities

For extra activities, games, and helpful videos on these math concepts, refer to **generations.org/math3**.

God gives us tools like math to measure and study His world.

CHAPTER 6

Doing Math in Real Life

DAY 66 Many Ways to Measure

Much of God's world is **measurable**—we are *able* to measure it. We are able to count time in seconds. We can measure a trip to the grocery store in miles. We can measure your growth using a scale or a measuring tape.

Yet, much of God's world is **immeasurable**. That means there are some things you cannot measure. How would you measure a thought? How about an emotion? How about measuring love? How can we measure the love Jesus had for us when He gave Himself up on the cross?

Knowing what you can and cannot measure helps you appreciate God's greatness and the richness of the world He made. Science measures physical things. But, as this verse says, science cannot measure God's wisdom.

> Great is our Lord, and abundant in power; his understanding is beyond measure. (Psalm 147:5 ESV)

The purpose of this chapter is to learn how to connect the world of math with the world of stuff you see around you. Measuring is a big part of making that connection.

Three Dimensions

Let's pretend that you are hiking in a long, straight canyon with steep sides you can't climb. You can only walk in a line, back and forth in the bottom of the canyon. In math, we say a line has 1 dimension. When we measure that 1 dimension, we can only measure **1** thing—its **length**. In the canyon, you hike in 1 dimension.

Now pretend you are on a flat field. You can still walk in a line, but you can also walk any direction to the side, or in a circle, or to any other place in the field. We have added another dimension. In math, we say something has 2 dimensions if it's flat and wider than a line. Rectangles, circles, squares, and triangles all have 2 dimensions. When we measure something with 2 dimensions, we measure **2** things—its **length** and its **width** to find its area. In the field, you can walk in 2 dimensions.

Now let's say you are walking in the field and you come across some wild geese that are also walking. When you approach the geese, they fly away. You can't fly. You must stay on the flat, 2-dimensional field. But the geese can use all 3 dimensions. They can walk in a line. They can walk around in the flat field. And they can fly up in the air. Balls, blocks, cups, and cubes all have 3 dimensions because they are not flat. They use space on the ground and space in the air. To measure one of these objects, we measure **3** things—its **length**, **width**, and **height**. From these 3 measurements, we can find the object's volume.

One Dimension

Measure this blue line in centimeters to get a one-dimensional measurement. The answer will be in decimals.

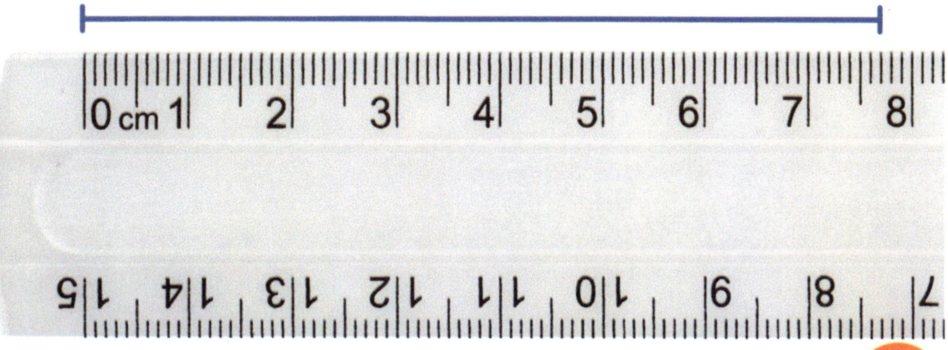

1. ______ cm

Two Dimensions

Let's say that your family plans to build a deck outside your back door. It will be shaped like the design below. Your parents ask you to practice measuring in 2 dimensions by finding the area of this design. Measure the area of each section in square inches by multiplying its length by its width and writing its area inside the section. Then add up the area of all the sections.

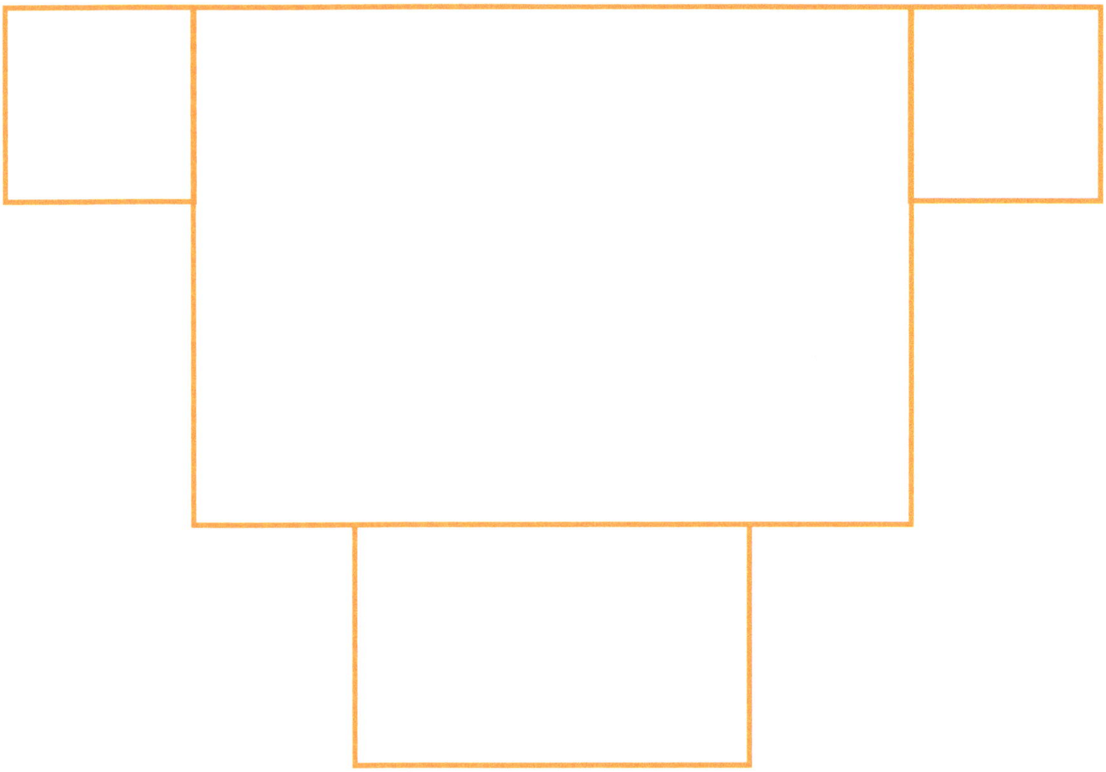

2. _______ total square inches = the area of this 2-dimensional deck design

Three Dimensions

Imagine that this cube is 2 inches wide, 2 inches long, and 2 inches in height. Next to it, you see another cube that is 1 inch wide, 1 inch long, and 1 inch tall. Now, can you guess how many of these 1-inch cubes will fit into the 2-inch cube?

One way to do this would be to make a 2-inch cube out of cardboard, and then make a few 1-inch cubes, and see how many 1-inch cubes would fit into the the 2-inch cube.

But let's use math to figure this out a quicker way. First we will find the larger cube's volume. Multiply its length times its width. Then multiply that answer times its height:

Length x Width x Height = Volume

2 x 2 x 2 = 8

2 inches x 2 inches x 2 inches = 8 cubic inches

We say the volume is measured in cubic inches because 8 1-inch cubes would fit inside the 2-inch cube.

Let's try another one. How many 1-inch cubes could fit inside a 3-inch cube? This cube would look like a big Rubik's Cube.

3. 3 x 3 x 3 = ☐ cubic

You can measure water bottles in cubic inches or cubic centimeters. Can you imagine how many cubic inches or cubic centimeters could fit into a water bottle? (A cubic centimeter is about $\frac{1}{16}$ the size of a cubic inch.) Sometimes we refer to cubic centimeters as "cc's" for short. How many cc's do you think the water bottle in the picture holds? It would be hard to fill a bottle with little centimeter blocks to figure this out. Someday you will learn how to measure the volume of bottles using math. Then you will find that this water bottle holds 500 cc's.

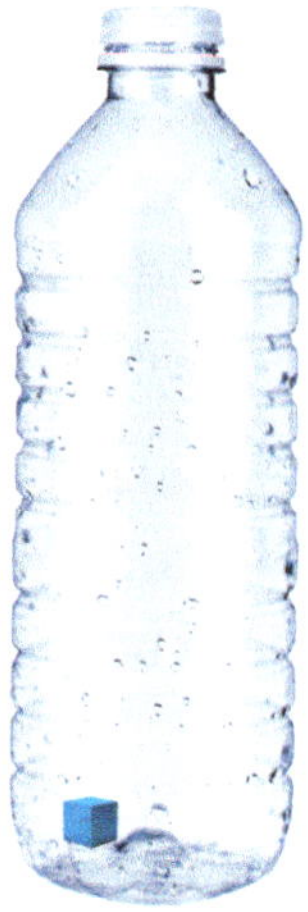

 ________ cc

The following Bible passage tells a measurement of volume.

> Now there were set there six waterpots of stone, according to the manner of purification of the Jews, containing twenty or thirty gallons apiece. Jesus said to them, "Fill the waterpots with water." And they filled them up to the brim. And He said to them, "Draw some out now, and take it to the master of the feast." And they took it. When the master of the feast had tasted the water that was made wine, and did not know where it came from (but the servants who had drawn the water knew), the master of the feast called the bridegroom. And he said to him, "Every man at the beginning sets out the good wine, and when the guests have well drunk, then the inferior. You have kept the good wine until now!" (John 2:6-10)

4. How many water pots did Jesus fill up with the wine?

____

5. How many gallons would each water pot hold?

____

6. How many gallons of wine did Jesus make?

____

Now let's see if you can figure out dimensions. For each statement below, think about how you would measure. Would you make 1 measurement (length) for 1 dimension, 2 measurements (length and width) for 2 dimensions, or 3 measurements (length, width, height) for 3 dimensions? Write each dimension in its box—1D, 2D, or 3D.

7. The amount of gas that will fit in your car's gas tank ☐

8. The distance from your home to grandma's house ☐

9. The area of your house in square feet or square meters ☐

10. The wingspan of an eagle (distance between wingtips) ☐

Student Exercises

Use your wisdom of long addition and long subtraction to figure out how God makes numbers come together to make new amounts.

1.

	1	5
+	1	8

2.

	3	7
+	2	1

3.

	3	5
−	1	6

4.

	1	4	6
+	2	8	7

5.

	7	8
−	4	2

6.

	2	5	3
−	1	6	0

Fill in the place values for each of these numbers. The first one has been filled in for you.

	NUMBER	hundreds	tens	ones	decimal	tenths	hundredths
7.	345.58	3	4	5	.	5	8
8.	587.83						
9.	145.87						
10.	999.99						
11.	100.01						
12.	68.43						
13.	45.38						
14.	199						

Fast add the following money by counting it up in your head, beginning with the largest bills, and then adding in the coins (the decimals).

15. $20 $20 $10 $5 0.25 0.25 $______.______

16. $10 $10 $10 $5 $5 $1 $1 0.25 $_______.____

17. $50 $10 $10 $10 $1 $1 $1 $1 0.25 0.10 $_______.____

18. $20 $20 $20 $5 $1 $1 $1 0.25 .05 .01 $_______.____

19. Fast add the following money by adding up the coins first. Then count the bills in your head and add their total to the coin total.

Coins	3	.25 (quarters)	________	Bills	2	$20	________
	8	.10 (dimes)	________		3	$10	________
	10	.05 (nickels)	________		4	$5	________
	10	.01 (pennies)	________		7	$1	________
		Total	________		Total		________

Add these together:__________

DAY 67 Practice

Check out each of these numbers and answer the questions. Add in commas to make the numbers easier to read.

1. 7 1 6 0 3 4 5 1 0 4

   How many 10s? ________ How many 1,000s? ________ How many 10,000s? ________

2. 8 4 3 6 . 9 7

   How many 1s? ________

   How many 100s? ________

   How many hundredths? ________

3. 4 7 2 0 8 7 4

   How many 100s? ________

   How many 1,000s? ________

   How many 1,000,000s? ________

What is the volume of the following 3-dimensional objects? Remember, the volume is the width times the length times the height. The name for the shape of these objects is *rectangular prism*.

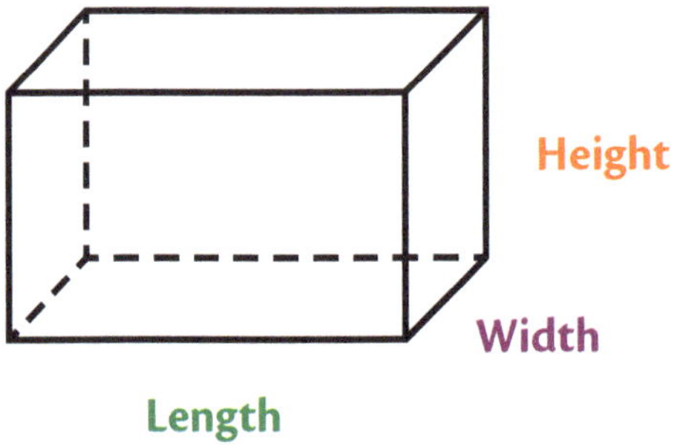

4. This fish tank is 2 feet long, 1 foot wide, and 1 foot high.

   _____ x _____ x _____ = ☐ cubic feet

5. This dog carrier is 3 feet long, 2 feet wide, and 2 feet high.

_____ x _____ x _____ = ☐ cubic feet

6. This storage shed is 2 meters wide, 4 meters long, and 4 meters high.

_____ x _____ x _____ = ☐ cubic meters

7. This garbage bin is 5 feet high, 3 feet wide and 3 feet long. How much garbage can it hold?

_____ x _____ x _____ = ☐ cubic feet

Practice multiplying and dividing from memory.

9. $7 \times 8 =$ ☐

10. $10 \div 5 =$ ☐

11. $9 \div 3 =$ ☐

12. $6 \times 4 =$ ☐

13. $3 \times 9 =$ ☐

14. $12 \div 2 =$ ☐

15. $\frac{15}{3} =$ ☐

16. $4 \times 3 =$ ☐

17. $4 \times 9 =$ ☐

18. $\frac{12}{4} =$ ☐

19. $4 \times 12 =$ ☐

20. $16 \div 4 =$ ☐

Measure the volume of a few 3-dimensional shapes at your home. Find three things that are shaped like rectangular prisms. For each item, choose whether you will measure in centimeters, inches, feet, yards, or meters. Measure its length, width, and height. Round each of these measurements to the nearest whole number. Then multiply these three measurements together. You may use a calculator if the multiplication numbers are larger than you know the facts for. Things you could measure:

- The main area of a dog house
- A refrigerator or freezer
- The inside of the back of a van
- Inside a drawer or cupboard
- A jewelry box
- A toy box

Use the area below to record your measurements.

21. What did you measure? ____________________

____ x ____ x ____ = ☐ cubic ____________________

22. What did you measure? ____________________

____ x ____ x ____ = ☐ cubic ____________________

23. What did you measure? ____________________

____ x ____ x ____ = ☐ cubic ____________________

Adding & Subtracting in Real Life DAY 68

You've already learned to apply math to "story/word exercises" up to this point. But let's see if you can find ways to tell your own math stories and set up your own exercises.

Today, let's learn how to connect the words people use with adding and subtracting in God's world.

Here are some of the addition clue words you need to look for:

Add, Sum, Total, Together, Come, Join

Outside of a sentence, these words don't make much sense. But when you see them used in a sentence, it often means there is some adding going on. Here are a few examples.

• This Sunday two of my friends are joining my family for an afternoon in the park!

• I've just bought a set of ten new screwdrivers to add to my collection in the garage.

Here are some subtraction clue words to look for:

Subtract, Difference, Split, Separate, Leave, Remove

These are words that might mean subtraction is happening. They don't always mean subtraction, but you'll learn to recognize when they do. Can you see the subtraction going on in these sentences?

• Our church used to have 150 members, but 25 people left to plant another church in a different town.

• I bought 5 pounds of meat from the butcher, but I had to cut almost 2 pounds of fat off before I cooked it.

• When we checked out at the grocery store, $20 dollars was subtracted from the bill of $135 because we used coupons to save money.

Activity

Read the following sentences and decide whether there is addition or subtraction going on. In some of these sentences, there is both. Watch for that and describe what is happening in your own words as in the example.

Example:

> So [Rebekah's family] sent away Rebekah their sister and her nurse, and Abraham's servant and his men. (Genesis 24:59)

A certain number of people were at Rebekah's family home. Three people plus a number of other men were sent away.

Addition or subtraction?

Subtraction

1. In summer, we have no sandhill cranes in our area. But in spring and fall, the cranes come by the thousands and gather together in the lakes and cornfields. They stop to eat and rest before continuing on their migration.

Addition or subtraction?

2. "Look at the birds of the air, for they neither sow nor reap nor gather into barns; yet your heavenly Father feeds them. Are you not of more value than they? Which of you by worrying can add one cubit to his stature?" (Matthew 6:26-27)

In your own words:

____

____

____

Addition or subtraction?

____

3. I bought 100 trees in order to plant a forest. I've already planted most of them.

In your own words:

____

____

____

Addition or subtraction?

____

4. Then the children of Israel did so and gathered [the bread], some more, some less. (Exodus 16:17)

Addition or subtraction?

5. Jesus prayed, "While I was with [my disciples] in the world, I kept them in Your name. Those whom You gave Me I have kept; and none of them is lost except the son of perdition, that the Scripture might be fulfilled." (John 17:12)

Addition or subtraction?

6. I spent some time eating breakfast, reading a book, and getting dressed. In total, it took me about one hour.

Addition or subtraction?

7. Now, create one addition and one subtraction story/word exercises yourself. You might think of stories from normal life, or you can use your imagination to create some interesting situations.

Addition Story/Word Exercise:

Subtraction Story/Word Exercise:

Practice DAY 69

Here is some practice with adding and subtracting. Use your wisdom of long addition and long subtraction to figure out how God makes these numbers come together to make a new amount.

1. $\begin{array}{r} 34 \\ +\ 18 \\ \hline \end{array}$

2. $\begin{array}{r} 45 \\ -\ 27 \\ \hline \end{array}$

3. $\begin{array}{r} 75 \\ -\ 52 \\ \hline \end{array}$

4. $\begin{array}{r} 62 \\ +\ 31 \\ \hline \end{array}$

5. $\begin{array}{r} 116 \\ +\ 586 \\ \hline \end{array}$

6. $\begin{array}{r} 506 \\ -\ 250 \\ \hline \end{array}$

Find the right number to answer these math questions.

7. What plus 7 equals 15? ________ + 7 = 15
8. What plus 9 equals 15? ________ + 9 = 15
9. What minus 5 equals 9? ________ − 5 = 9
10. What minus 7 equals 9? ________ − 7 = 9
11. What times 2 equals 12? ________ × 2 = 12
12. What times 4 equals 16? ________ × 4 = 16
13. What divided by 2 equals 4? ________ ÷ 2 = 4
14. What divided by 3 equals 7? ________ ÷ 3 = 7

Add the fractions below. Color in the right number of pie slices for the answer by filling in whole pies first and then adding additional slices as needed. In the box, write the answer as a mixed fraction. The first one is completed for you. Remember, add the numerators, but don't add the denominators. Denominators stay the same when you add and subtract fractions.

$$\frac{2}{3} + \frac{2}{3} = 1\frac{1}{3}$$

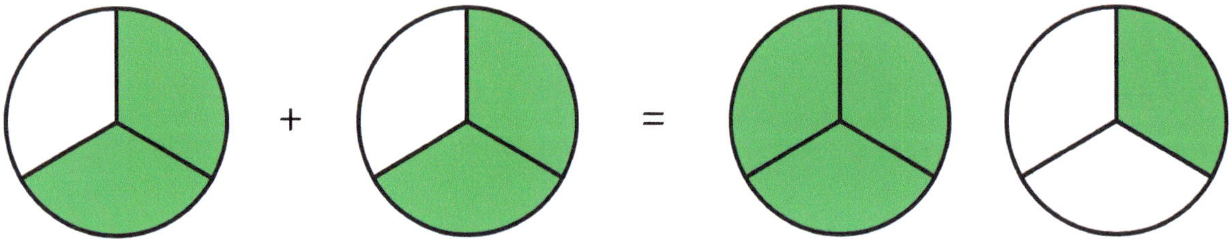

15. $\frac{3}{7} + \frac{5}{7} =$ ☐

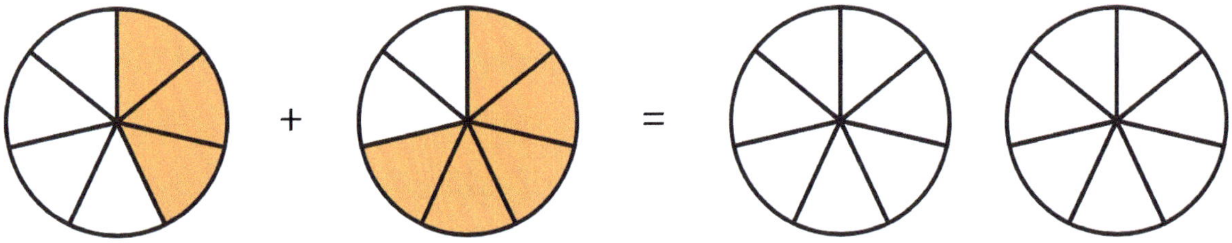

16. $\frac{4}{6} + \frac{5}{6} =$ ☐

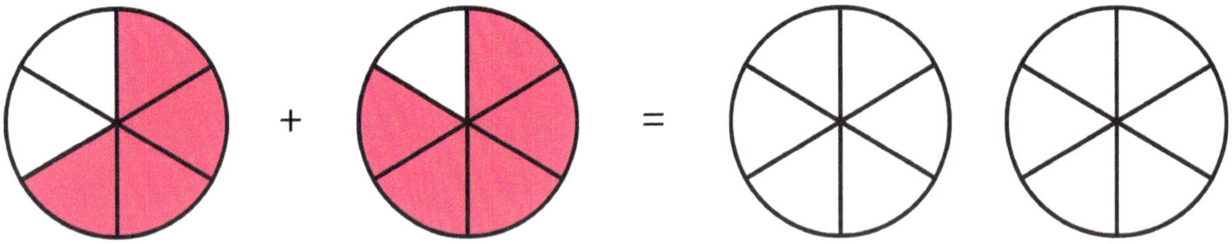

Decide whether to add or subtract when you find the answer to these word exercises.

17. Suppose it takes a total of 20 hours of driving to reach Grandma's house. You plan to drive 13 hours the first day. How many remaining hours will there be to drive on the second day? _________ If you start driving at 8:00 a.m. on the second day, what time will you arrive? _________

18. Suppose you plan to help with yard work for 3 days, earning $124 each day. When that job is finished, you would like to buy a bicycle that costs $398. How much more money will you need to buy the bike? _________

19. At a football game, 56 parents were sitting on the bleachers, and 25 parents were standing. How many parents were at the game to cheer on the kids? _________

20. The donut shop had made 60 donuts. That morning, they sold 41 of the donuts. How many donuts were left to sell in the afternoon? _________ If a customer calls and places an order for 36 donuts to be picked up in the evening, how many more donuts does the shop need to make? _________

DAY 70 Adding & Multiplying in Real Life

Now let's learn more about addition and multiplication in real life. Addition and multiplication both make amounts larger, but they make amounts larger in different ways.

Here are some clue words that are used to talk about making things larger. They are used for both addition and multiplication:

Add, Increase, Bigger, Higher, More, Faster, Heavier, Longer

Here are some clue words that are used to talk specifically about multiplication:

Multiply, Times, Each/Every

Check out this example from the Bible:

> And Joab said to the king, "Now may the LORD your God add to the people a hundred times more than there are…
> (2 Sam 24:3a)

The way people use language can be really tricky, but that's why you are doing this lesson! God will help you learn to deal with hard wording like this.

Here are a few examples of how multiplication can be captured in everyday words. Can you underline the clue words that mean there's multiplication going on?

- If I worked on math for 30 minutes each day, I wonder how long I would work on math in a week.
- At one point my father weighed ten times what I weighed.
- My family went up to the mountains to search for gemstones. We found so many, that we multiplied the size of our collection 4 times!

Activity

Read the following sentences and write whether there is addition or multiplication going on. Be careful. Some are challenging as to which one is meant.

1. Then he took servings to them from before him, but Benjamin's serving was five times as much as any of theirs. So they drank and were merry with him. (Genesis 43:34)

____

2. Three times you shall keep a feast to Me in the year (Exodus 23:14)

____

3. One day, I was watching a bird bounce around my backyard gathering twigs, one by one, to build a nest.

____

4. If riches increase,
Do not set your heart on them. (Psalm 62:10)

____

5. When parents have kids, and their kids have kids, and their kids have kids, families grow.

____

6. And I [God] will add to your days fifteen years. I will deliver you and this city from the hand of the king of Assyria; and I will defend this city for My own sake, and for the sake of My servant David. (2 Kings 20:6)

____

7. Then Nebuchadnezzar was full of fury, and the expression on his face changed toward Shadrach, Meshach, and Abed-Nego. He spoke and commanded that they heat the furnace seven times more than it was usually heated.
(Daniel 3:19)

____

Which is the larger fraction? Fill in the bubble with a greater than sign (>) or less than sign (<). The colored slices should help you get the right answer.

8. 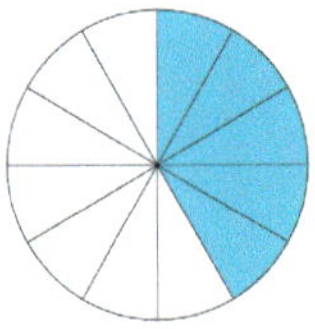 $\frac{5}{12}$ ◯ $\frac{7}{12}$ 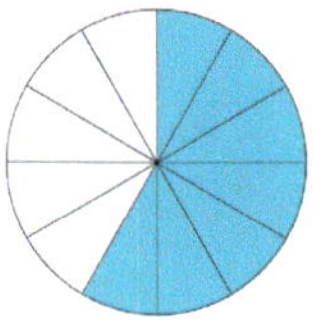

9. 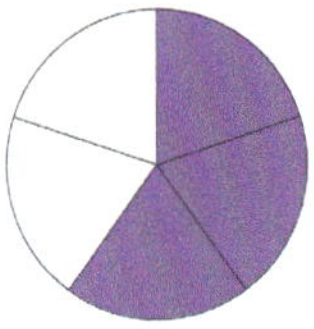 $\frac{3}{5}$ ◯ $\frac{1}{2}$ 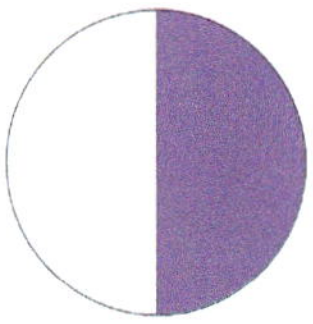

10. 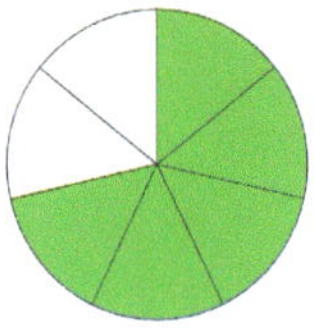 $\frac{5}{7}$ ◯ $\frac{2}{3}$ 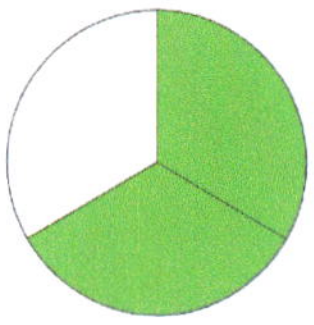

11. 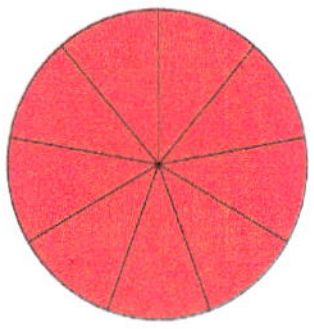 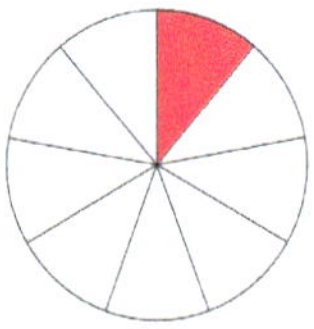 $\frac{10}{9}$ ◯ $\frac{7}{6}$ 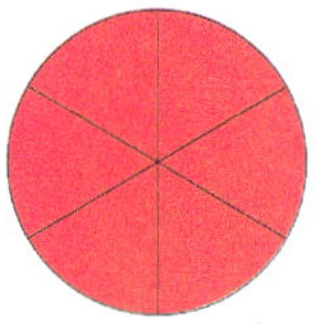 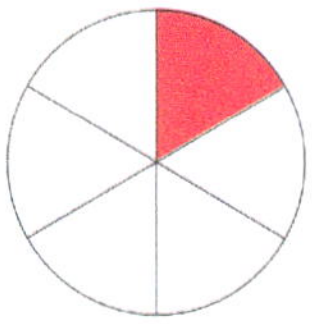

Fill in each bubble with a greater than (>), less than (<), or equal (=) sign.

12. 2 x 2 ◯ 2 + 2

13. 15 – 7 ◯ 2 x 5

14. 12 ÷ 4 ◯ 12 – 8

15. 4 x 3 ◯ 5 + 8

16. 6 x 6 ◯ 4 x 9

17. 3 + 6 ◯ 42 ÷ 6

Practice DAY 71

Practice multiplying and dividing with numbers that are a little bigger. You may be able to get these from memory, but there is a good chance you'll want to use your multiplication table in the back of the book for help. Don't forget to pray, too!

1. $5 \times 7 =$ ☐

2. $25 \div 5 =$ ☐

3. $21 \div 3 =$ ☐

4. $7 \times 8 =$ ☐

5. $6 \times 5 =$ ☐

6. $36 \div 9 =$ ☐

7. $\frac{30}{6} =$ ☐

8. $6 \times 6 =$ ☐

9. $5 \times 8 =$ ☐

10. $\frac{45}{5} =$ ☐

Let's practice measuring angles. First, guess a measurement for each angle below. Then measure to see how close your guess was.

11. Your guess:

______

The measurement:

______

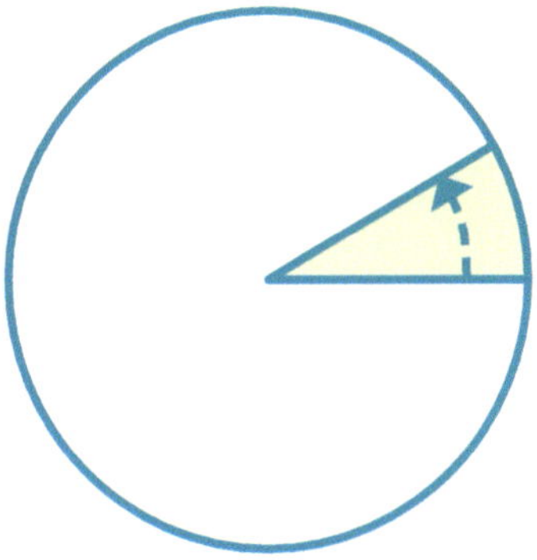

13. Your guess:

______

The measurement:

______

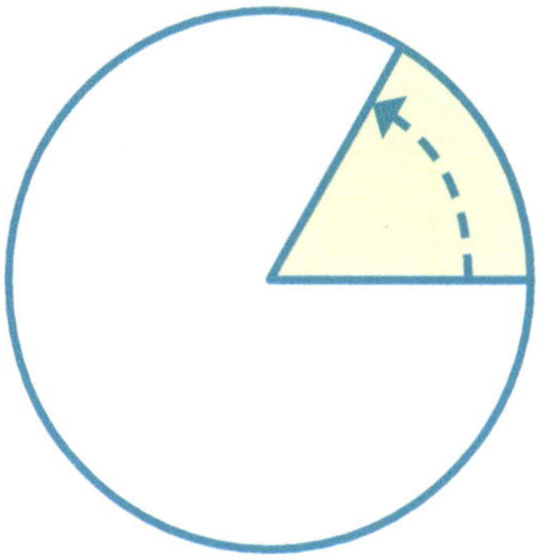

12. Your guess:

______

The measurement:

______

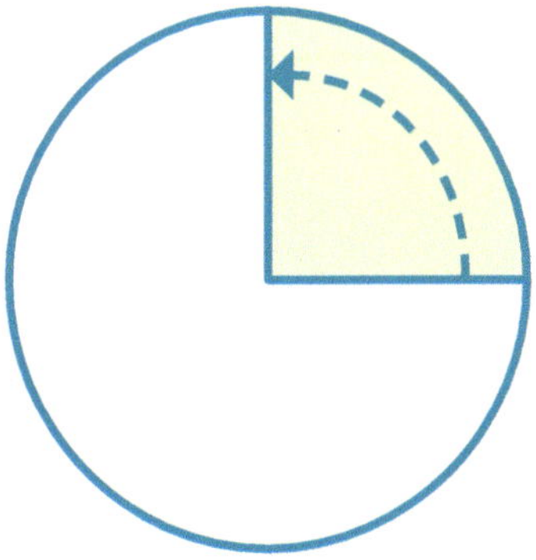

14. Your guess:

______

The measurement:

______

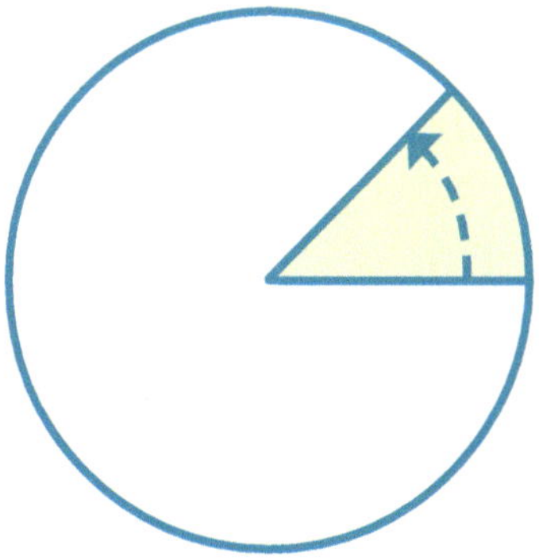

Practice rounding these numbers! Rounding helps us understand large numbers in God's world.

15. Round 6,824,053 to the nearest 100

____

16. Round 6,824,053 to the nearest 10,000

____

17. Round 6,824,053 to the nearest 100,000

____

18. Round 123,456 to the nearest 1,000

____

19. Round 123,456 to the nearest 10,000

____

20. Round 123,456 to the nearest 100,000

____

Pretend that the drawing below is a map of a house with many rooms. You are going to practice finding area by measuring on this map. Find the area of each of these squares and rectangles in square inches. Write the area of each inside its shape. Then add up the areas to get the total area.

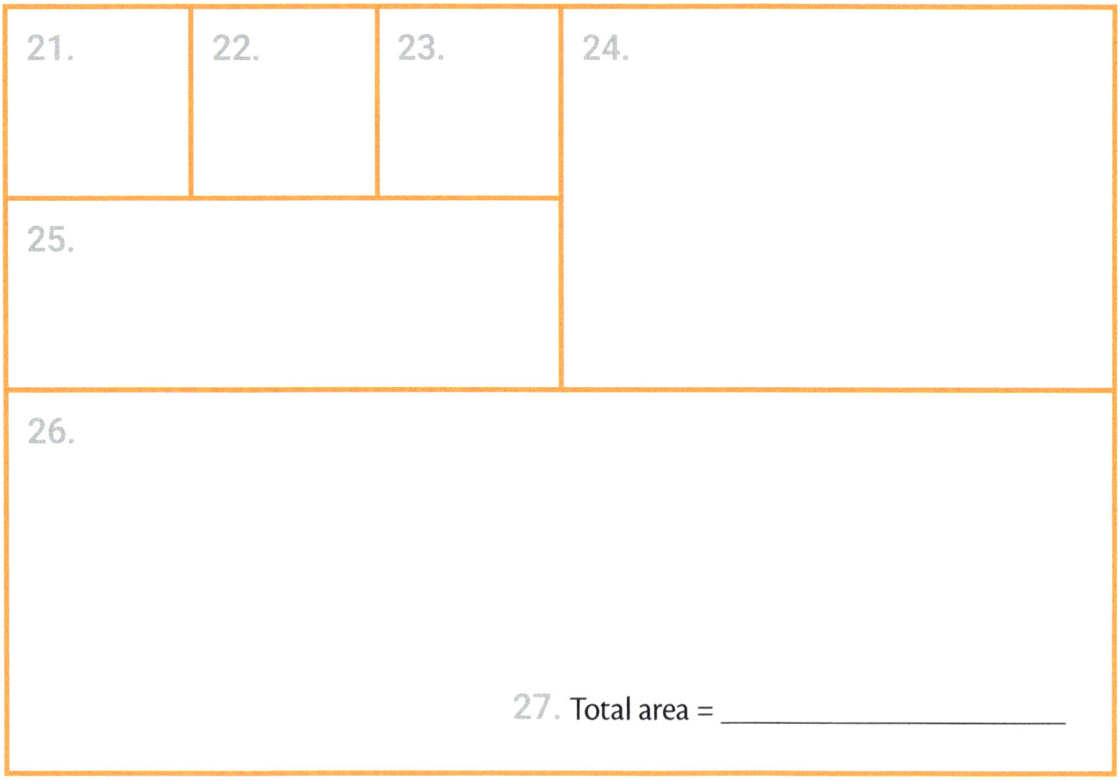

Practice DAY 72

Subtract these fractions. Color in the right number of pie slices for the answer. Just as with adding fractions, when you subtract fractions, you only subtract the numerators. The denominators stay the same.

1. $\frac{5}{6} - \frac{3}{6} =$ ☐

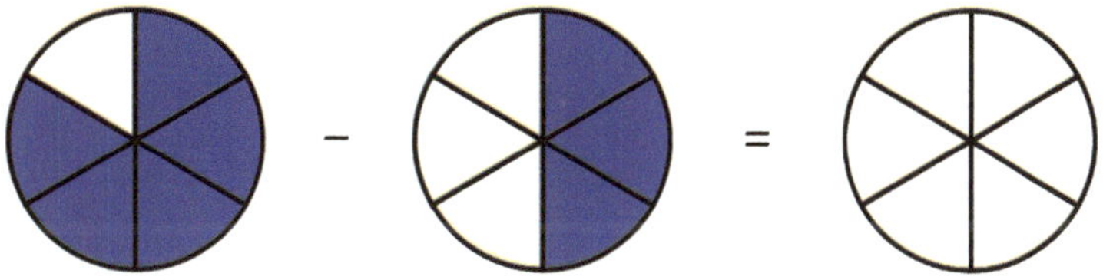

2. $\frac{11}{12} - \frac{7}{12} =$ ☐

3. $\frac{7}{5} - \frac{3}{5} =$ ☐

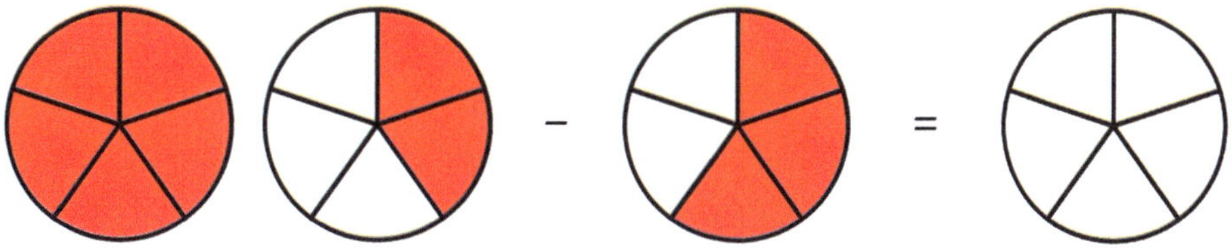

4. $\frac{5}{6} - \frac{4}{6} =$ ☐

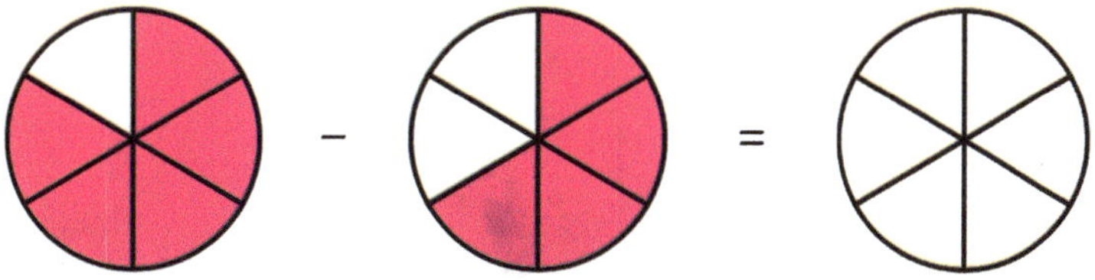

5. Suppose you have 15 biography books about missionaries and scientists. You notice that 8 of the books are about missionaries. What fraction of the books are about scientists?

____

6. You have $\frac{2}{5}$ of a tube of blue paint and $\frac{4}{5}$ of a tube of yellow paint. If you mix all this paint together to get a green color, what is the total amount of paint in the mix?

$\frac{2}{5} + \frac{4}{5} =$ __________

7. When you planted a new tree in the yard, it was only $\frac{2}{6}$ of a yard tall. Three months later, the tree was $\frac{5}{6}$ of a yard tall. How much did the tree grow over those months?

$\frac{5}{6} - \frac{2}{6} =$ __________

8. Draw the hands on each clock. Remember, the shorter hand points towards the hour number, and the longer hand points to the number of minutes. Count by 5s to place the minute hand on the clock's correct number.

Dividing in Real Life DAY 73

Let's learn about certain key words that show division is going on. Some of these division words are:

Divide, Per

Division means to split up equally. Underline the key words in the following division sentences.

1. Apples are one of my mom's favorite foods. She has 8 in the fridge right now and will probably eat two per day.

2. When I have money to give for helping others, I sometimes divide it equally. Other times I give people different amounts because some needs require more money than others.

3. We often walk to the library. It takes about 60 minutes to walk the three miles. I think 3 miles per hour is a pretty good speed.

Activity

Read the following sentences and decide whether there is multiplication or division going on. Using math language explain in your own words what is happening. An example is shown.

> Divide the plunder into two parts between the warriors who went out to battle and all the congregation. (Numbers 31:27 ESV)

The total amount of plunder was divided by two for the two groups of people.

4. This year for Christmas, I plan to spend the same amount of money on a gift for my dad as I spend for my mom. I think I will spend $60 dollars for dad's gift. How much will I spend for both gifts together?

____

____

5. Micah said to him, "Stay with me, and be to me a father and a priest, and I will give you ten pieces of silver a year and a suit of clothes and your living." And the Levite went in. (Judges 17:10 ESV)

How long would it take before the priest collected 40 pieces of silver?

____

____

6. My grandfather just came by and gave us 50 small snack bags of homemade apple chips. We liked them so much that we ate 5 bags as soon as he arrived. Then we decided to make the rest of them last a long time. Some of us want to make a rule that we can only eat 5 bags per week. Others want to allow us to eat 9 bags per week. How long will the apple chips last if we eat 5 bags per week?

____

____

7. Let's pick up 10 kilograms of apples at the store. If it's $2.00 for each kilogram, how much money should we take to buy the apples?

____

____

Now create a few division word exercises yourself. Since subtraction and division both make smaller numbers, think about how your division stories will be different from your subtraction stories.

8. ______________________________

____

____

____

9. ______________________________

____

____

____

____

Add this money. Be sure to include the decimal point in the correct place.

10.

	1.	5	0
+	1.	7	5

11.

	2	3.	4	5
+	2	8.	8	4

12.

	1.	8	9
+	1.	5	4

13.

	3	4.	4	5
+	4	7.	8	3

14.

	2	5.	7	5
+	2	7.	8	2

15.

	8	7.	2	3
+	2	3.	4	5

DAY 74 Practice

Today let's practice finding degrees and lengths. First, guess a measurement for each of the shapes. Then measure to see how close your guess was.

1. Guess the size of this angle. _____

Measure the size of the angle. _____

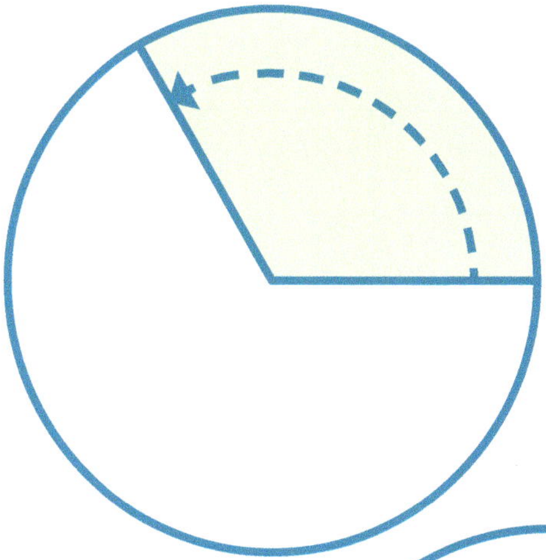

2. Guess the size of this angle. _____

Measure the size of the angle. _____

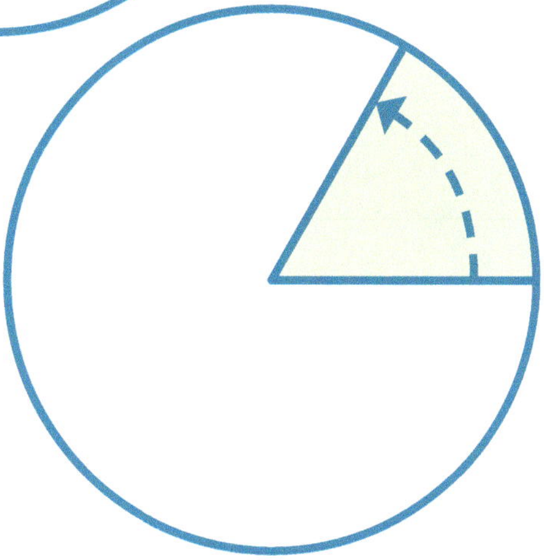

3. Guess the length of this line in centimeters. _____ Measure the length of the line: _____

4. Guess the length of this line in inches. _____ Measure the length of the line: _____

5. Guess the length of this line in centimeters. _____ Measure the length of the line: _____

6. Guess the length of this line in inches. _____ Measure the length of the line: _____

Everyday Math

7. Peter and Jamie want to buy Grandma and Grandpa a Christmas present. Peter has $12.35 and Jamie has $11.75. If they combine their money, how much will they be able to spend for the gift?

____

8. The children's choir contained 28 singers. If 12 of the singers were boys, how many were girls?

____

9. The six children in the Jones family each collected 14 flowers. How many total flowers was the family able to collect? (Hint: For this exercise you will need to multiply 14 x 6. But you haven't memorized multiplication math facts for 2-digit numbers like 14. Later in this book, you will learn a special way to multiply large numbers like this, but for now, let's try a different way. Break 14 into one 10 and 4 ones. Then you can separately multiply 10 x 6 and 4 x 6. Then add those two answers together to find the total number of flowers.)

____

10. Suppose your mom has asked you and your brother to write addresses on the 26 Christmas cards that she wants to mail. If you and your brother decide to split the job evenly, how many cards will each of you need to address? (Hint: Just as with the last exercise, you have not memorized math facts for the number 26. But you can do this exercise too, by splitting 26 into 10s and 1s. Decide whether to add, subtract, multiply, or divide those 10s and 1s. Then add up the two answers to see how many cards you will each address.)

____

Speed Test

Ask your parent/teacher to time your speed on these multiplication exercises. You can practice this page every now and then to help you memorize these facts. Your goal (by the time you have completed Level 3 Math) should be to answer the exercises correctly in 90 seconds.

Average Time: 120 seconds

Target Time: 90 seconds

Best Student Time: 60 seconds

Expert/Teacher Time: 45 seconds

Your Time ________

1. 6 x 5 =
2. 9 x 9 =
3. 9 x 8 =
4. 10 x 3 =
5. 9 x 7 =
6. 5 x 7 =
7. 3 x 7 =
8. 6 x 0 =
9. 9 x 10 =
10. 3 x 5 =
11. 6 x 9 =
12. 8 x 5 =
13. 5 x 0 =
14. 7 x 4 =
15. 10 x 1 =
16. 8 x 8 =
17. 4 x 3 =
18. 3 x 6 =
19. 6 x 6 =
20. 2 x 9 =
21. 6 x 8 =
22. 1 x 5 =
23. 6 x 7 =
24. 1 x 8 =
25. 2 x 8 =
26. 1 x 2 =
27. 6 x 0 =
28. 4 x 6 =
29. 2 x 6 =
30. 5 x 5 =
31. 5 x 9 =
32. 3 x 9 =
33. 3 x 2 =
34. 5 x 2 =
35. 8 x 4 =
36. 8 x 7 =
37. 10 x 1 =
38. 5 x 4 =
39. 8 x 8 =
40. 4 x 9 =
41. 5 x 8 =
42. 4 x 4 =
43. 4 x 6 =
44. 7 x 7 =
45. 9 x 9 =

Solving Problems with Math DAY 75

One of the main reasons for learning math is to solve problems or to answer questions. Hopefully, you are learning to use your math to help your family and others to solve problems. Suppose your family has a garden bench that has become splintery from the rain and sun. A good solution to this problem would be to cover the bench with stone tiles. Each tile is 1 square foot, and costs $2.

Here is a diagram of the bench. The width (2 feet) and the length (4 feet) are written out next to lines with two arrows. These measurements are called *dimensions*. Architects and engineers use drawings like this to plan out, or design, what they will set out to build.

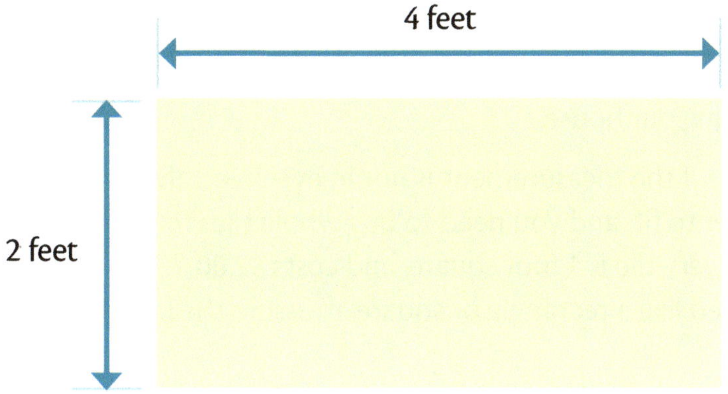

1. $2 \times 4 =$ ☐ sq. ft.

Each square tile that your family wants to buy has an area of 1 square foot. Because each tile is a square, its length is the same as its width. The four sides are each 1 foot long.

If each tile costs $2, how much will it cost to buy all the tiles needed for the bench?

2. $2 \times$ ☐ $=$ ☐

Now, suppose your family decides to use smaller square tiles that measure ½ foot on each side. These would be 6-inch by 6-inch tiles. How many ½-foot red tiles (See picture.) would it take to fill in 1 square foot? It would take 4 of these to equal 1 square foot.

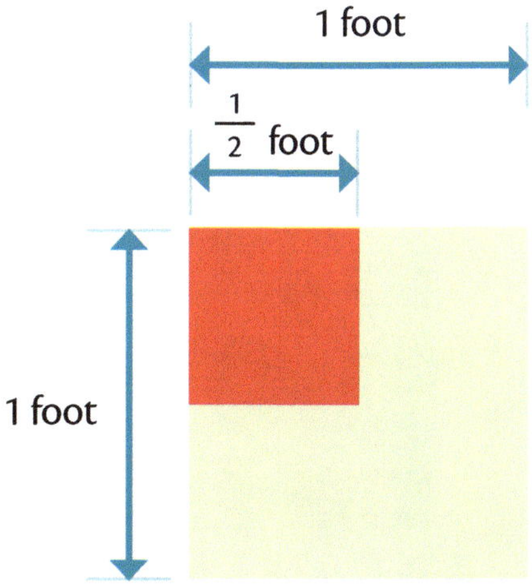

Architects and carpenters have a shorter way to say the size of rectangles and squares. Instead of saying your bench is "4 feet long and 2 feet wide," they would say it's "2 feet by 4 feet." They would write this as "2ft x 4ft."

Since these are smaller tiles, it would take 4 tiles for every square foot of the bench. Let's find out how many small tiles it would take to cover the bench.

8 square feet x 4 tiles per square foot = 32

3. If each of these small tiles cost $1, how much will all the small tiles cost?

☐ x ☐ = ☐

Activity

Now plan the flooring for a bathroom in your house.

- Take the measurements in feet. If the measurement is not in even feet, always round up. Some tiles must be cut in order to fit, and you need to buy whole tiles to cut.
- For this activity, assume that every tile is 1 foot square, and costs $2.00.
- If your bathroom floor is shaped like a rectangle or square, measure the length and width of that shape.
- If the floor is made up of several rectangles or squares, measure each shape separately. Their areas can be added together later.
- You may use a calculator for the multiplication of 2-digit numbers.

Note for parent/teacher: The student may take measurements in decimeters or meters instead of feet. Assume that a 1-square-decimeter tile is $1.

If your bathroom floor is one rectangle, take your measurements and fill in the dimensions on the drawing. If not, you may make your own drawing to write your measurements on.

4. How many 1-square-foot tiles do you need to complete this project? Multiply the width and the length for each rectangular area.

☐ x ☐ = ☐ Sq. Ft.

5. Write additional equations for additional rectangular areas if necessary. Add up the areas and write the total area here: ________________ square feet

How much would this project cost if tiles are $2 each? ☐ x ☐ = $ ☐

How much would this project cost if your tiles were $3 each? ☐ x ☐ = $ ☐

Student Exercises

1. Sometimes, outdoor wooden surfaces become rotten and dangerous to step on. Below is a drawing for the surface of a wooden porch floor that needs to be replaced. Let's figure out how many boards the job will take. Each board is 6 inches (½ foot) wide and 20 feet long. How many boards will fit in each foot of the porch's width? __________

How many boards will it take to replace the whole porch floor? __________ Draw each board if that helps.

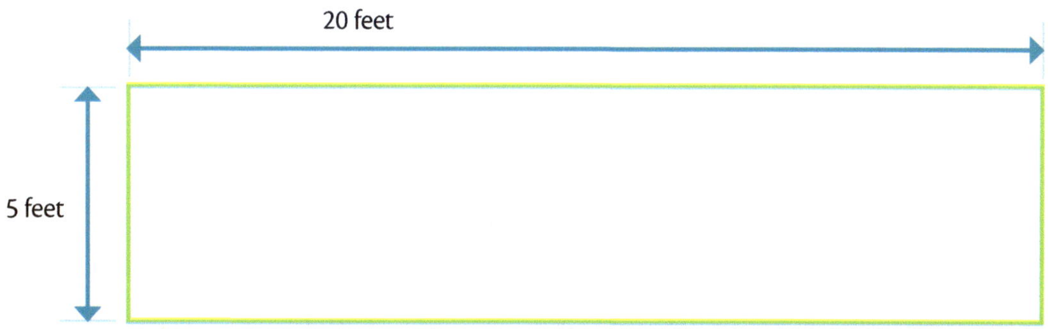

If each board costs $10, how much will it cost to build this porch? ________________

2. Dad was watching some construction workers make the outer wall of a new grocery store. The workers had extra concrete blocks that measured 1 foot, by 1 foot, by 1 foot. Dad checked with the owner of the blocks and asked if he could buy the extras. He wanted to build a short wall in the yard to keep the little ones away from their bee hives. The wall would need to be 60 inches (5 feet) long and 36 inches (3 feet) high. How many 1-cubic-foot concrete blocks should Dad ask to buy? ___________

☐ x ☐ = ☐

3. How many concrete blocks would be needed for this wall if they were 6 inches high, 1 foot long, and 1 foot wide? ___________

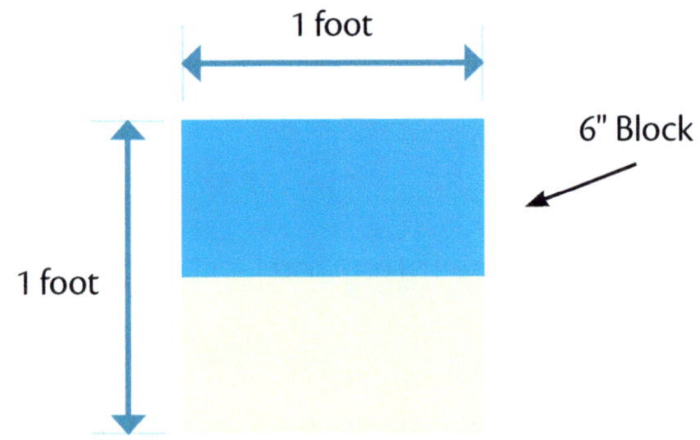

4. How many square feet of wallpaper would be needed to paper these four walls of a room? Add them up!

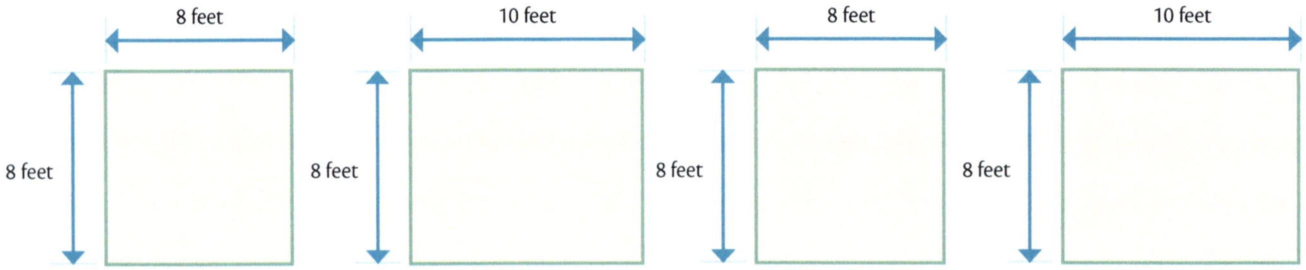

☐ + ☐ + ☐ + ☐ = ☐ Sq. Ft.

Fill in the answer for each of these numbers.

	How many **hundreds**?	How many **tens**?	How many **ones**?	How many **tenths**?	How many **hundredths**?
5. 123.45	_____	_____	_____	_____	_____
6. 345.56	_____	_____	_____	_____	_____
7. 567.89	_____	_____	_____	_____	_____
8. 90.09	_____	_____	_____	_____	_____
9. 909.09	_____	_____	_____	_____	_____
10. 999.9	_____	_____	_____	_____	_____

11. Draw the hands on each clock. Remember, the shorter hand points towards the hour number, and the longer hand points to the number of minutes. Count by 5s to place the minute hand at one of the clock's numbers. If the time is not exactly at one of the numbers, draw the minute hand where you think that time should be between two clock numbers.

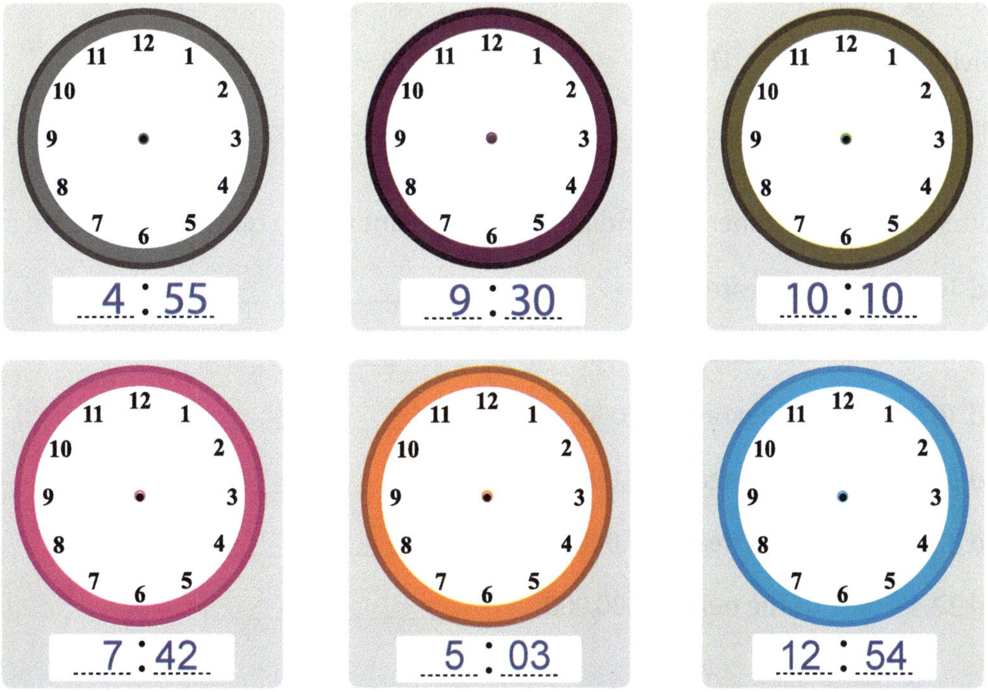

DAY 76 Practice

You have probably heard the word *operation* before. It comes from the word *operate* which means to make something do what it's meant to do. A cow's digestion operates using its four-part stomach. A doctor does an operation on a patient, working inside the patient's body to make it function better. Operation Christmas Child is a ministry made up of many people working to help people around the world while they share the Gospel of Christ with them.

We use the word operation in math too. A math operation is the process we use in an equation to show how the numbers will work together to find an answer. The most common math operations are addition, subtraction, multiplication and division (+, −, ×, ÷).

Write the answer to these math questions. You will use all four operations.

1. What plus 6 equals 19? _______ + 6 = 19
2. What plus 11 equals 13? _______ + 11 = 13
3. What minus 10 equals 8? _______ − 10 = 8
4. What minus 4 equals 9? _______ − 4 = 9
5. What times 5 equals 15? _______ × 5 = 15
6. What times 3 equals 9? _______ × 3 = 9
7. What divided by 13 equals 1? _______ ÷ 13 = 1
8. What divided by 3 equals 6? _______ ÷ 3 = 6

Practice rounding these numbers. Rounding is an important way to quickly understand amounts.

9. Round 167,285 to the nearest 10. _______________
10. Round 167,285 to the nearest 10,000. _______________
11. Round 167,285 to the nearest 100,000. _______________
12. Round 32,485,882 to the nearest 1,000 _______________
13. Round 32,485,882 to the nearest 1,000,000 _______________
14. Round 32,485,882 to the nearest 100,000 _______________

15. Each of the family's 5 students needs a new notebook for Bible class, math class, and English class. How many notebooks does their mom need to buy?

____

16. Mom wants everyone to look nice for family photos. She buys herself a dress for $25.99. She also buys Dad a shirt for $18.00 and shirts for you and your brother that cost $12.99 each. How much does she spend on new clothes for the family?

____

17. Dad decides to take you to a professional baseball game. The stadium is an hour's drive away. Dad wants to allow an extra half hour to park, walk, and find your seats before the game starts. If the game starts at 6:00pm, what time should you leave to be at the game in time?

____

Solve these exercises.

18. $7 \times 9 =$ ☐

19. $10 \times 4 =$ ☐

20. $31 - 16 =$ ☐

21. $106 - 28 =$ ☐

22. $\frac{9}{3} - \frac{7}{3} =$ ☐

23. $\frac{1}{4} + \frac{1}{4} =$ ☐

24. $\frac{2}{7} + \frac{6}{7} =$ ☐

25. $\frac{8}{9} - \frac{3}{9} =$ ☐

DAY 77 Go Take Care of God's World

This lesson integrates math into everyday life. With this activity, the student will apply God's patterns and wisdom in meaningful ways to home and community. Take a break from memory work and academic exercises to make math part of everyday life. The following are suggestions or examples. Other ideas may be added to the list.

> Then the Lord God took the man and put him in the garden of Eden to tend and keep it. (Genesis 2:15)

> And the Lord God said, "It is not good that man should be alone; I will make a helper [Eve] comparable to him." (Genesis 2:18)

1. Just as God gave Adam and Eve the garden of Eden to tend and watch over, God still gives families land to tend. Everything belongs to God, but He lets people be the earthly owners of land. He lets them use their land and take care of it. God's Word has laws against stealing someone's land.

Now, let's pretend that you were given 7,000 square feet of land with a house on it. The house takes up 1,500 square feet of the land. What would you do with the rest of the land? Would you plant a grassy yard? Would you grow a garden or raise a few sheep? Maybe you would build a dog run so you could breed dogs and sell them.

Use the drawing on the next page to plan out your land. Decide how much land you will allow for the things you want to produce. Remember, you are responsible for tending all of this property as God's steward (the one who takes care) of the land.

- Use a ruler to measure and draw the outlines (fences) of the different spaces. Make each space's border in even inches. Each inch of length on the plan will stand for 10 feet of length on your land. (If you want a space for a garden, first do the "If you want to plan a garden" activity that follows.)
- Color the spaces you've outlined with different colors.
- Use a calculator to multiply and find the area of each space. Write the square feet within each space. Add up the square feet of all the spaces to make sure they add up to 7000 square feet.

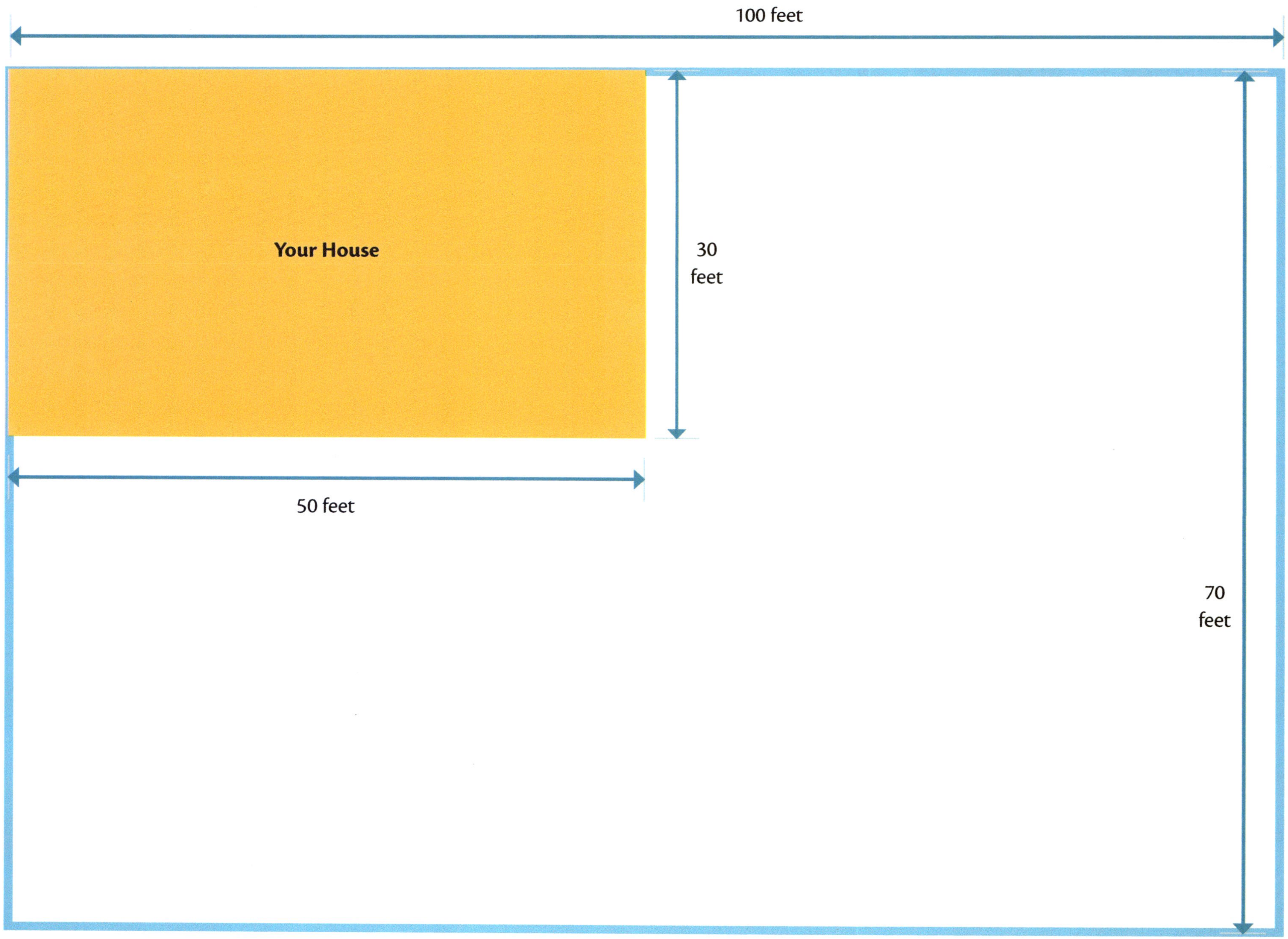
100 feet
Your House
30
feet
50 feet
70
feet

2. Let's suppose you would like to grow a garden on some of your land. You can use the "Garden Crop Yields" chart to help you decide what crops to plant. The chart shows the yield (the amount of food produced) for each crop during its growing season. For each crop in the chart, you can see the yield for a certain number of plants in a 30 foot row.

You can use a calculator and ask for help in figuring out a plan for your garden that will fit a space. Garden rows should usually be planted 3 feet apart. This means that you could plant 10 rows across a space that's 30 feet wide. If your rows are all 30 feet long, you would have a square garden that's 30 ft x 30 ft. (Say, "30 feet by 30 feet.") You could make your garden larger if you would like to grow more food. If your plan for the imagined 7,000 square feet of land includes a garden, draw and color your garden plan in its space.

Of course, growing food takes a lot of work. Remember, God designed the plants, water, sunshine, and soil to work together to make our food, so give Him thanks!

> He has filled the hungry with good things. (Luke 1:53)

GARDEN CROP YIELDS (PER 30 FOOT ROW)

GARDEN CROP	HARVEST WEIGHT (LBS)	PLANTS PER ROW
Asparagus	10	20
Basil	10	30
Snap Beans (bush)	15	120
Snap Beans (pole)	15	60
Beets	30	120
Broccoli	15	20
Brussel Sprouts	15	20
Cabbage	45	15
Carrots	30	150
Cauliflower	30	20
Celery	30	45
Corn	25	45
Cucumbers	30	40

GARDEN CROP	HARVEST WEIGHT (LBS)	PLANTS PER ROW
Dill	15	30
Eggplant	30	20
Garlic	10	60
Kale	20	30
Lettuce	20	30
Melons	50	10
Onions	30	90
Peas	10	90
Peppers	30	20
Potatoes	45	30
Pumpkins	50	15
Radishes	12	150
Rhubarb	30	6
Spinach	15	60
Summer Squash	150	14
Winter Squash	100	10
Strawberries	20	20
Sweet Potatoes	30	30
Tomatoes	80	10

3. Make a plan to wallpaper a room in your house. Measure the length and width of the four walls to the next largest foot. This will give you a little extra paper. Multiply to find the area of each wall. Add up the areas of the four walls.

Additional Activities

For extra activities, games, and helpful videos on these math concepts, refer to **generations.org/math3**.

God will
probably send
rain today.

CHAPTER 7

Probability

DAY 78 Coin Flipping

And Elijah said to Ahab, "Go up, eat and drink, for there is a sound of the rushing of rain." So Ahab went up to eat and to drink. And Elijah went up to the top of Mount Carmel. And he bowed himself down on the earth and put his face between his knees. And he said to his servant, "Go up now, look toward the sea." And he went up and looked and said, "There is nothing." And he said, "Go again," seven times. And at the seventh time he said, "Behold, a little cloud like a man's hand is rising from the sea." And he said, "Go up, say to Ahab, 'Prepare your chariot and go down, lest the rain stop you.'" And in a little while the heavens grew black with clouds and wind, and there was a great rain. (1 Kings 18:41-45)

Elijah prayed for rain. Then, in faith, he predicted it would rain before he even saw a cloud. When there is a drought going on, or even in slightly dry weather, we can pray for rain too.

Often, scientists study God's creation and then make predictions according to what they find. Their predictions say how likely a certain thing might happen. We also can try to predict what will happen by knowing God's creation. We say something like this: "When dark clouds have come over us in the past, it usually begins to rain. There are dark clouds coming over us now, so we will probably see some rain soon."

Science deals with probability. We use the word *probability* to talk about how likely (how probable) it is for a certain thing to happen. Elijah knew for sure that it would rain. He could have said that there was a high probability it would rain.

If the weather report says there is a low probability for rain today, it means that we don't usually get rain when the weather conditions are as they are today. Past weather conditions are used to figure out the probability for rain.

Let's see how we can use numbers to talk about probability.

Activity

Most coins are well balanced. This means that one side of the coin weighs about the same as the other side. Many coins have a picture of a person's head on one side. We call this side *heads*. We call the side without the person's head *tails*. If you flip a coin into the air, you don't know if it will it land with heads facing upwards or tails facing upwards. You've probably guessed that about half the time the coin will land heads up, and about half the time it will land tails up.

Now let's flip a coin 50 times. Record your results in the chart. Use tally marks like this 𝍸 to keep track of how many times you get "heads" (coin landing heads up) and "tails" (coin landing tails up).

HEADS	TAILS

Now let's put your results into fractions. Count the tally marks to see how many times you got heads. Count how many times you got tails. Write these numbers in the numerators below.

Heads: $\frac{\_\_\_}{50}$ *Tails*: $\frac{\_\_\_}{50}$

How close to $\frac{25}{50}$ did you get for each? 25 is half of 50, so $\frac{25}{50}$ is the same fraction as $\frac{1}{2}$.

Next, flip the coin 10 times. Record the results on the chart using tally marks. Fill in the first fraction to show how many of the 10 flips were heads. Fill in the second fraction to show tails.

HEADS	TAILS
___	___

You have just completed an experiment (a test used to discover something). We could say that, based on this experiment,

- We have a _____ out of 10 probability of getting heads.
- We have a _____ out of 10 probability of getting tails.

This time, flip two coins at the same time. (Make sure they are different kinds of coins so you can tell the difference between them.) Instead of 2 possible outcomes (heads and tails), you will see 4 different outcomes: heads-heads, tails-heads, heads-tails, and tails-tails. Do 20 flips with both coins at the same time. Record your results below. Use tally marks to keep track of how many times you get each combination. Then count the number of tally marks for each and write them as numerators in the fractions.

HEADS-HEADS	HEADS-TAILS	TAILS-HEADS	TAILS-TAILS

Heads-Heads: ___ / 50

Heads-Tails: ___ / 50

Tails-Heads: ___ / 50

Tails-Tails: ___ / 50

Coin tosses like this should end up with almost equal probabilities of 5/20 for each combination. Did this happen for you?

> The lot is cast into the lap, but its every decision is from the Lord. (Prov. 16:33)

Sometimes we flip a coin to help us make a choice. Or we roll dice in a board game to tell us how many spaces to move our piece. Let's remember that God is in control of each roll of the dice or flip of the coin. He has already determined what will happen. If He has planned ahead that you will flip the coin five times and get heads each time, then that's what you'll get. But in His ordinary purposes for the world, He usually makes sure that the coin will land on heads about half the time and tails the other half. Remember, God is always in control of everything!

Play It!

Let's play a coin toss game. This time, we won't pay attention to whether the coins land as heads or tails. Place a large basket on the floor or ground. Label four small pieces of paper with numbers 1 thru 4. Place Paper 1 about 4 feet from the bowl. Place the other papers, counting up, at 4-foot intervals farther away from Paper 1. Gather a handful of coins. Stand on Paper 1 and practice tossing the coins into the bowl one at a time. After you master one spot, step back to the next farthest paper and try to toss the coins again.

Once you have tried the coin toss from all four distances, let's keep score. Use only 5 coins to make 5 tosses at each position. This time, write fractions that show how many tosses landed in the basket from each place. Write numerators to show the number of tosses out of 5 that made it into the basket.

	Paper 1	Paper 2	Paper 3	Paper 4
$\frac{\text{Successful tosses}}{\text{Total tosses}}$	$\frac{\square}{50}$	$\frac{\square}{50}$	$\frac{\square}{50}$	$\frac{\square}{50}$

Talk with your parent/teacher about the probability of tossing coins into the basket from Paper 1 as compared with Papers 2, 3, and 4.

Decide which number is larger and write the correct symbol (>, <, or =).

1. 329 ☐ 239
2. $\frac{3}{4}$ ☐ $\frac{3}{9}$
3. 3.4 ☐ 2.1
4. 2,345 ☐ 2,346
5. 2.5 ☐ $2\frac{3}{4}$
6. $\frac{1}{10}$ ☐ 0.2

Put these numbers in order from least to greatest.

437	734	347

_______, _______, _______
7. 8. 9.

DAY 79 Practice

Here is some practice with adding and subtracting. Use your wisdom of long addition and long subtraction to figure out how God makes numbers come together to make different amounts. Some of the carries and borrows have been provided for you.

1.
```
   1
   3 7
 + 1 8
 -----
```

2.
```
   3 13
   4̸ 3̸
 - 2  9
 ------
```

3.
```
   1 3 2
 -     8
 -------
```

4.
```
   5 3 4
 + 8 8 5
 -------
```

5.
```
   5.2
 + 3.4
 -----
```

6.
```
   5.9
 - 4.7
 -----
```

7.
```
   1 4 5,5 4 2,4 3 6
 +                 0
 -------------------
```

8.
```
   1,2 3 4,3 4 5
 -             0
 ---------------
```

Today let's see how fun it is to count by 11s. Counting chunks of 11 is easy because of the pattern it makes:

0, 11, 22, 33, 44, 55, 66, 77, 88, 99, 110, 121, 132

Practice counting by chunks. For each list of numbers, figure out if the list is counting up or counting down. Then fill in the blanks to continue the pattern.

9. Count by 4s: 0, 4, ______, ______, ______, ______

10. Count by 4s: 40, 36, ______, ______, ______, ______

11. Count by 6s: 0, 6, ______, ______, ______, ______

12. Count by 6s: 60, 54, ______, ______, ______, ______

13. Count by 8s: 0, 8, ______, ______, ______, ______

14. Count by 11s: 0, 11, ______, ______, ______, ______

Activity

Do your own experiment. Flip a bottle cap in the air 10 times. Count how many times the flat side lands facing up and how many times the open side lands facing up. Use tally marks like this 𝍸 to record your data.

Fill in the fractions below to show the results.

FLAT SIDE UP	OPEN SIDE UP

Flat side up: $\frac{\_\_}{10}$

Open side up: $\frac{\_\_}{10}$

DAY 80 Rolling a Die

When you flip a coin, you only have 2 possible outcomes—heads or tails. The coin isn't going to land on its edge like this!

When you throw or "roll" a six-sided die, it's different. The die can land with any one of the 6 sides facing up. So there are 6 possible outcomes to the roll of a die—numbers 1-6, shown by the number of dots on top. Since dice are well balanced, with all sides equal, the probability of tossing any one number is the same as tossing any other number.

What is the probability of rolling the number 3? The probability of rolling that 1 number would be 1 out of 6 or $\frac{1}{6}$.

What is the probability of rolling a 3 or a 1? The probability of rolling either of these 2 numbers would be 2 out of 6 or $\frac{2}{6}$. What is the probability of rolling a 4, 5, or 6? The probability of rolling any of these 3 numbers would be 3 out of 6 or $\frac{3}{6}$. 3 is half of 6, so the probability is that half of the time, you would roll a 4, a 5, or a 6. Now, let's experiment with a die. Roll a die 36 times and record your results below. Use tally marks to keep track of how many times you get each of the six numbers. Then count the tally marks and write the probability fraction for each number. Since the probability is the same for any of the six numbers, you should expect to get about a $\frac{6}{36}$ result for rolling each of them. Were you close to 6/36 for each number? __________

1	2	3	4	5	6

___/36	___/36	___/36	___/36	___/36	___/36

Student Exercises

Practice multiplying and dividing these numbers. They are a bit larger, so you might need to use your multiplication/division table. The more you use it, the more familiar you will become with how God made numbers to combine in these ways.

1. $6 \times 5 =$ ☐
2. $36 \div 6 =$ ☐
3. $60 \div 6 =$ ☐
4. $6 \times 7 =$ ☐
5. $8 \times 5 =$ ☐
6. $40 \div 8 =$ ☐
7. $\frac{80}{8} =$ ☐
8. $9 \times 8 =$ ☐
9. $4 \times 7 =$ ☐
10. $\frac{49}{7} =$ ☐

Everyday Math

11. Let's pretend that your mother has sent you to the store to buy apples, oranges, and a pineapple to make a fruit salad for the church fellowship meal. She needs 5 pounds of apples which cost $1.00 per pound. She also needs 3 pounds of oranges, which cost $2.00 per pound. The pineapple will cost $3.59. How much money should you take?

$1 x 5 = __________

$2 x 3 = __________

	$	.		
	$	.		
+	$	.		
	$	.		

12. Now suppose that you are baking bread for the neighbors. They pay $8.00 for an order of two loaves. You sold 6 orders the first week. You sold another 8 orders on the second week. Then, you sold 7 more orders of bread on the third week. How much did you earn selling the bread during these three weeks?

$8 x 6 = __________

$8 x 8 = __________

$8 x 7 = __________

	$			
	$			
+	$			
	$			

DAY 81 Practice

Figure out the correct number to answer each math question. Write it in the blank. Then write an equation that includes your answer. The first one is done for you.

1. What plus 4 equals 13? 9 + 4 = 13
2. What plus 10 equals 13? ☐ + ☐ = ☐
3. What minus 8 equals 8? ☐ − ☐ = ☐
4. What minus 7 equals 8? ☐ − ☐ = ☐
5. What times 3 equals 12? ☐ x ☐ = ☐
6. What times 3 equals 15? ☐ x ☐ = ☐
7. What divided by 2 equals 4? ☐ ÷ ☐ = ☐
8. What divided by 2 equals 6? ☐ ÷ ☐ = ☐

Write these multiplication facts and use the chart if you need to.

9. 7 × 8= ☐
10. 5 × 6 = ☐
11. 7 × 6 = ☐
12. 10 × 7 = ☐
13. 8 × 5 = ☐
14. 8 × 9 = ☐

X	1	2	3	4	5	6	7	8	9	10
1	1	2	3	4	5	6	7	8	9	10
2	2	4	6	8	10	12	14	16	18	20
3	3	6	9	12	15	18	21	24	27	30
4	4	8	12	16	20	24	28	32	36	40
5	5	10	15	20	25	30	35	40	45	50
6	6	12	18	24	30	36	42	48	54	60
7	7	14	21	28	35	42	49	56	63	70
8	8	16	24	32	40	48	56	64	72	80
9	9	18	27	36	45	54	63	72	81	90
10	10	20	30	40	50	60	70	80	90	100

Today you are going to practice finding degrees, lengths, and areas. First, guess a measurement for each one. Then measure to see how close your guess was.

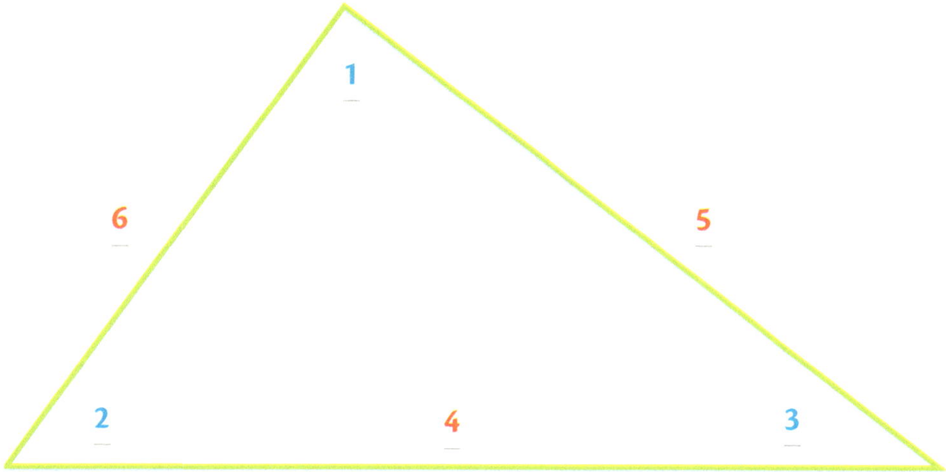

15. Guess the size of angle 1 _____ Measure the size of the angle _____

16. Guess the size of angle 2 _____ Measure the size of the angle _____

17. Guess the size of angle 3 _____ Measure the size of the angle _____

18. Guess the length of line 4 in inches _____ Measure the length of the line _____

19. Guess the length of line 5 in inches _____ Measure the length of the line _____

20. Guess the length of line 6 in inches _____ Measure the length of the line _____

21. Guess the area of rectangle 7 in square centimeters _____
Measure the length and width of the rectangle and figure out its area
_____

22. Guess the area of rectangle 8 in square centimeters _____
Measure the length and width of the rectangle and figure out its area
_____

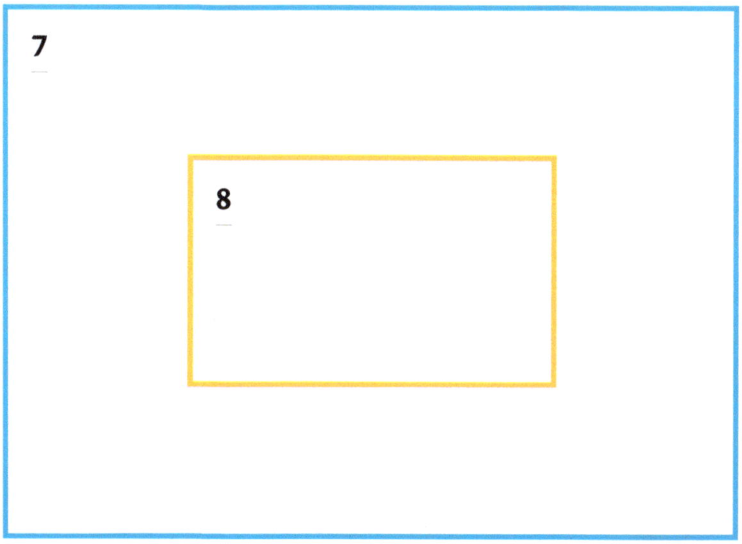

Let's learn about temperatures. We can communicate how hot or cold something is by saying its temperature in a certain number of degrees. Most thermometers in America show temperatures using the number of *Fahrenheit* degrees. Most thermometers in the rest of the world use *Celsius* degrees. Fahrenheit and Celsius are two different ways of dividing up hotness and coldness. Study this table that compares Fahrenheit and Celsius.

TEMPERATURE FOR:	FAHRENHEIT	CELSIUS
Freezing Point of Water	32°	0°
Hot Day	99°	37°
Boiling Point of Water	212°	100°

Read the temperatures on the Celsius thermometers below. Write the number of degrees in the box below each thermometer.

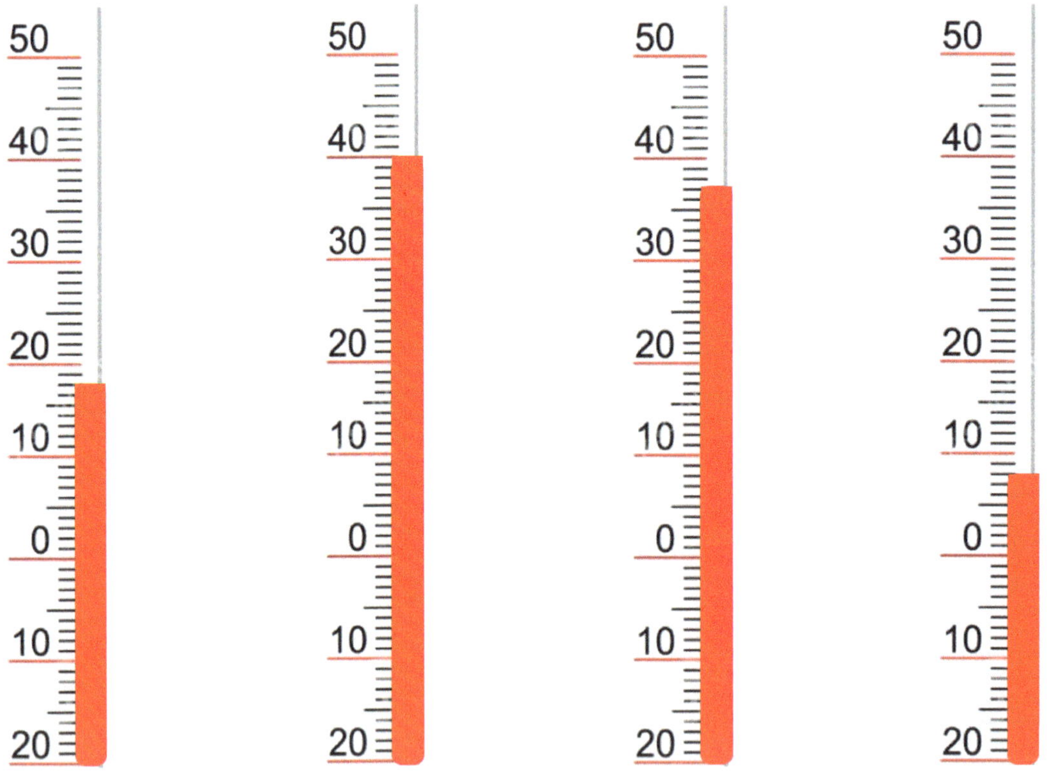

23. ☐ °C

24. ☐ °C

25. ☐ °C

26. ☐ °C

Rolling Two Dice DAY 82

When you flipped one coin, you had 2 possible outcomes: heads or tails.

When you flipped two coins, you had 4 possible outcomes: heads-heads, tails-tails, heads-tails, or tails-heads.

When you rolled one die, you had 6 possible outcomes: 1,2,3,4,5, or 6.

Now, how many possible outcomes might you get if you rolled two dice? To figure this out, let's fill in the possible pairs, using this table. A few are filled in for you.

	⚀	⚁	⚂	⚃	⚄	⚅
⚀	1,1	1,2				
⚁	2,1					
⚂						
⚃						
⚄						
⚅						

1. How many possible outcomes do you see? ______

Let's learn a simpler way to figure this out. The first die has 6 possible outcomes. The second die also has 6 possible outcomes. Let's multiply those together.

2. 6 x 6 =_____

This simpler way shows that there are 36 possible outcomes when rolling two dice.

3. In games where you roll two dice, you have to add up their dots. Let's pretend we're playing a game that uses two dice. Add each number from the top of this table to each of the numbers from the side of the table to see what number was rolled. A few spaces have been filled in for you.

TWO-DICE TOTAL

DIE B ↓ / DIE A →	1	2	3	4	5	6
1	2	3				
2	3	4	5			
3			6			
4						
5						
6						

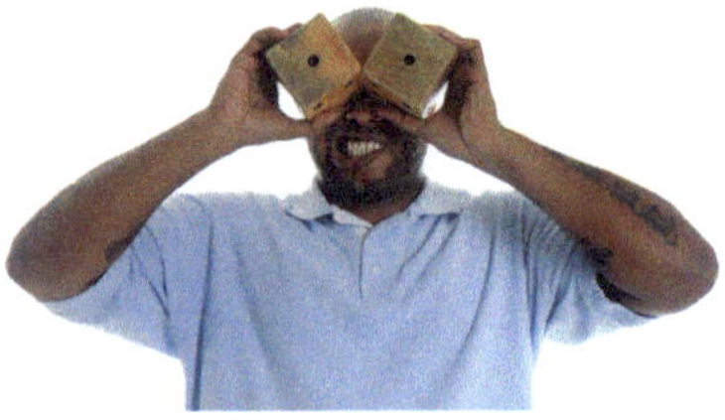

4. Do you see any patterns in this table? What's the biggest number you can possibly get if you roll two 6-sided dice? ______

Student Exercises

What is the probability of rolling a 2? We only get a 2 if we roll two 1s. (When this happens we say we've rolled snake eyes.) Count the 2s on the "Two-Dice Total" table (previous page). There is only one 2, so the probability of rolling a 2 is 1 in 36. Use the same table to answer these probability questions.

1. What is the probability of rolling a 2? 1 in 36
2. What is the probability of rolling a 3? ☐ in 36
3. What is the probability of rolling a 4? ☐ in 36
4. What is the probability of rolling a 5? ☐ in 36
5. What is the probability of rolling a 6? ☐ in 36
6. What is the probability of rolling a 7? ☐ in 36
7. What is the probability of rolling an 8? ☐ in 36
8. What is the probability of rolling a 9? ☐ in 36
9. What is the probability of rolling a 10? ☐ in 36
10. What is the probability of rolling an 11? ☐ in 36
11. What is the probability of rolling a 12? ☐ in 36

Round these numbers. Rounding is an important way to understand large numbers in God's world!

12. Round 6,790,768 to the nearest 1,000. ________________

13. Round 6,790,768 to the nearest 10,000. ________________

14. Round 6,790,768 to the nearest 100,000. ________________

15. Round 64,083 to the nearest 10 ________________

16. Round 64,083 to the nearest 100 ________________

17. Round 64,083 to the nearest 1000 ________________

Activity

18. Let's do an experiment with two dice. Roll both dice 36 times and add their dots to see what you get. Use tally marks to keep track of how many times you get each dice total.

TWO-DICE RESULTS

1	2	3	4	5	6	7	8	9	10	11	12

Compare your results with the probabilities you recorded in the questions on the last page. Your results are probably a little different.

Practice DAY 83

Read the following sentences and decide how the numbers in the story are combining—addition, subtraction, multiplication, or division. Circle the correct operation.

1. I wonder how long my 15 bags of chips will last if I give away 3 bags per day. (+ – × ÷)

2. I left 6 bananas on the counter too long, and 4 have already become rotten. (+ – × ÷)

3. My day is split up into 3 parts that are almost equal amounts of time: work, sleep, and play. (+ – × ÷)

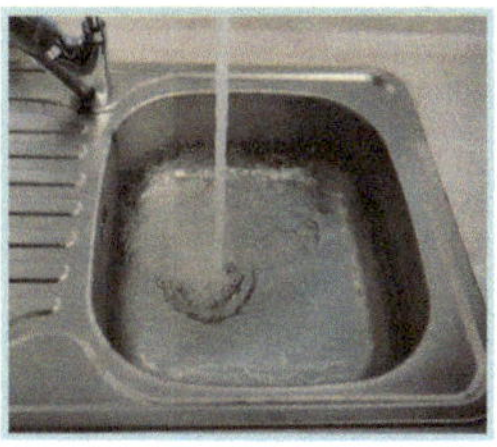

4. When you plug the sink but keep the water running, the sink fills up. (+ – × ÷)

5. The number of dogs increased a lot when each of them had 5 new puppies. (+ – × ÷)

6. For Christmas this year, Grandma gave me 2 books, Dad gave me 1 book, and my cousin gave me 3 books. (+ – × ÷)

7. I water my tree 2 times a day, 7 days a week. (+ – × ÷)

8. In some sports, the referees will take players out of the game because of penalties. (+ – × ÷)

Decide which number is larger and write the correct symbol in the circle.

9.	$\frac{4}{5}$ ◯ $\frac{1}{5}$	11.	10.1 ◯ 1.01
10.	100 ◯ 101	12.	$\frac{4}{6}$ ◯ $\frac{4}{2}$

Put these numbers in order from least to greatest.

573	375	537

__________, __________, __________
13. 14. 15.

Here is a fun multiplication puzzle. Find three numbers in the order that would make a correct multiplication equation. The three numbers can go in any direction but they must be in a straight line, horizontally, vertically, or diagonally. Find at least 20 equations.

32	5	6	30	4	5	20	2	6
4	12	10	32	18	9	2	6	9
8	3	8	18	5	1	36	10	54
16	4	4	45	2	9	18	2	0
2	5	5	25	3	9	6	20	24
8	20	18	2	4	8	4	5	4
24	6	4	5	3	15	24	4	6
3	2	5	10	7	5	35	12	3
2	16	20	4	40	6	6	36	2
1	2	2	4	7	28	49	9	6
10	8	10	16	8	2	3	4	12

Probability in God's Creation

DAY 84

Today we will look at probability in God's creation.

As you have seen, we can use numbers to show how likely it is that something will happen. We can be like scientists and observe the probability of something happening. For example, how likely is it that the sun will rise tomorrow? It is certain that the sun will rise. Why are we so sure? Because if you are 9 years old, you have observed the sun rise 3,300 times! Someone who is 30 years old has seen it rise 11,000 times. People have observed the sun rise about 2,200,000 times since God created the world. That's why we are certain the sun will rise.

Now, what is the probability of getting a 0 or a 7 in the throw of one die? That would be impossible. There is no probability of this happening because dice only have numbers 1 through 6.

Let's figure out more probabilities.

> When evening had come, He sat down with the twelve. Now as they were eating, Jesus said, "Assuredly, I say to you, one of you will betray Me." (Matthew 26:20,21)

The probability that one of the disciples would betray Jesus was certain because Jesus said it would happen. But the disciples didn't know which one of them would betray Jesus.

1. What was the probability that it was Matthew or any one of the twelve disciples? The probability was 1 out of ______.

2. On Saturday, 6 straws will be given out, one to each of 6 children. But 1 of the straws is a short straw. The other 5 are long. What is the probability that a certain child will get the short straw?

The probability is 1 out of _______.

3. What if 1 straw is given to each of 8 children? This time, 2 of the straws are short straws. The other 6 straws are long. What is the probability that a certain child will get a short straw.

The probability is 2 out of 8. Write this as a fraction.

$\frac{\square}{\square}$

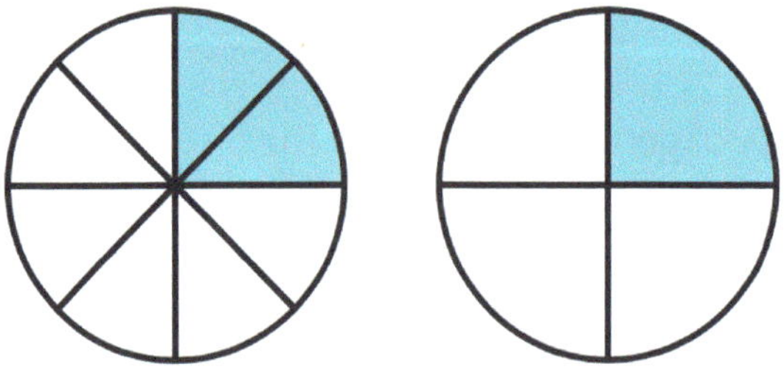

$$\frac{2}{8} = \frac{1}{4}$$

As you can see from the pie figures, $\frac{2}{8}$ is the same as $\frac{1}{4}$. So each child has a 2 out of 8 (1 out of 4) probability of picking a short straw.

Student Exercises

1. 3 + 4 = ______	4. 4 + 7 = ______	7. 9 + 9 = ______	10. 22 + 11 = ______
2. 8 − 7 = ______	5. 8 + 6 = ______	8. 18 − 11 = ______	11. 22 − 11 = ______
3. 15 − 9 = ______	6. 15 − 7 = ______	9. 14 + 12 = ______	12. 14 − 12 = ______

Add or subtract these fractions.

13. $\frac{2}{3} - \frac{2}{3} = \frac{\quad}{3}$

14. $\frac{1}{2} + \frac{1}{2} = \frac{\quad}{2}$

15. $\frac{5}{4} + \frac{2}{4} = \frac{\quad}{4}$

16. $\frac{6}{9} - \frac{3}{9} = \frac{\quad}{9}$

17. $\frac{3}{8} + \frac{4}{8} = \frac{\quad}{\quad}$

18. $\frac{8}{10} - \frac{3}{10} = \frac{\quad}{\quad}$

Practice DAY 85

It's handy to know how to count by 12s. Counting by 12 helps us do things like changing feet to inches. And since donuts and eggs are often packaged in boxes of 12, we can count by 12s to find their total number.

Here's how we count by chunks of 12:

0, 12, 24, 36, 48, 60, 72, 84, 96, 108, 120, 132, 144

Do you see a pattern in this list? The 1s place is increasing by two — 0, 2, 4, 6, 8 . . .

We often say "one dozen" when referring to 12 items. Here are 1 dozen donuts.

Count these eggs by 12:

____

____

____

____

____

____

____

____

____

____

____

____

Let's practice counting by chunks. For each list of numbers, see if the list is counting up or counting down. Then fill in the blanks to continue the pattern.

1. Count by 3s: 60, 57, ____, ____, ____, ____
2. Count by 6s: 24, 30, ____, ____, ____, ____
3. Count by 8s: 40, 48, ____, ____, ____, ____
4. Count by 10s: 670, 660, ____, ____, ____, ____
5. Count by 11s: 132, 121, ____, ____, ____, ____
6. Count by 12s: 24, 36, ____, ____, ____, ____

Practice multiplying and dividing these numbers using your memory. If you get stuck, your multiplication/division table will help you.

7. $2 \times 3 =$ ☐
8. $10 \div 5 =$ ☐
9. $9 \div 3 =$ ☐
10. $6 \times 2 =$ ☐
11. $3 \times 12 =$ ☐
12. $12 \div 2 =$ ☐
13. $\frac{15}{3} =$ ☐
14. $4 \times 3 =$ ☐
15. $2 \times 2 =$ ☐
16. $\frac{12}{4} =$ ☐

Everyday Math

17. Suppose you shared a pizza pie with your friends and they ate $\frac{2}{3}$ of it. How much of the pizza is left for you? Fill in the fraction here:

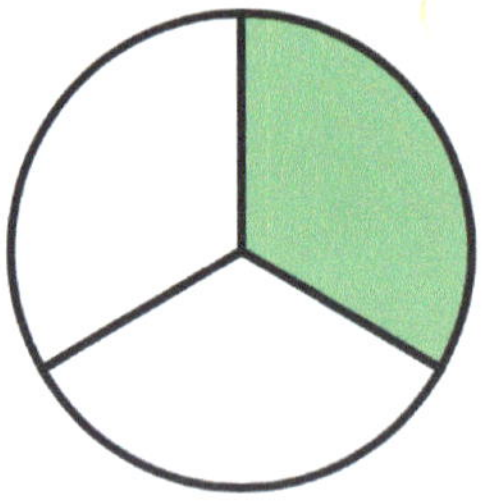

$\frac{\square}{3}$

18. Let's say you started a cookie business. For your first batch, you spent $96 on flour, sugar, chocolate chips, and other ingredients and made 450 cookies. If you sold 342 cookies, how many cookies would you have left? ___________ How many dozen cookies is the amount you have left? ___________

$$\begin{array}{r} \square \\ - \ \square \\ \hline \end{array}$$

Cookies Left Over: $\square$

19. From number 17, if you received $346 from your customers, how much money would you have earned? ___________

$$\begin{array}{r} \square \\ - \ \square \\ \hline \end{array}$$

Money Earned: $\square$

20. Suppose your grandma baked 48 mini muffins for your soccer team. If your team has 12 players, how many muffins will each player get? ___________

Extra Challenge

21. Add the following. Color in the answer. What is the answer in fourths? What is the answer in halves?

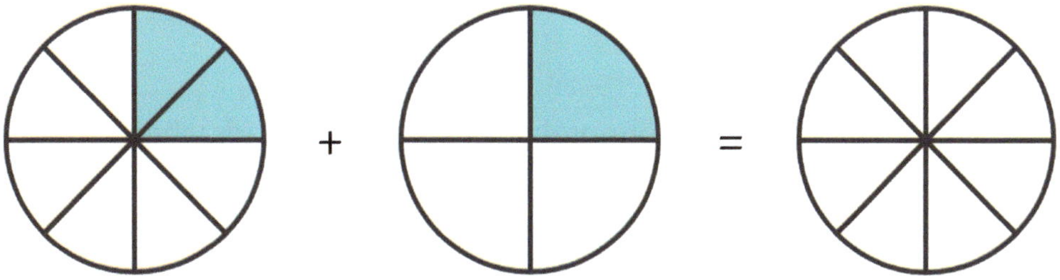

$$\frac{2}{8} + \frac{1}{4} = \frac{\square}{4} \text{ or } \frac{\square}{2}$$

DAY 86 Percent, Fractions, Decimals, and Probability

You have been learning how fractions and decimals show parts of whole amounts. Today you will learn another way to talk about part of a whole amount—percent! Percent means a certain amount out of 100. Notice that the word percent has the word *cent* in it. Since there are 100 cents in a dollar, that will help you remember that percent has something to do with the number 100.

Percent also has the word *per* in it. You will soon understand that, instead of saying "out of 100," we can say "per cent." We write percent with this symbol after the number: %.

Now let's see how to use percent. If there are 100 school-age children in your church, and all 100 of them brought their Bibles last Sunday, that's 100 out of 100. We would say that 100% of them brought their Bibles. If only 50 children brought their Bibles, we would say that 50% of them did.

What percent of children brought their Bibles if 85 children brought them? ________%

Practice writing the percent symbol five times:

__________ __________ __________ __________ __________

People often use the word *percent* to talk about a fraction of a whole amount even if that whole amount is not 100 things. For example, when you say you have finished 100% of the work, it means you've finished all of it. If you have eaten 50% of the pie, it means you have eaten half of it. This is because 50 is half of 100. If you have finished 25% of today's math, it means you've finished 1/4 of it because 25 is 1/4 of 100. These are important percents to memorize:

100% is the same as 1.

50% is the same as $\frac{1}{2}$.

25% is the same as $\frac{1}{4}$.

You can also use percent for probability. Your probability of getting heads when you toss a coin is 1 out of 2 or $\frac{1}{2}$. If you want to say this as a percent, you would say that the probability of getting heads is 50%.

You've seen number lines from 0 to 10 and maybe from 0 to 100. But let's look at a number line from 0 to 1. Between 0 and 1, you will see decimals that show tenths. Below the decimals, you will see two lines of fractions that equal the decimals above them. At the very bottom, you will see percents that equal the amounts above each of them. These percents are parts of the whole number 1 on the number line.

0	0.1	0.2	0.3	0.4	0.5	0.6	0.7	0.8	0.9	1.0
$\frac{0}{10}$	$\frac{1}{10}$	$\frac{2}{10}$	$\frac{3}{10}$	$\frac{4}{10}$	$\frac{5}{10}$	$\frac{6}{10}$	$\frac{7}{10}$	$\frac{8}{10}$	$\frac{9}{10}$	$\frac{10}{10}$
$\frac{0}{100}$	$\frac{10}{100}$	$\frac{20}{100}$	$\frac{30}{100}$	$\frac{40}{100}$	$\frac{50}{100}$	$\frac{60}{100}$	$\frac{70}{100}$	$\frac{80}{100}$	$\frac{90}{100}$	$\frac{100}{100}$
0%	10%	20%	30%	40%	50%	60%	70%	80%	90%	100%

Study the number line and the fractions and percents. Do you see that 90% is close to 1? Do you see that 10% is close to 0? 99% is almost 100%. When we say that something will happen with 99% probability, it almost certainly will happen. When we say that something will happen with 1% probability, it probably won't happen. Let's look at a few examples:

1. The weather report said that there is a 1% chance of rain tomorrow. Circle the right answer.

It probably will rain tomorrow.

It probably will not rain tomorrow.

2. The weather report said there is a 95% chance of snow tomorrow. Circle the right answer.

It probably will snow tomorrow.

It probably will not snow tomorrow.

3. There is a 3% probability that you will get snake eyes on your next roll of these two dice.

You probably will not roll snake eyes

You probably will roll snake eyes.

Each Bible passage below contains a word that's bold. In each box, write the answer to the question about the bold word.

> He will again have compassion on us, and will subdue our iniquities. You will cast **all** our sins into the depths of the sea. (Micah 7:19)

4. What percent of our sins will God cast into the depths of the sea?

%

> Blessed be God Most High, who has delivered your enemies into your hand!" And Abram gave him a **tenth** of everything. (Genesis 14:20, ESV)

5. What percent of his goods did Abram give to God? %

> Moses also had given an inheritance to **half** the tribe of Manasseh; it was for half the tribe of the children of Manasseh according to their families. (Joshua 13:29)

6. What percent of the tribe of Manasseh received an inheritance? %

> He gives rain on the earth, And sends waters on the fields. (Job 5:10)

> For He says to the snow, 'Fall on the earth'; Likewise to the gentle rain and the heavy rain of His strength. (Job 37:6)

> Who can number the clouds by wisdom? Or who can pour out the bottles of heaven? (Job 38:37)

Activity

With a little help from your parent/teacher, check out your area's weather forecast for today or tomorrow. Looking at a weather website, what is the percent probability it will rain or snow?

____

How will you dress for the weather?

____

____

Student Exercises

Find the right number to answer these math questions. Write your own equations.

1. What plus 50 equals 60? ☐ + ☐ = ☐
2. What times 8 equals 56? ☐ x ☐ = ☐
3. What divided by 8 equals 7? ☐ ÷ ☐ = ☐
4. What plus 54 equals 60? ☐ + ☐ = ☐
5. What times 6 equals 54? ☐ x ☐ = ☐
6. What divided by 6 equals 9? ☐ ÷ ☐ = ☐
7. What minus 6 equals 54? ☐ – ☐ = ☐
8. 1 plus what equals 2? ☐ + ☐ = ☐

DAY 87 Practice

Today let's practice finding degrees, lengths, and areas. First, guess a measurement for each. Then measure to see how close your guess was.

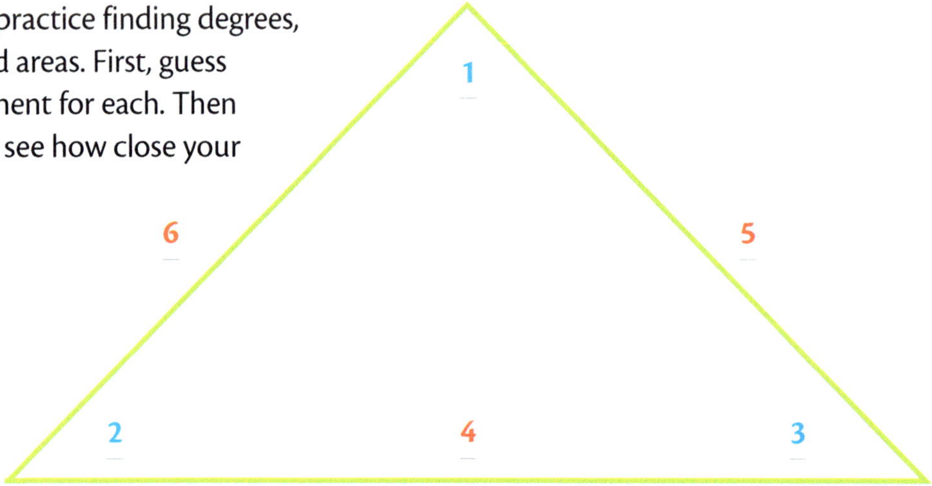

1. Guess the size of angle 1 _____ Measure the size of the angle _____

2. Guess the size of angle 2 _____ Measure the size of the angle _____

3. Guess the size of angle 3 _____ Measure the size of the angle _____

4. Guess the length of line 4 to the nearest centimeter. _____ Measure the length of the line to the nearest tenth of a centimeter. _____

5. Guess the length of line 5 to the nearest centimeter. _____ Measure the length of the line to the nearest tenth of a centimeter. _____

6. Guess the length of line 6 to the nearest centimeter. _____ Measure the length of the line to the nearest tenth of a centimeter. _____

7. Guess the area of rectangle 7 in square centimeters. _____
Measure the length and width of the rectangle and figure out its area.
_____

8. Guess the area of rectangle 8 in square centimeters. _____
Measure the length and width of the rectangle and figure out its area.
_____

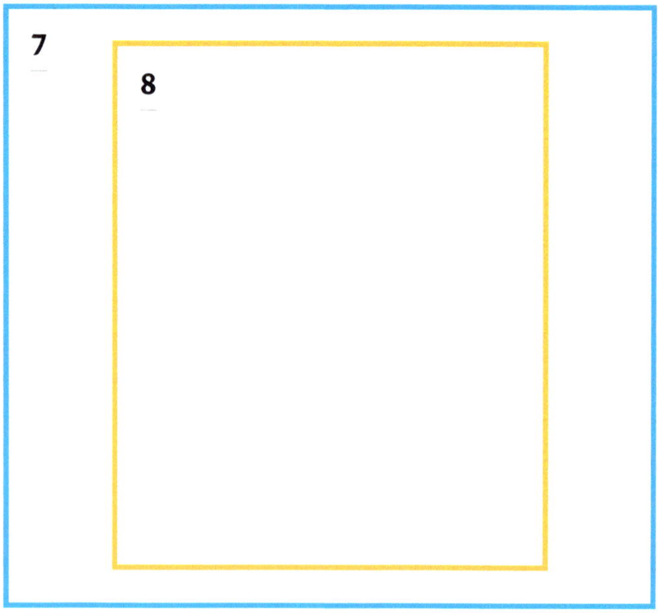

Practice rounding these numbers. Rounding is an important way to understand amounts in God's world.

9. Round 87.39 to the nearest 0.1. ______________

10. Round 87.39 to the nearest 1. ______________

11. Round 87.39 to the nearest 10. ______________

12. Round 56,132 to the nearest 100 ______________

13. Round 56,132 to the nearest 1,000 ______________

14. Round 56,132 to the nearest 10,000 ______________

Count up by 11s:

15. 11, _____ _____ _____ _____ _____ _____ _____ _____ _____ _____

16.
$$\begin{array}{r} 64 \\ +\ 78 \\ \hline \end{array}$$

17.
$$\begin{array}{r} 85 \\ -\ 47 \\ \hline \end{array}$$

18.
$$\begin{array}{r} 11 \\ \times\ \ 5 \\ \hline \end{array}$$

19.
$$\begin{array}{r} 12 \\ \times\ \ 8 \\ \hline \end{array}$$

20. $\frac{3}{4} + \frac{4}{4} = \square$

21. $\frac{2}{6} - \frac{1}{6} = \square$

DAY 88 Applying Probability Every Day

This lesson integrates math into everyday life. With this activity, the student will apply God's patterns and wisdom in meaningful ways to home and community. Take a break from memory work and academic exercises to make math part of everyday life. The following are suggestions or examples. Other ideas may be added to the list.

Here are some everyday ways in which you can use probability!

1. Have fun with a family member or friend playing this probability game. This dice game is called Pig. Here's how it works:

Use one die. On your turn, you can roll as many times as you like. Each time you roll, add the number you rolled to your score. BUT, if you get a 1 on any roll, your total score goes back to zero and it becomes the other person's turn. If you stop rolling before you get a 1, you can keep your score. The first player to reach 100 wins.

Or you can use two dice. For each turn, you can roll both dice as many times as you like. But if one of the two dice ends up as a 1, your total score goes to zero and it becomes the next person's turn. Two 1s (snake eyes) are fine though. If you get two 1s, you get 25 points!

Keep in mind that with one die, your probability of getting a 1 is one out of 6 rolls. But when your roll two dice, your probability of getting a 1 is about 1 out of 3 rolls. When you roll two dice, your probability of getting snake eyes is 1 out of every 36 rolls. But always remember, God is in control of the roll of the dice.

2. Help plan a vacation. Find a place you would like to go, and check the weather. Look at the forecast for the next week. What is the probability each day for rain or snow?

Day 1 ______%

Day 2 ______%

Day 3 ______%

Day 4 ______%

Day 5 ______%

Day 6 ______%

Day 7 ______%

3. God can overcome situations where the probability is 0%. Read 1 Samuel 17. Explain why David had a 0% chance to conquer Goliath with his own strength.

____

____

____

Who really won the battle?

____

Why?

____

____

4. From verse 4, what was the height of Goliath in Bible measurements?

____

____

5. Look up to see how tall that would be in measurements you use.

____

Find a place in your home where you can stretch out a tape measure to see that length.

Additional Activities

For extra activities, games, and helpful videos on these math concepts, refer to generations.org/math3.

God has given us amazing shapes. Man uses them to construct beautiful buildings.

CHAPTER 8

Geometry

DAY 89 More Flat Shapes (in Two Dimensions)

Let's learn about polygons. A **polygon** is a many-sided shape. It could have 5 sides, 10 sides, or even 20 sides.

The buds of the porcelain flower are shaped like pentagons.

Pentagon (5 sides)

Hexagon (6 sides)

Heptagon (7 sides)

Octagon (8 sides)

Let's think about the perimeter, or distance around, each of these polygons. There are two ways to figure out a perimeter for these shapes. If the sides of the shape are different lengths, you can add them up. If the sides are all the same length, you can do it the fast way and multiply the length of one side by the number of sides. For example, the perimeter of a pentagon would be either:

Pentagon Perimeter = Side + Side + Side + Side + Side

Or:

Pentagon Perimeter = 5 x Side

If a hexagon has equal sides, its perimeter would be:

Hexagon Perimeter = 6 x Side

If an octagon has equal sides, its perimeter would be:

Octagon Perimeter = 8 x Side

Now suppose each side of these three shapes is 6 cm. What is the perimeter of each?

1. Pentagon Perimeter = 5 X 6 = ____
2. Hexagon Perimeter = 6 X 6 = ____
3. Octagon Perimeter = 8 X 6 = ____

Something to know:

Two straight lines are **parallel** if there is the same distance between them along their whole length. Parallel lines never cross each other even if they go on forever.

Here are a few more shapes you should know about:

A **parallelogram** is made of two sets of parallel lines that are of equal length and opposite each other.

A **rhombus** is a parallelogram where all four sides are of equal length.

A **trapezoid** has one set of parallel lines. The other lines can go any which way.

Parallelogram

Rhombus

Trapezoid

Did You Know?

A rectangle is a special type of parallelogram. It has square corners that are 90° angles.

A square is a special type of rhombus. It has square corners that are 90° angles.

A rhombus is a special type of parallelogram. Its sides are all the same length.

Student Exercises

Find the perimeter of these strange shapes. The lengths of some sides are labeled. You should be able to figure out the lengths of the other sides by looking at the definitions on the previous page. Add up the sides of each shape and write its perimeter in the box.

Parallelogram

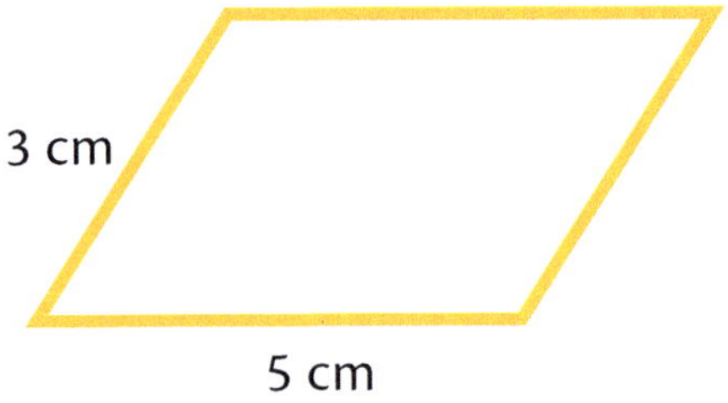

1. _____ + _____ + _____ + _____ = _____

Rhombus

2. _____ + _____ + _____ + _____ = _____

Trapezoid

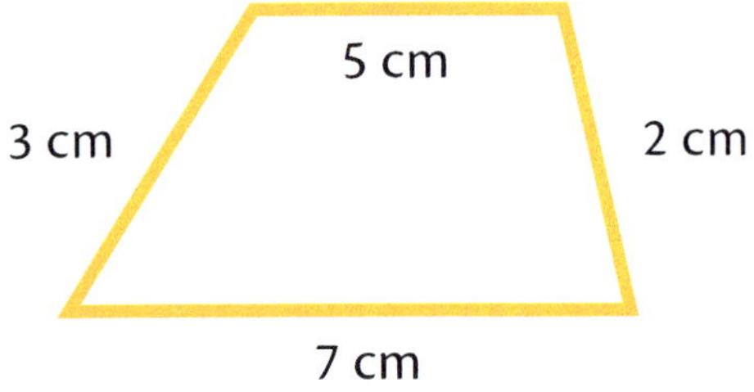

3. _____ + _____ + _____ + _____ = _____

Decide which number is larger and write the correct symbol in the circle.

4.	45	◯	54	6.	$\frac{3}{4}$	◯	$\frac{5}{4}$
5.	3.06	◯	3.6	7.	$\frac{1}{2}$	◯	1

Put these numbers in order from least to greatest.

981	189	198

__________, __________, __________
8. 9. 10.

Everyday Math

Find the perimeters of these familiar things:

11. A professional baseball diamond is a square measuring 90 feet between each base. If you hit a home run, how far would you have to run to go all the way around the diamond?

12. A stop sign looks like a(n) ________________. If each side is 30 centimeters long, what is the sign's perimeter?

13. This umbrella looks like a(n) ____________________. If each side is 12 inches long, what is the umbrella's perimeter?

God we praise You in all weather!

14. This nut from the swing set looks like a(n) ______________________. If each side is 4 millimeters long, what is the nut's perimeter?

$6 \times 4 =$ ☐

15. Find at least three more perimeters of shapes around your house or yard. Ideas: a square of chocolate, a window, your bed, a backyard shed, or your whole home.

___________________________________ ☐

___________________________________ ☐

___________________________________ ☐

Use your wisdom of long addition and subtraction to figure out how God made these numbers come together and make something new. Some carries and borrows have been provided as hints.

16.

	6	5
+	7	5

17.

	1	1	
	4	3	5
+		6	7

18.

	1	3,	2	4	5
+		2,	6	0	1

19.

	8	7
−	5	7

20.

	3.	4
+	4.	7

21.

		2	11
	1	~~3~~	~~1~~
−		1	2

22.

	1	3.	9
−		1.	2

23.

	3	4	5
−	2	9	7

Try It!

Find an area of your home that has tiles on the floor or wall. Use painter's tape to create shapes with the tile lines and corner points. Can you make a hexagon and an octagon? Measure the length of a tile to the nearest inch or centimeter, and use it to figure out the perimeter of the shapes you create.

Practice DAY 90

For each list of numbers, figure out if the numbers are counting up or counting down. Fill in the blanks to continue the pattern.

1. Count by 2s: 100, 98, ______, ______, ______, ______, ______, ______

2. Count by 4s: 100, 96, ______, ______, ______, ______

Circle the answers in number 1 that you also see in number 2.

3. Count by 7s: 70, 63, ______, ______, ______, ______

4. Count by 7s: 0, 7, ______, ______, ______, ______

5. Count by 100s: 100, 200, ______, ______, ______, ______

Practice multiplying and dividing. You might need to use your multiplication/division table. The more you use it, the more familiar you will become with how God made numbers to combine.

6. $4 \times 6 =$ ☐

7. $24 \div 4 =$ ☐

8. $36 \div 4 =$ ☐

9. $8 \times 7 =$ ☐

10. $7 \times 5 =$ ☐

11. $63 \div 7 =$ ☐

12. $\frac{81}{9} =$ ☐

13. $9 \times 5 =$ ☐

14. $9 \times 7 =$ ☐

15. $\frac{100}{10} =$ ☐

Everyday Math

16. A rectangular walkway measures 14.2 feet long and 3.9 feet wide. Round the measurements to the nearest foot and find the perimeter of the walkway.

______ + ______ + ______ + ______ = ______

17. Each carton contains about 20 strawberries. Figure out about how many strawberries are in this whole box.

_____ + _____ + _____ + _____ + _____ +

_____ + _____ + _____ + _____ + _____ =

_____

When we have more than one of something, we can say we have multiple things. Practice these multiple-operation exercises and use more than one kind of operation. Do each exercise, from its left to its right, in your head or on paper.

18. 3 + 5 + 6 = ☐

19. 3 x 2 x 1 = ☐

20. 1 + 10 + 4 = ☐

21. 2 x 3 x 2 = ☐

22. 4 + 6 + 3 + 7 = ☐

23. 3 x 2 x 2 ÷ 2 = ☐

24. 4 x 3 ÷ 2 = ☐

25. 5 + 3 + 1 + 1 + 1 + 1 = ☐

Three-Dimensional Shapes Pop Out of the Page

DAY 91

Let's learn the names of some three-dimensional shapes.

Cube	A three-dimensional shape with 6 square surfaces, all the same size.	
Sphere	A ball-like shape whose outline looks like a circle no matter how it's turned.	
Cylinder	A can-like shape that looks round from the top and like a rectangle from the side. If we took slices straight across in several places, the slices would all be the same-sized circle.	
Cone	A three-dimensional shape whose bottom is a circle, and whose top is a single point. The sides are sloped. Looks like a triangle from the side.	
Pyramid	A three-dimensional shape whose bottom is a square or rectangle, and whose top is a single point. The sides are sloped. Each side surface is a triangle.	
Rectangular Prism	A three-dimensional shape with 6 square or rectangular sides. The rectangles/squares opposite each other are the same size and shape.	

With three-dimensional shapes, we can measure two things:

Surface area: Finding a shape's surface area is like finding out how much paper we would need to cover the shape. We figure out the area of each of the shape's surfaces and add them together.

Volume: finding the volume of a shape is like finding how much it can hold. You already know how to find the volume of a cube or a rectangular prism — Length x Width x Height = Volume

Origami Cube

Let's find the surface area and volume for a cube whose edges are all 2 inches long.

1. **Surface area**: How much paper would it take to cover the cube?

First we find the area of one side of the cube.

Area of one side = 2 x 2 = 4 square inches

Next we multiply that area by 6 because cubes have 6 equal sides.

Total surface area of cube = 4 x 6 = 24 square inches

Formula:
Surface Area of a cube =Area of one side of the cube x 6

2. **Volume**: What's the volume of this cube?

Formula:
Volume of Cube = Length x Width x Height

$$\begin{aligned}\text{Volume} &= \underbrace{2 \times 2} \times 2 \\ &= 4 \times 2 \\ &= 8 \text{ cubic inches}\end{aligned}$$

A **formula** is a rule for how to do something. Math formulas look like equations. For example, the formula for finding the area of a rectangle is Area = Length x Width.

Student Exercises

1. How much paper would it take to cover a cube that's 3 inches long, 3 inches wide, and 3 inches high?

Area of one surface = 3 x 3 = ☐

Total surface area = ☐ x 6 = ☐

2. What is the volume of the cube?

Volume = 3 x 3 x 3

= ☐ x 3

= ☐ cubic inches

Match the shapes with their names drawing lines between them. Use the most exact name for each shape.

3. 

4. 

5. 

6. 

A. Parallelograms

B. Hexagons

C. Trapezoids

D. Rhombuses

Activity

Look around the house for 3-dimensional shapes from today's lesson. Make a list of the ones you find. If you can find a cube (like a Rubik's Cube™), measure its edges and find its surface area and volume.

Craft It!

Create three-dimensional shapes using foldable paper patterns from this website. Here is an example of instructions on folding an origami cube: https://www.wikihow.com/Fold-an-Origami-Cube

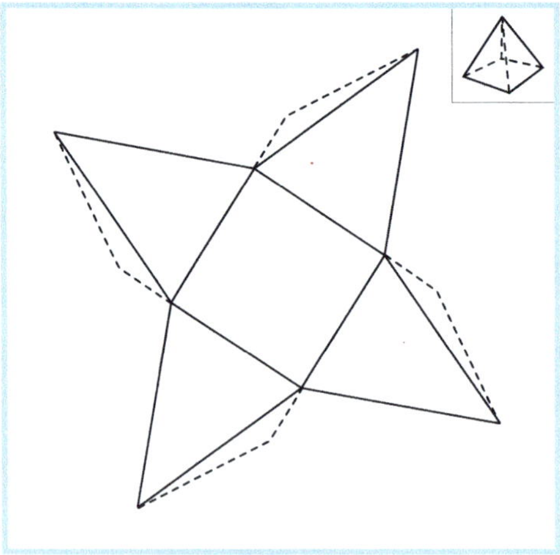

Practice DAY 92

Finish each list of shapes. Find the pattern, and then draw a picture that keeps that pattern going.

1. 

2. 

3. 

4. 

> And they stood up in their place and read from the Book of the Law of the Lord their God for one-fourth of the day; and for another fourth they confessed and worshipped their God. (Nehemiah 9:3)

5. What made the people see their need to confess to God?

____

____

6. How much time altogether was spent in the Word and worshipping God?

____

For each exercise, count the shaded parts to see the total number of pieces. Write that number in the blank numerator.

7. $1\frac{2}{4} = \frac{}{4}$

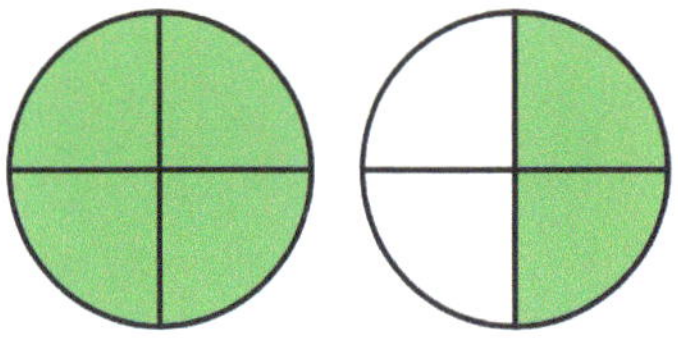

8. $2\frac{1}{3} = \frac{}{3}$

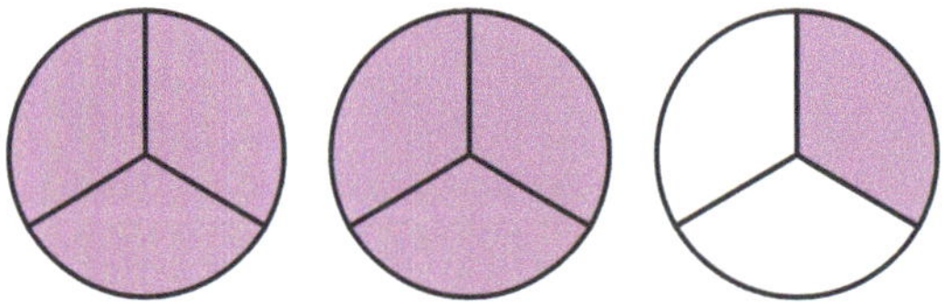

9. $4\frac{1}{2} = \frac{}{2}$

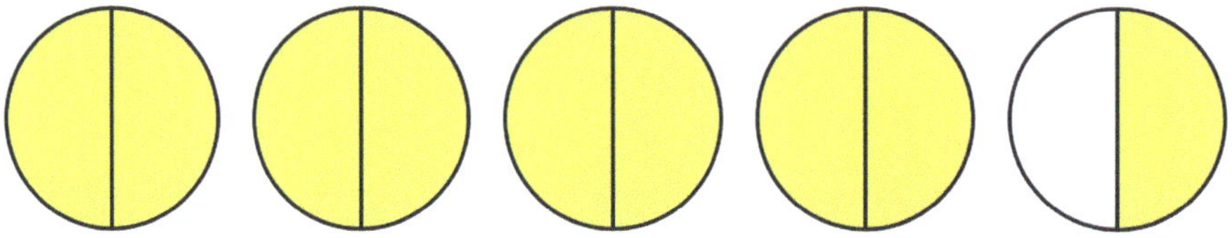

Split these numbers up! How many whole numbers are there and how many pieces are left over?

10. $\frac{8}{3} = \square\frac{}{3}$

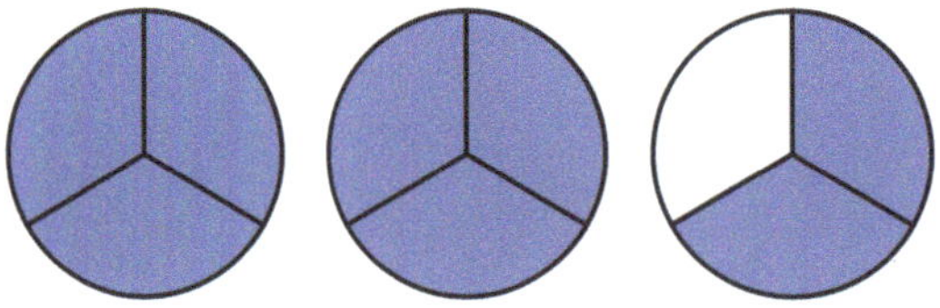

11. $\frac{7}{4} = \square\frac{}{4}$

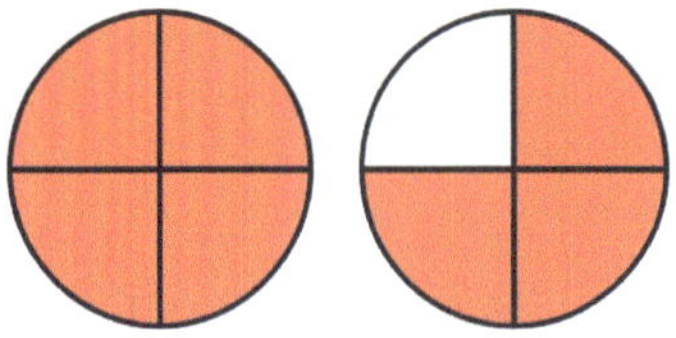

Color in one-half of each pie below and write the numerator of the fraction. The first is finished for you.

12. 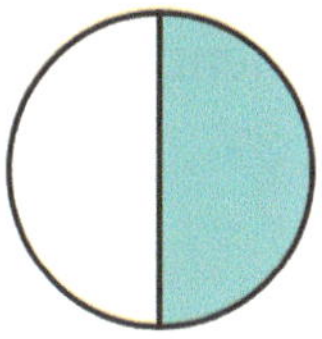 $\frac{\quad}{2}$

13. 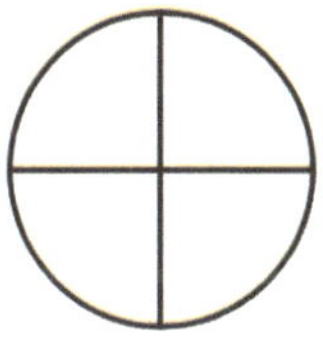 $\frac{\quad}{4}$

14.  $\frac{\quad}{6}$

15.  $\frac{\quad}{8}$

16.  $\frac{\quad}{10}$

17.  $\frac{\quad}{12}$

Figure out the correct number to answer each math question. Write it in the blank. Then write an equation that includes your answer. The first one is done for you.

18. What plus 8 equals 12? 4 $4 + 8 = 12$

19. What plus 7 equals 14? ______ ______________

20. What minus 3 equals 13? ______ ______________

21. What minus 5 equals 9? ______ ______________

22. What times 4 equals 16? ______ ______________

23. What times 4 equals 0? ______ ______________

24. What times 11 equals 121? ______ ______________

25. What divided by 1 equals 15? ______ ______________

26. What divided by 3 equals 3? ______ ______________

27. What divided by 12 equals 12? ______ ______________

DAY 93 Measuring Three-Sided Shapes

Suppose you have two friends that live across a large park from you—one toward the right and one toward the left. When you visit each other, you are able to ride bikes in a straight line between any of the three homes. If you bicycle from your home to Right Friend, then to Left Friend, and then home again, you would have ridden the perimeter of a triangle.

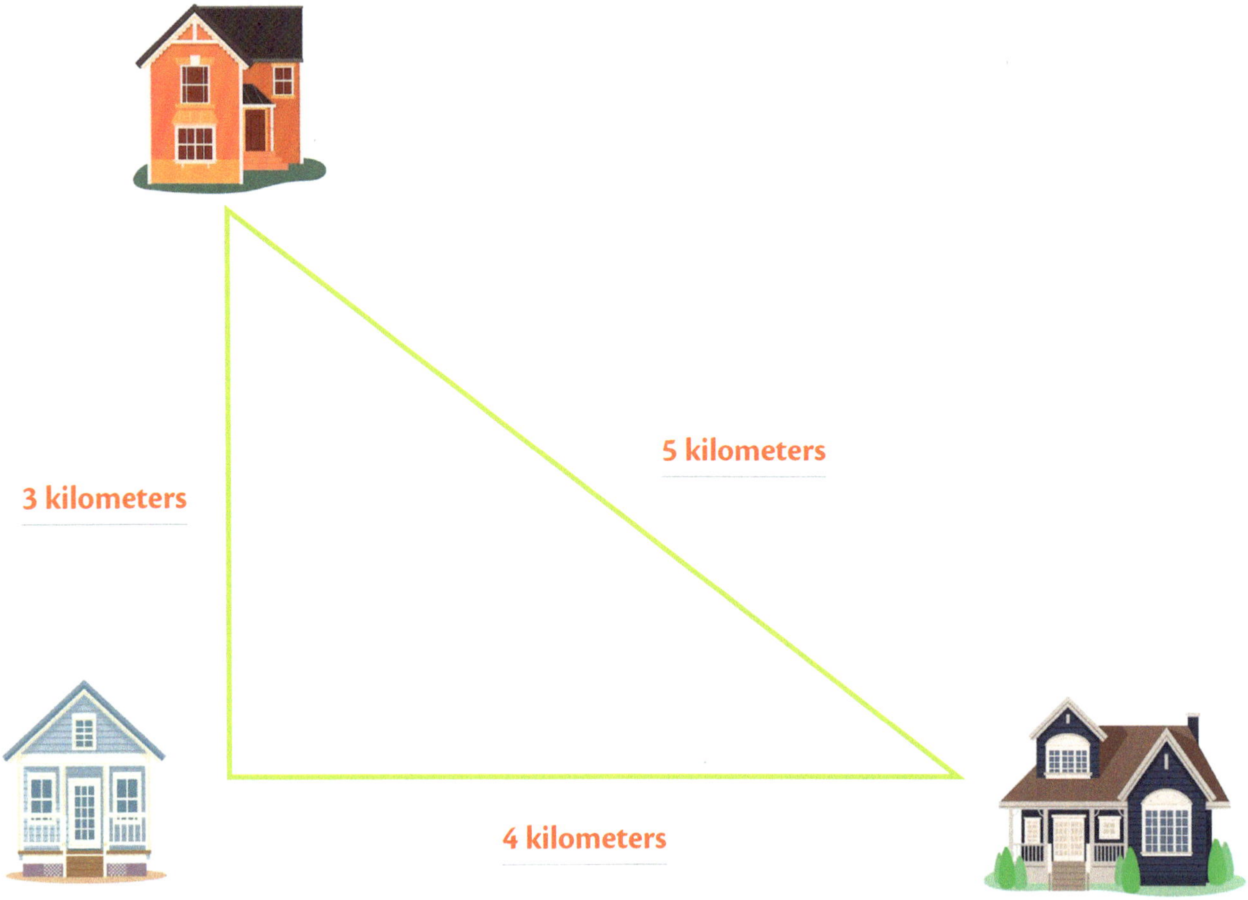

How far would you have ridden? You can figure out the perimeter of a triangle the same way you figure out other shapes that are made of straight lines—by adding the sides. To visit both friends, you would bicycle 5 + 4 + 3 = 12 kilometers.

Finding the perimeter of a triangle is easy. Now let's learn how to find the area of a triangle.

Suppose you have a choice between a piece of peppermint bark candy shaped like this rectangle and one shaped like the triangle.

Which piece of candy do you think is larger (has more area)? Let's learn how to figure that out. Of course, you would choose the smaller piece for yourself and let someone else eat the larger one.

You already know how to find the area of a rectangle. You multiply its length by its width. Now, take a look at the blue triangle on the right. It has a square corner like the rectangle does. Do you see that the blue triangle is exactly half of the whole rectangle?

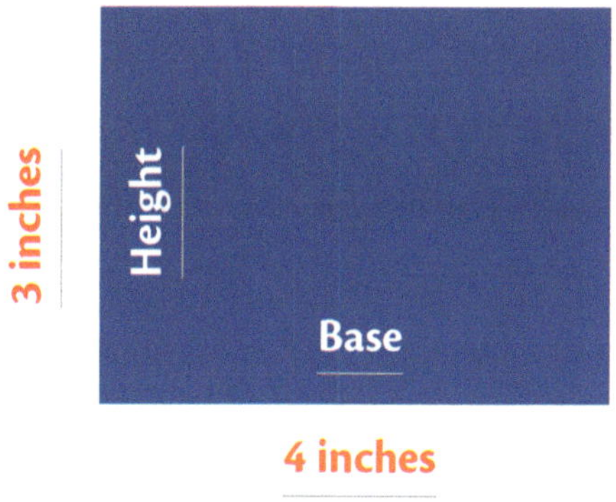

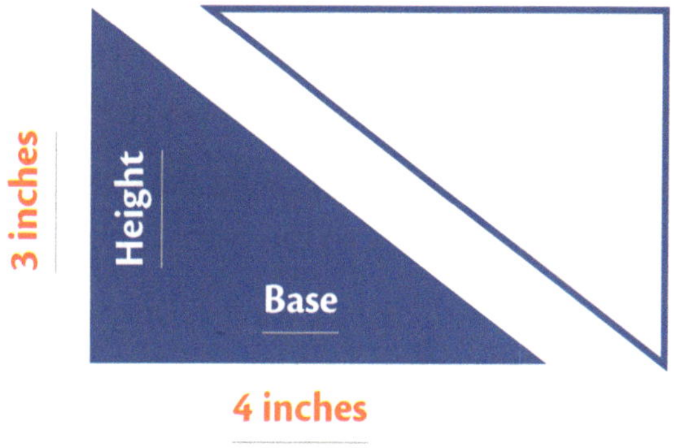

So, to find the area of a triangle like this, we start with a whole rectangle and multiply its length and width together to find its area. Since the triangle is half of the rectangle, we divide the rectangle's area by 2. The blue rectangle above is 4 inches by 5 inches. Its area = 4 x 5 = 20 square inches. The triangle will be half that size. We divide 20 by 2, and get 10 square inches.

$$20 \div 2 = 10$$

Now let's compare the pieces of peppermint bark to see which one has the most area. Fill in the boxes.

2 inches

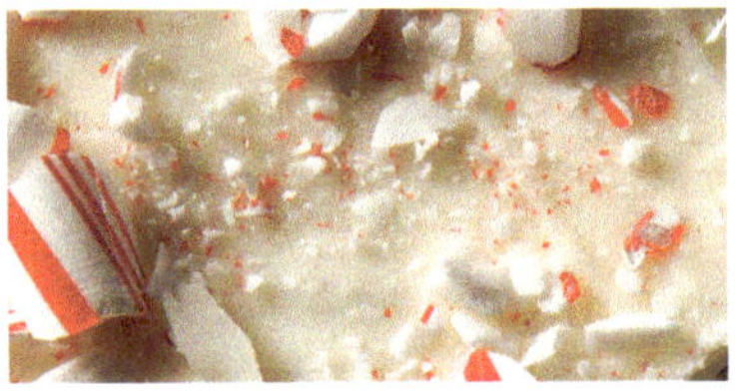

4 inches

3 inches

4 inches

Area of Rectangle = 2 x 4 = ☐

Area of Triangle = 3 x 4 ÷ 2

= 12 ÷ 2 = ☐

Which piece of candy is larger: the rectangle or the triangle? ____________________________

When we talk about triangles, we don't say "length" and "width." We call one of the sides its **base**. It doesn't matter which side we use as the base, but it's usually easiest to use the bottom side.

To find the height of the triangle, we set the triangle on its base. The **height** of the triangle is the distance from the base to the highest point of the triangle.

Amazingly, in God's world, there is one way to find the area of all triangles. This applies to **right triangles** with a square corner and to all other triangles.

The area of a triangle is the length of the base times the height divided by two.

Area of Triangle = Base x Height ÷ 2

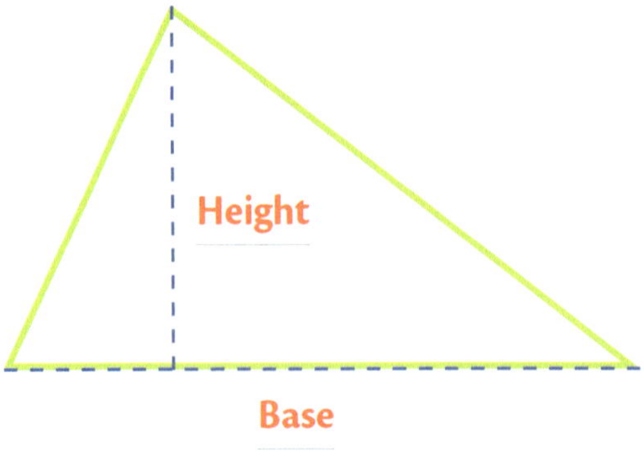

Student Exercises

Find the area of these triangles.

1. Area of Triangle = Base x Height ÷ 2

Area = 6 x 2 ÷ 2

= ☐ ÷ 2

= ☐ square inches

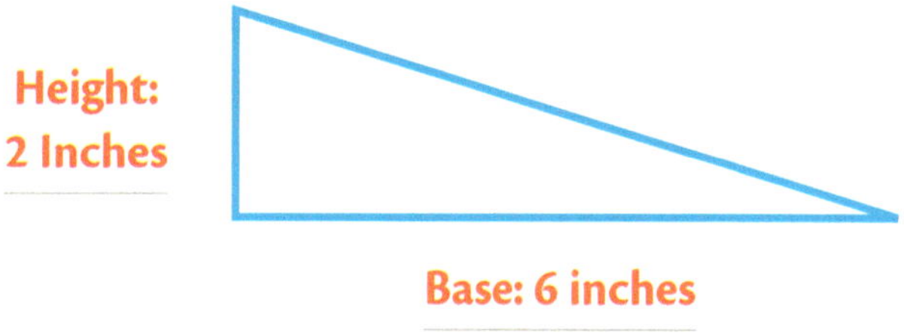

2. Area of Triangle = Base x Height ÷ 2

Area = 4 x 5 ÷ 2

= ☐ ÷ 2

= ☐ square inches

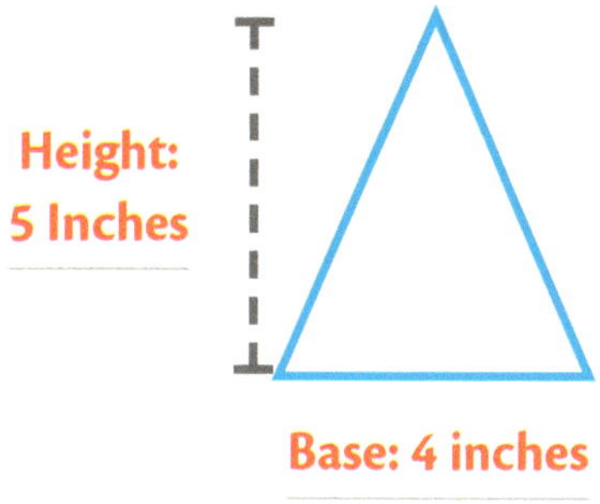

3. Area of Triangle = Base x Height ÷ 2

Area = 8 x 1 ÷ 2

= ☐ ÷ 2

= ☐ square inches

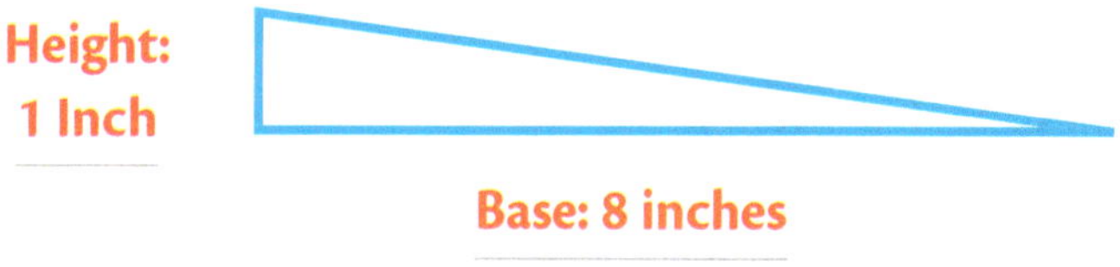

4. Area of Triangle = Base x Height ÷ 2

Area = 4 x 3 ÷ 2

= ☐ ÷ 2

= ☐ square inches

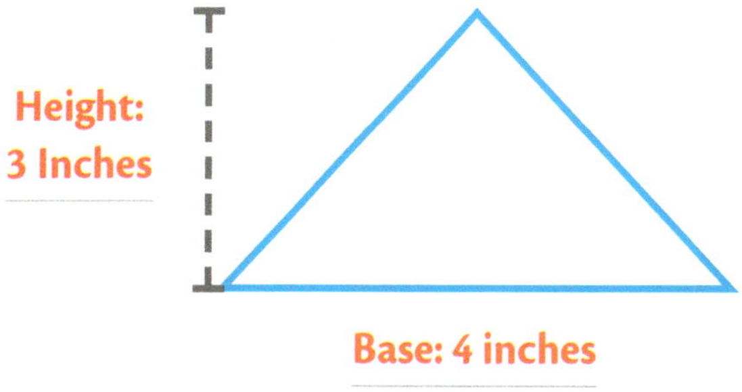

Read the following sentences and decide how the numbers in the story are combining–addition, subtraction, multiplication, or division. Circle the correct operation.

5. This week, 7 new children came to our Bible club. (+ − × ÷)

6. I gave 2 chocolates to each of my cousins. (+ − × ÷)

7. My Bible study guide has 200 pages grouped into 5 sections that are all the same size. (+ − × ÷)

8. When I walk on the beach I pick up every interesting thing that God has created on the shore. (+ − × ÷)

9. Every aisle at the grocery store has 3 different kinds of foods. (+ − × ÷)

10. I spent 10 dollars of my spending money at a garage sale. (+ − × ÷)

11. Dividing chores equally gives everyone a chance to be a blessing. (+ − × ÷)

12. Sometimes when you take away notes from a song it actually sounds better. (+ − × ÷)

Practice DAY 94

Decide which number is bigger and write the correct symbol in the circle.

1. 2003 ◯ 3002

2. $\frac{3}{100}$ ◯ $\frac{4}{10}$

3. $\frac{4}{6}$ ◯ $\frac{1}{6}$

4. 4.58 ◯ 4.56

Put these numbers in order from least to greatest.

456	645	564

5. ________, 6. ________, 7. ________

Complete these long addition and long subtraction exercises. The numbers to carry and to borrow are included for you.

8.

Carry		1	
	3	3	4
+	1	5	9

9.

Carry	1		
	4	9	2
+	2	1	3

10.

Carry	1	1	
	4	9	9
+	5	1	9

11.

Borrow		5	13
	3	~~6~~	~~3~~
−	1	5	9

12.

Borrow	0	12	
	~~1~~	~~2~~	5
−		9	3

13.

Borrow	4	10	
	~~5~~	~~0~~	5
−	3	4	4

Find the areas of these three shapes to see which is the largest piece of chocolate. The heights of the triangles are written to the side.

14. Area of Triangle = Base x Height ÷ 2

Area = 5 x 8 ÷ 2

= ☐ ÷ 2 = ☐ square inches

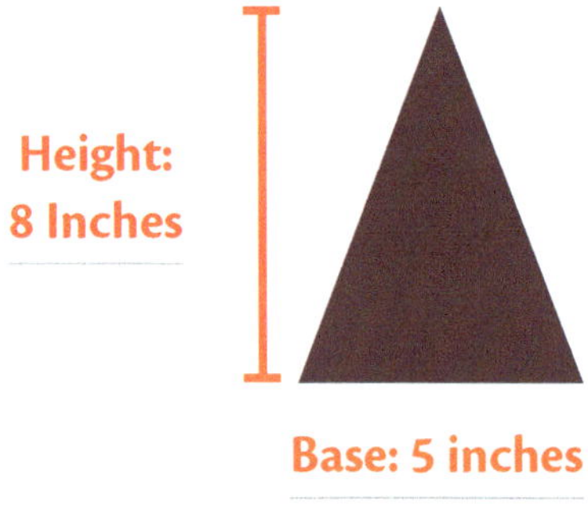

15. Area of Triangle = Base x Height ÷ 2

Area = 7 x 2 ÷ 2

= ☐ ÷ 2 = ☐ square inches

16. Area of Rectangle = Width x Length

Area = ☐ x ☐ = ☐

Extra Challenge

When you flip a coin three times, there are 8 different outcomes you could possibly get.

HHH (three heads in a row)

HHT (two heads first, then tails)

HTT

THT

HTH

THH

TTH

TTT

Use the list to figure out the probability for the following outcomes. Each time, the coin will be flipped three times—no more and no less.

17. All heads (HHH) $\frac{\quad}{8}$

18. Two tails (TT) in a row $\frac{\quad}{8}$

19. 4 tails (TTTT) $\frac{\quad}{8}$

20. All heads or all tails (TTT or HHH) $\frac{\quad}{8}$

21. One head (H) $\frac{\quad}{8}$

22. One tail (T) $\frac{\quad}{8}$

DAY 95 Measuring Four-Sided Shapes

Let's try a hands-on exercise to find areas of squares, rectangles, triangles, and parallelograms. You will need scissors and three pieces of notebook-sized paper of different colors.

But first, let's review these area formulas.

Formulas

Area of a Rectangle = **length x width**

Area of a Square = **length of any side x length of any side**

Area of a Triangle = **base x height ÷ 2**

Activity

1. Cut out one 5 inch x 5 inch square. What is its area?

   Area of Square = ____ x ____ = ____ square inches

2. Cut out two 3 inch x 6 inch rectangles. What is the area of each?

   Area of Rectangle = ____ x ____ = ____ square inches

3. Cut each square and rectangle diagonally into two triangles as shown on the next page. What is the area of each triangle? (These areas would be exactly half the area of the square or rectangle they came from.) Take the areas of the square or rectangle and divide it by 2.

   Area of Triangle (from the Square)

   = ____ ÷ 2 = ____ square inches

   Area of Triangle (from the Rectangle)

   = ____ ÷ 2 = ____ square inches

Put the same-colored triangles together a different way to make parallelograms as shown.

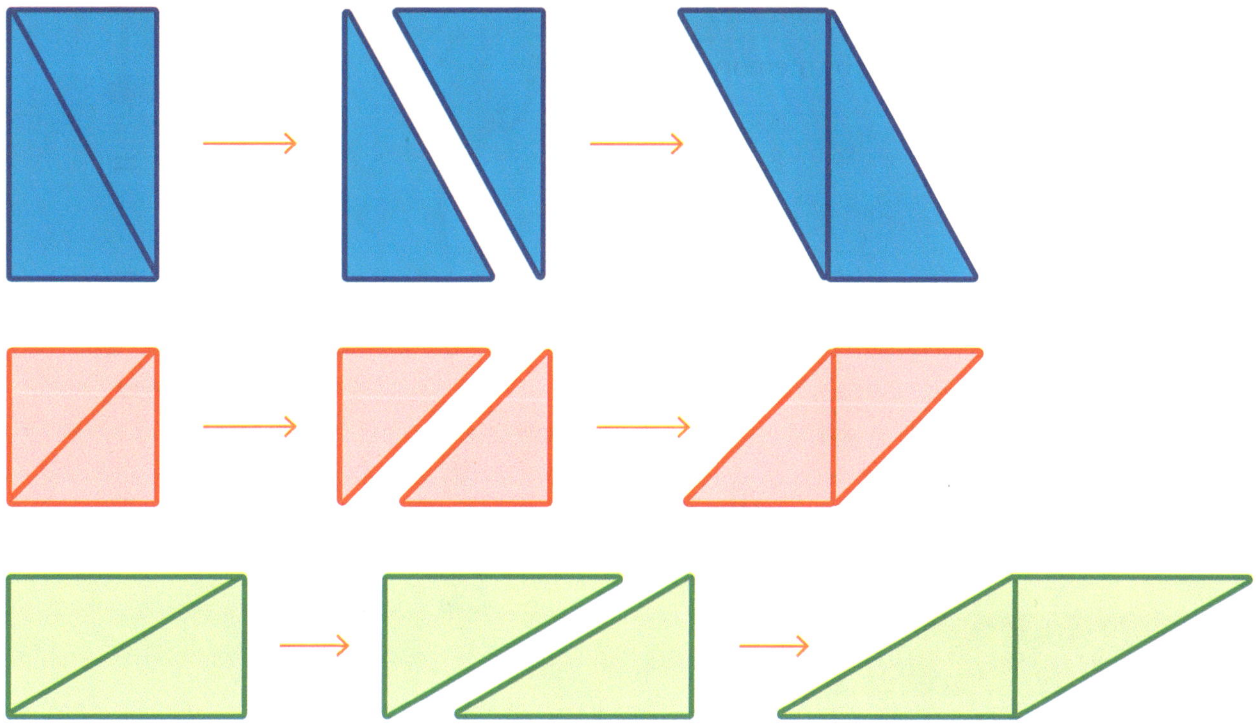

Now you know the area of each parallelogram! The area is exactly the same as the square or rectangle you started with.

When you find the area of a parallelogram, be careful that you use the correct measurements. You can't measure its sides and multiply them together like you do when you find a square's area or a rectangle's area. Remember, two of the parallelogram's sides were formed by the diagonal cut you made across the original shape, so it would be too long. Instead, look at the place where you taped your triangles together. Those two edges would be the correct place to measure. We call that measurement the parallelogram's *height*. The side that makes a square corner with the parallelogram's height is called its *base*.

Area of Parallelogram = Base x Height

4. If this parallelogram's height is 5 inches and its base is 9 inches, what is its area?

☐ x ☐ = ☐

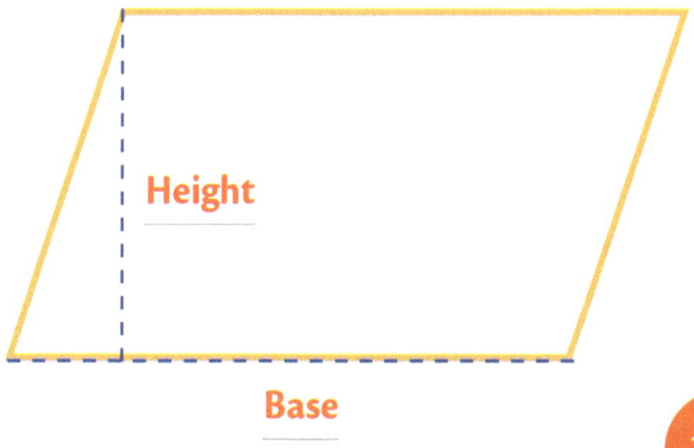

Student Exercises

Practice adding and subtracting by memory.

5. $5 + 5 =$ ☐

6. $18 - 9 =$ ☐

7. $16 - 8 =$ ☐

8. $7 + 7 =$ ☐

9. $8 + 8 =$ ☐

10. $12 - 6 =$ ☐

11. $6 + 6 =$ ☐

12. $20 - 10 =$ ☐

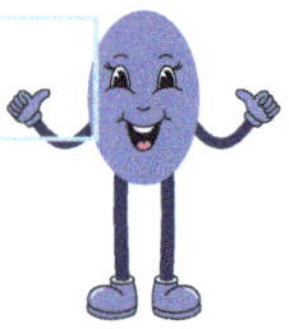

13. $\frac{2}{4} - \frac{1}{4} = \frac{\square}{\square}$

14. $\frac{2}{3} + \frac{2}{3} = \frac{\square}{\square}$

15. $\frac{1}{8} + \frac{1}{8} = \frac{\square}{\square}$

16. $\frac{6}{8} - \frac{1}{8} = \frac{\square}{\square}$

Everyday Math

13. Your father ate ¼ of the pizza pie. You ate ¼ of the pizza pie. How much was left of the pie? Shade the fraction of the pie that has already been eaten. Then fill in the answer.

$\frac{4}{4} - \frac{1}{4} - \frac{1}{4} = \frac{\square}{\square}$

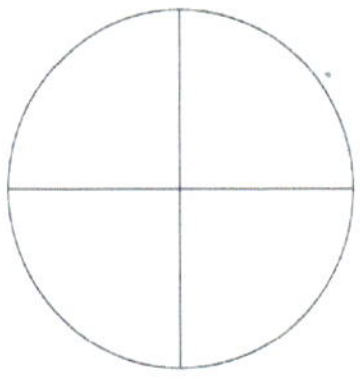

14. Your mother bought 3 pizzas and each pizza was cut into 8 pieces. Your family ate one and a half pizzas (12 pieces) for dinner. Then you ate six more pieces (6/8 of a pizza) for lunch. How much is left over for the next meal? Shade the fraction of the pies that have already been eaten. Then fill in the answer.

$\frac{24}{8} - \frac{\_\_}{8} - \frac{\_\_}{8} = \frac{\square}{\square}$

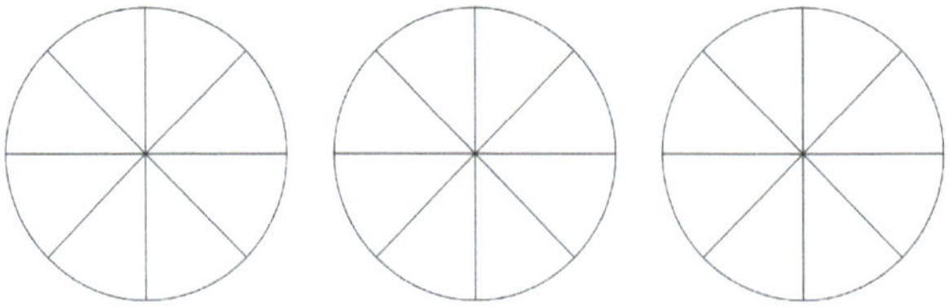

15. You have 24 chocolate candies, and you want to split them up equally with yourself and three friends. How many does each child receive?

☐ ÷ ☐ = ☐

Practice DAY 96

Practice counting by chunks. For each list of numbers, figure out if the list is counting up or counting down. Then fill in the blanks to continue the pattern.

1. Count by 3s: 90, 87, ______, ______, ______, ______, ______, ______
2. Count by 6s: 90, 84, ______, ______, ______, ______

Circle the answers in number 1 that you also see in number 2.

3. Count by 9s: 0, 9, ______, ______, ______, ______
4. Count by 9s: 90, 81, ______, ______, ______, ______
5. Count by 100s: 1000, 900, ______, ______, ______, ______

Practice multiplying and dividing these numbers using your memory. If you get stuck, use your multiplication/division table.

6. 4 × 3 x 2 = ☐
7. 10 ÷ 5 = ☐
8. 9 ÷ 1 = ☐
9. 3 × 2 x 5 = ☐
10. 8 x 9 = ☐
11. 12 ÷ 6 = ☐
12. $\frac{15}{15}$ = ☐
13. $\frac{8}{4}$ = ☐
14. 7 × 2 x 1 = ☐
15. 4 × 2 x 2 = ☐
16. 11 x 8 = ☐
17. 60 ÷ 12 = ☐

Remember, when you take a fraction of something, you are dividing it up. Let's divide up these apples. When you take ½ of 6, you are dividing 6 into two smaller groups. When you take ⅓ of 6, you are dividing 6 into three smaller groups. The first example is completed for you.

18. What is $\frac{1}{2}$ of 6? 3

19. What is $\frac{1}{2}$ of 12? ☐

20. What is $\frac{1}{2}$ of 18? ☐

21. What is $\frac{1}{3}$ of 6? ☐

22. What is $\frac{1}{3}$ of 12? ☐

23. What is $\frac{1}{3}$ of 18? ☐

Walking Around the Circle DAY 97

> To whom then will you liken God? Or what likeness will you compare to Him? . . . It is He who sits above the circle of the earth, and its inhabitants are like grasshoppers, Who stretches out the heavens like a curtain, and spreads them out like a tent to dwell in. (Isaiah 40:18,22)

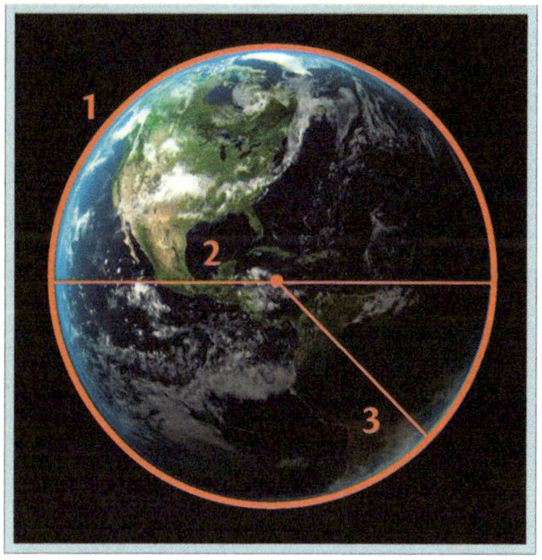

This verse talks about the circle of the earth. The earth is a three-dimensional circle that we call a *sphere* or a *globe*. If you look at a picture of the earth from space, you will see a round object. God sees this circle of the earth. Its inhabitants seem as tiny as grasshoppers from His view. God sits on a throne above the heavens and the earth. He is King over all.

Let's think about flat shapes and their numbers of sides. Triangles have three sides. Squares have four. But how many sides make up a circle? The simplest answer is to say that a circle doesn't have any sides.

The numbered lines on the picture will help you learn these new words:

1. The perimeter of a circle is called the **circumference**.
2. A line across a circle, right through the middle and from edge to edge, is called its **diameter**.
3. A line from the center of a circle to its edge is called its **radius**.

Activity

Let's measure the circumference of a circle.

You will need:

- a ruler
- a 12-inch piece of string and a 20-inch piece of string
- a piece of cardboard at least 8 inches x 8 inches
- 16 push pins

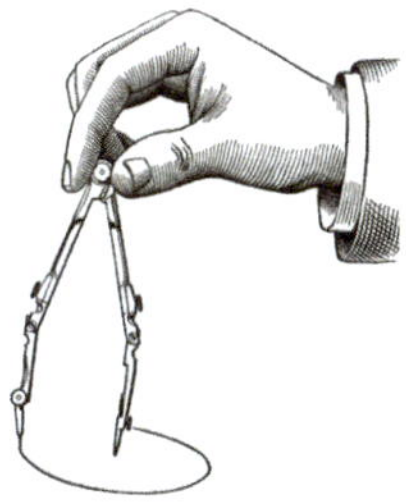

1. On the cardboard, draw a circle with a **diameter** of 6 inches: Tie one end of the 12-inch string to a pencil. You will be stretching the pencil out from the middle of the cardboard to make a circle around a point. Use a pin to fasten the loose part of the string in the middle of the cardboard so that the length of string between the pin and the pencil is 3 inches. If you have a compass, you may use it to draw the circle instead of using string.

The distance between the point where the pin is and the circle made by the pencil is the measure of the circle's radius. A circle's radius is half the length of its diameter.

2. Now let's try to figure out the circle's circumference. Poke four push pins into the edge of the circle, about the same distance apart. Stretch the 20-inch string around the outside of the pins, making a square (first picture). Make a mark on the string to show how much of it was used for the square. Remove the string.

Now attach four more stick pins between the first four (second picture). Stretch the string around the pins, and mark the length. Then attach eight more stick pins around the circle. Stretch the string around the pins and mark this length. Measure these lengths by laying your string along a ruler.

__________ __________ __________

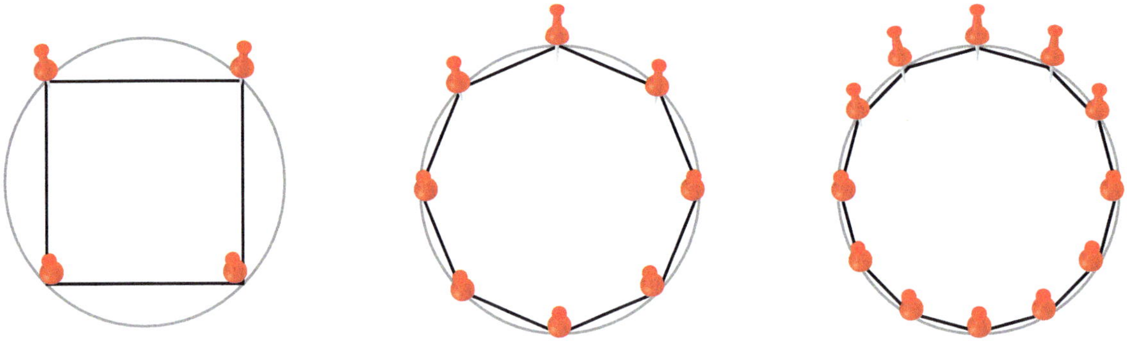

3. Next, lay the string around the circumference of the circle. Mark its length and measure it with a ruler. This is your most accurate measurement. Round it to the nearest inch and write it below.

Circumference of Circle: ________

Is this measurement more or less than your earlier string measurements? ____________________

4. Now, measure the diameter of the circle to the nearest inch. Be sure that the ruler is going through the center of the circle.

Diameter of Circle: ________ (Keep your cardboard circle on hand for another lesson.)

5. Now you get to discover something amazing. You will learn another way that God's math shows His wisdom and reminds us that He doesn't change.

Divide the circumference of your circle by its diameter. For example, your circumference was probably around 18 inches and your diameter was 6 inches. Find the answer.

Circumference ÷ Diameter

18 ÷ 6 = ☐

God made circles so unchangeable that every time we divide any circle's circumference by its diameter, we get the same number. That number is about 3. No matter what size circle we have, it takes about 3 of its diameters to equal its circumference. We use this "about-3" number so often in circle math, that we have made a name for it. We call it pi (Say "pie.") and write it like this:

$$\pi$$

Pi is a mysterious number. Pi is a little bigger than 3. Usually, people doing circle math will use the number 3.14 for π. But its decimal part is actually longer than that. Someday you will learn how π reminds us of God's eternity. You will discover how its decimal goes on and on without end.

If you know the diameter of a circle, you can find its circumference!

Formula

Circumference of a Circle = π x diameter.

In this book, we will use 3 for π:

Circumference of a Circle = 3 x diameter

Using 3 for π, figure out some circumferences using the equations below.

π x 2 =_______

π x 5 =_______

π x 10 = _______

π x 12 = _______

Extra Challenge

Imagine rolling a die once. What is the probability of rolling the following numbers? Write the answer as a fraction. The first one is shown for you.

20. 4 or 6 $\frac{2}{6}$

21. 4 ______

22. An even number ______

23. 7 ______

24. A number smaller than 10 ______

25. An odd number ______

DAY 98 Practice

1. Guess the size of this angle. _____

Measure the size of the angle. _____

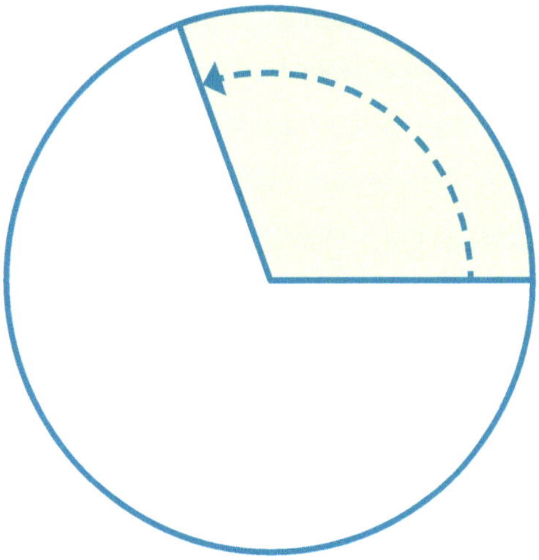

2. Guess the size of this angle. _____

Measure the size of the angle. _____

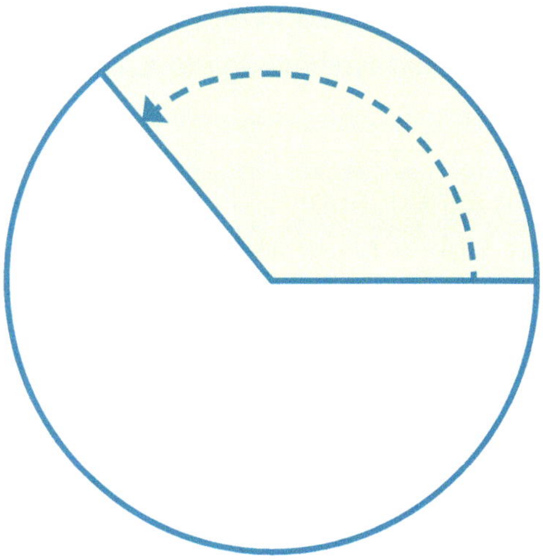

3. Guess the size of this angle. _____

Measure the size of the angle. _____

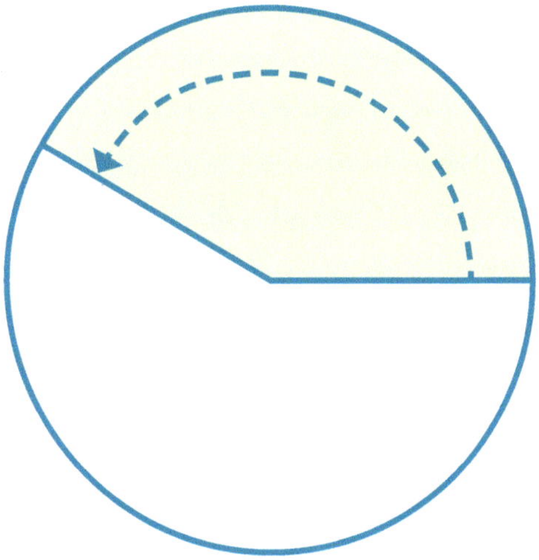

4. Guess the size of this angle. _____

Measure the size of the angle. _____

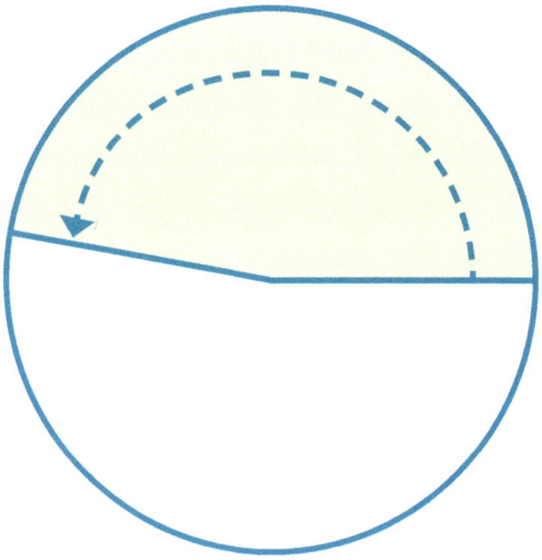

5. Guess the area of rectangle 5 in square inches.

______

Measure the length and width of the rectangle and figure out its area. ______

6. Guess the area of rectangle 6 in square inches.

______

Measure the length and width of the rectangle and figure out its area. ______

7. Guess the area of rectangle 7 in square inches.

______

Measure the length and width of the rectangle and figure out its area. ______

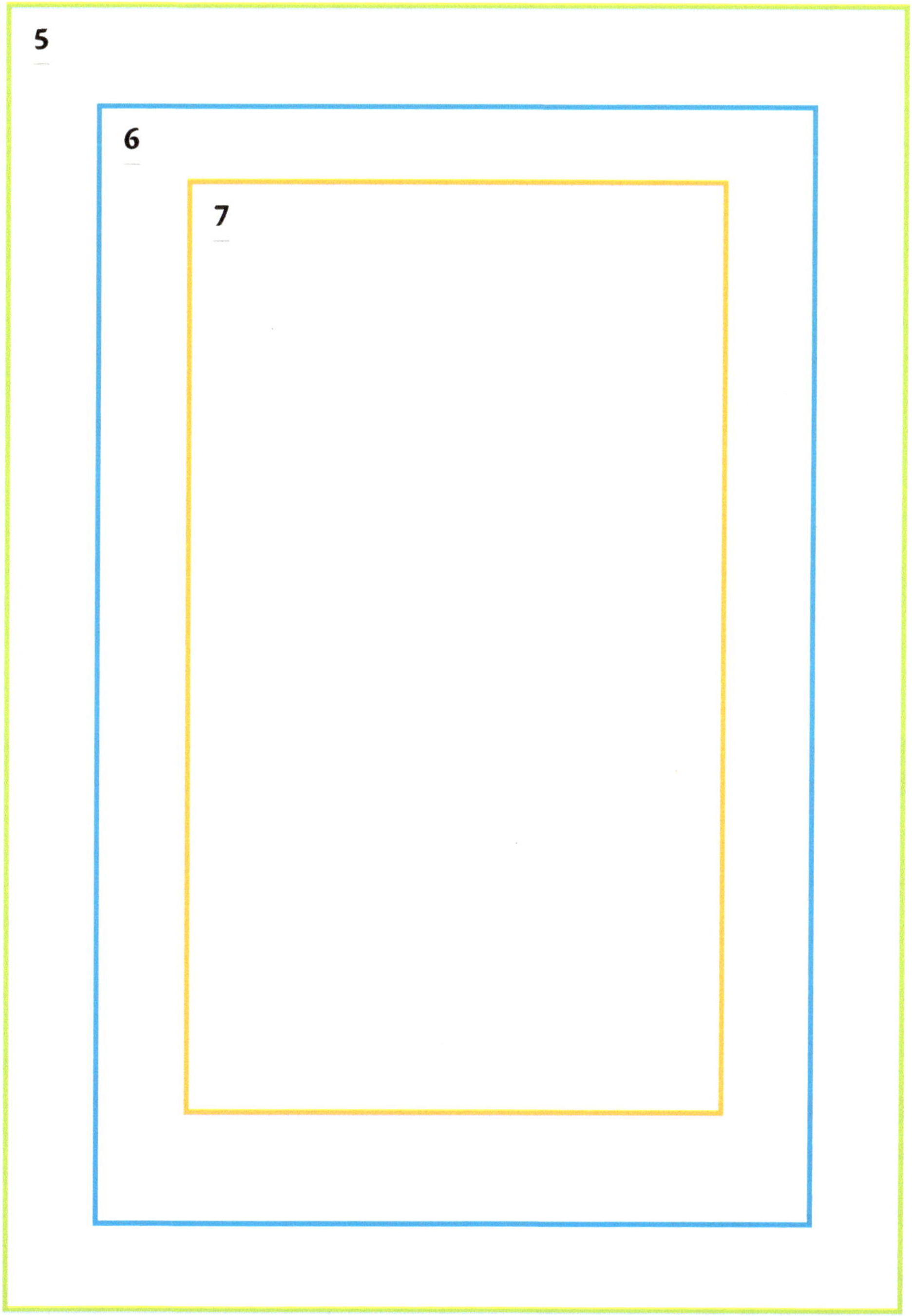

Finish each of these lists of shapes. Find the pattern, and then draw a picture that keeps that pattern going.

8. 

9. 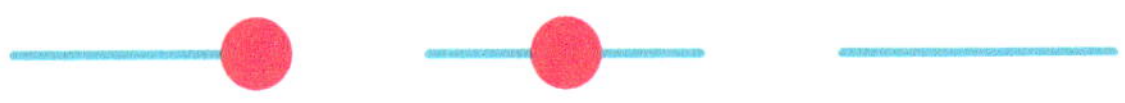

10. 

11. 

12. 

13. 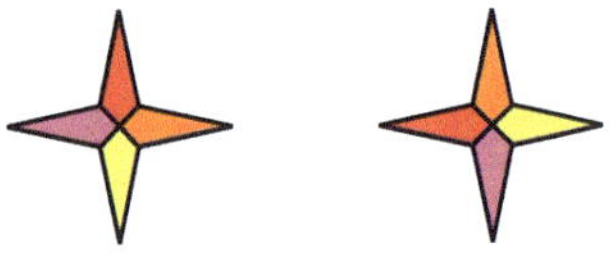

Figure out the correct number to answer each math question. Write it in the blank. Then write an equation that includes your answer. The first one is done for you.

14. What plus 5 equals 15? ___10___ ___10 + 5 = 15___
15. What plus 3 equals 13? ______ ______________
16. What minus 5 equals 9? ______ ______________
17. What minus 7 equals 3? ______ ______________
18. What times 5 equals 5? ______ ______________
19. What times 3 equals 18? ______ ______________
20. What divided by 5 equals 4? ______ ______________
21. What divided by 3 equals 2? ______ ______________

22. The length of the free-throw line on a basketball court is 6 feet. What is the circumference of the circle around that line? Remember the formula for the circumference of a circle is 3 times the diameter.

☐ x ☐ = ☐

23. Let's say you want to run 2 miles. You know that 4 laps around the track equals a mile. How many laps will you need to run in order to accomplish your goal?

☐ x ☐ = ☐

24. Continue the tessellation pattern. Draw more arrows to fill in all the blank empty areas.

DAY 99 Living Inside the Circle

How would you like to live in this round house? If you did, you might like to figure out its area. You already know how to find the area of square rooms and rectangler houses. Today you will learn to measure the area of a circle.

Let's guess the area of your cardboard circle. To help you guess, use your push pins and string to make a square inside the circle again. Measure the sides of your square, rounding up to the next highest inch. Then multiply the length of two sides together to get the area.

Area of a Square = Length of Any Side x Length of Any Side

☐ x ☐ = ☐

The area of this square is only a guess at the circle's area. You can see that there is quite a bit of space in the circle that's not included in your guess.

As God arranged it, the area of a circle can be figured out using the mysterious number π. In our forumula, we will also use the length of the radius which is half the length of the diameter.

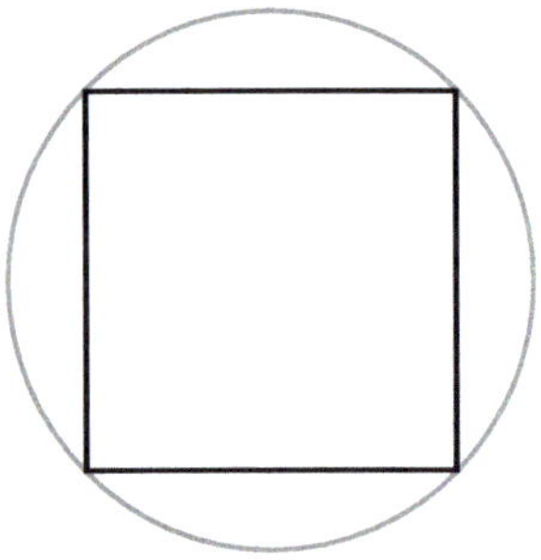

Formula:

Area of a Circle = π x Radius x Radius

In this book, we will use:

Area of a Circle = 3 x Radius x Radius

Use this formula to figure out the area of your circle. Compare your answer to the guess you made using the square inside your circle.

Area of Circle = 3 x 3 inches x 3 inches = ☐ square inches

Student Exercises

Write the mixed numbers in exercises 1-3 as fractions. Write the fractions in exercises 4-6 as mixed numbers.

1. $2\frac{3}{4} \Rightarrow \frac{\square}{\square}$
2. $1\frac{1}{2} \Rightarrow \frac{\square}{\square}$
3. $3\frac{2}{3} \Rightarrow \frac{\square}{\square}$
4. $\frac{7}{3} \Rightarrow \frac{\square}{\square}$
5. $\frac{12}{5} \Rightarrow \frac{\square}{\square}$
6. $\frac{9}{2} \Rightarrow \frac{\square}{\square}$

7. Identify the location of each piece of wood in the forest picture by writing its coordinates in the table. You will write each location with its column letter first and its row number second.

WOOD	PLACE
	F5

Find the perimeters of the following polygons. Each shape has sides of equal length. Remember you will find the perimeter by multiplying the number of sides by the length of each side.

1 inch

8. 1 × ☐ = ☐

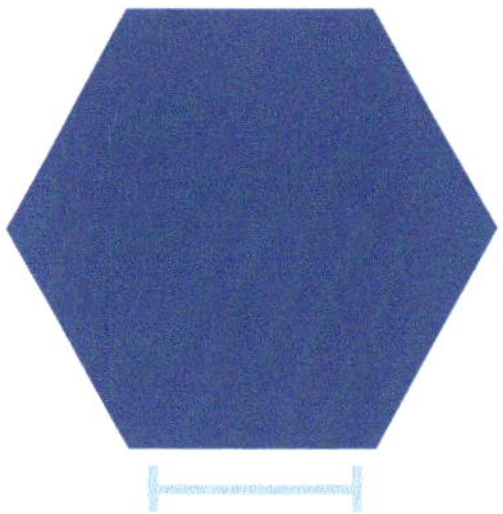

3 inches

10. 3 × ☐ = ☐

2 inches

9. 2 × ☐ = ☐

4 inches

11. 4 × ☐ = ☐

Complete these long addition and long subtraction exercises. Some of the carries and borrows are included for you.

12.

Borrow		4	11	
		~~5~~	~~1~~	9
	−	1	5	9

13.

Carry		1	1	
		9	3	8
	+		9	3

14.

Carry		1	1	
		6	4	3
	+		5	9

15.

Borrow				
		9	3	8
	−		9	3

16.

Borrow		6	14	16
		~~7~~	~~5~~	~~6~~
	−	1	7	8

17.

Carry				
		4	5	6
	+	4	5	6

Practice DAY 100

Read the following sentences and decide how the numbers are combining—addition, subtraction, multiplication, or division. Circle the correct operation.

1. My pocket had a hole in it, and I lost 2 quarters. (+ − × ÷)

2. I split my candy bar into 3 equal pieces to eat over three days. (+ − × ÷)

3. There's going to be a soccer game a few hours from now. (+ − × ÷)

4. And the LORD said to Moses, "I will bring one more plague on Pharaoh and on Egypt. Afterward he will let you go from here." (Exodus 11:1) (+ − × ÷)

5. For healthy teeth, brush 3 times a day, all 7 days of the week. (+ − × ÷)

6. In our dog-breeding business, 4 golden retrievers gave birth to 5 puppies each. (+ − × ÷)

7. I wonder how long my 30 bags of chips will last if I give away 5 per day. (+ − × ÷)

8. I left 6 avocados on the counter too long and 4 have already become rotten. (+ − × ÷)

Complete these long addition and long subtraction exercises. Some of the carries and borrows are shown for you.

9.

Borrow		5	11	
		~~6~~	~~1~~	4
	−	2	4	4

10.

Borrow			7	17
		6	~~8~~	~~7~~
	−	1	3	9

11.

Carry				
		2	7	5
	+	3	4	8

12.

Borrow		2	17	
		~~3~~	~~7~~	9
	−	2	9	9

13.

Carry		1	1	
		5	3	7
	+	1	9	6

14.

Carry				
		4	5	5
	+	1	2	5

15.

Carry		1	1	
		4	9	8
	+		7	9

16.

Carry		1	1	
		3	9	9
	+	2	9	9

17.

Carry				
		7	9	9
	+	7	9	9

Fill in the following circles with these two sets of numbers: circle A will contain 1, 2, 3, 4, 5. Circle B will contain 3, 5, 7, 9. Fill in the numbers for the union and intersection of these sets.

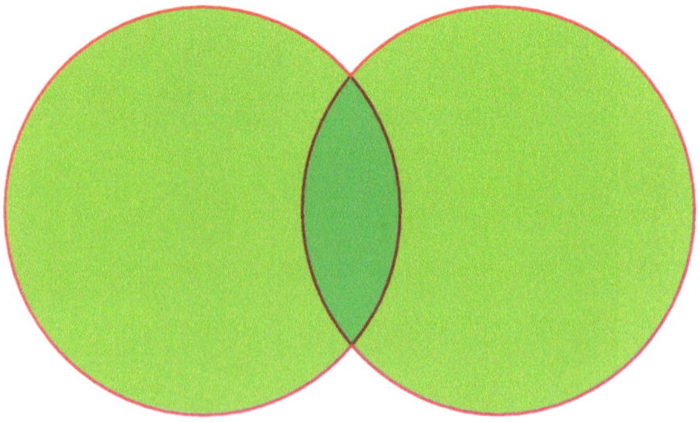

a. b

18. $A \cup B =$ ________________

19. $A \cap B =$ ________________

Go Enjoy God's Shapes

DAY 101

This lesson integrates math into everyday life. With this activity, the student will apply God's patterns and wisdom in meaningful ways to home and community. Take a break from memory work and academic exercises to make math part of everyday life. The following are suggestions or examples. Other ideas may be added to the list.

God made many beautiful structures in His world. Each wonderful structure, from a strong tree to a bird's sharp eye, God has made to do an amazing job.

God uses shapes in His structures. Here are several ways you can use shapes in your day-to-day life. Try one of these:

1. Design a bird house. Make sure everything fits together well. When two sides face each other, they should be exactly the same shape and size. Plan out all four sides with their measurements. Then plan the roof and the floor. How many rectangles, triangles, and circles will you use in the bird house? Be sure to make a way to open the birdhouse so you can clean it out before nesting time begins again.

2. Make cookies in the shape of rectangles, circles, stars, hearts, squares, and triangles. Decorate the cookies with different colors for a festive look!

3. Entertain some younger children with three-dimensional shapes. Find large boxes and join them to make tunnels and forts. Ask an adult's help to cut windows shaped like hexagons or circles.

Additional Activities

For extra activities, games, and helpful videos on these math concepts, refer to **generations.org/math3**.

Two horses are two times as strong as a single horse to pull the load.

CHAPTER 9

Long Multiplication

DAY 102 Multiplying by 11, 12, & More

> Two are better than one, because they have a good reward for their labor. For if they fall, one will lift up his companion. But woe to him who is alone when he falls, for he has no one to help him up. . . . Though one may be overpowered by another, two can withstand him. And a threefold cord is not quickly broken. (Ecclesiastes 4:9-12)

This verse shows an important biblical principle. You can see from the picture that two horses are stronger than one. Two horses provide two times the strength of one. Four horses would provide four times the strength. God has also made people able to team up and get more work done. Today we will learn more about God's wisdom of multiplication.

You have already learned multiplication facts for the numbers 1, 2, 3, 4, 5, 6, 7, 8, 9, 10 and maybe 11 and 12. Now let's multiply larger numbers. Remember, multiplying is a way to add the same number over and over again a certain number of times.

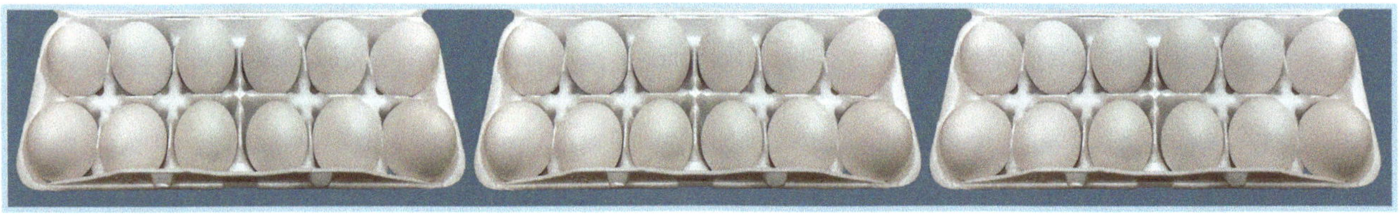

Here are 3 sets of 12 eggs. Let's see how many eggs there are altogether. If you haven't memorized the 3 x 12 multiplication fact, you can count the eggs. That would take a while. Or you can add them up like this: 12 + 12 + 12. But you can get the answer more quickly by multiplying 3 x 12.

Let's see how this works by multiplying with blocks. Here we have 3 groups of 12. Each group is shown as 1 chunk of 10 and two 1s. It will be easy to separately multiply the 10s and the 1s by 3.

Step 1. Multiply the 1s. 3 x 2 = 6

Step 2. Multiply the 10s. 3 x 10 = 30

Step 3. Add the answers together. 30 + 6 =36

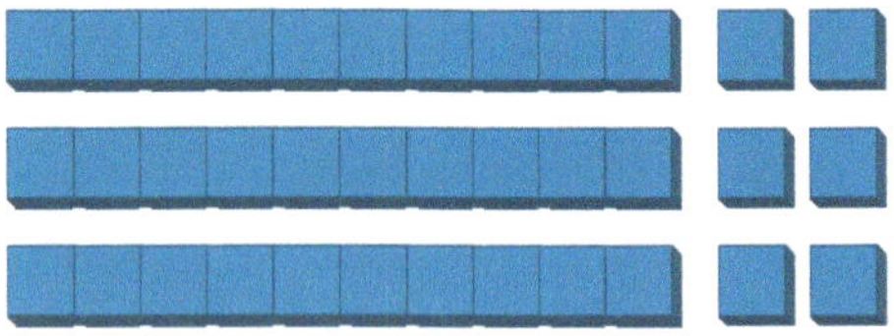

Let's try another one: 4 x 15. Fill in the blanks to get the final answer.

Step 1. 4 x 5 = ______

Step 2. 4 x 10 = ______

Step 3. _____ + ______ = ______

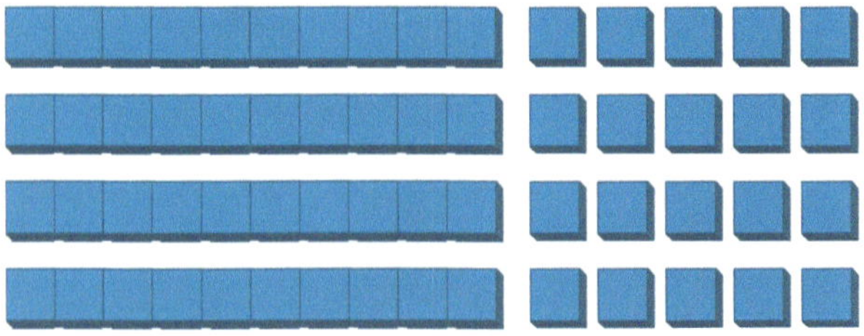

Student Exercises

Try a few three-step multiplications yourself. You can use your blocks or draw blocks if that helps you.

1.	2.	3.
3 x 11	2 x 12	4 x 13
Step 1	**Step 1**	**Step 1**
_____ x 1 = _____	_____ x _____ = _____	_____ x _____ = _____
Step 2	**Step 2**	**Step 2**
_____ x _____ = _____	_____ x 10 = _____	_____ x _____ = _____
Step 3	**Step 3**	**Step 3**
_____ + _____ = _____	_____ + _____ = _____	_____ + _____ = _____

Let's explore God's patterns with sets. Draw your own brackets around your answers.

4.	{1, 2, 3} U {a, b, c} =	7.	{1, 3, 7} ∩ {2, 4, 8} =
5.	{a, b, d} ∩ {a, c, d} =	8.	{4, 6, 8} U {8, 6, 4} =
6.	{x, y, z} U {w, y, z} =	9.	{y, x, z} ∩ {x, z, y} =

Everyday Math

10. You are planning to buy three presents for your family members for the upcoming holiday. The presents are priced $4.35, $2.43, and $3.11. How much will you spend for the three presents?

11. Your mother wants to make a vegetable dish using 1 pound of carrots at $1.00 per pound. She will also need 2 pounds of cherry tomatoes at $3.00 per pound. She will include 1 pound of yellow peppers for $1.50 per pound and 1 pound of cucumbers at $2.00 per pound. How much will she spend on the vegetable dish?

12. You would like to make your own trail mix. One pound of premade trail mix usually costs $9.90. Here is the cost of the items you will need for 1 pound of your trailmix:

Raisins: $ 0.50

Peanuts: $0.60

Chocolates: $1.00

Almonds: $1.20

Add up the cost of your mix. Then find out how much you will save by making it yourself.

Practice DAY 103

Write in commas to help you read these numbers easily. Answer the place-value questions about each number.

1. 2 3 5 7 8 4

   How many 10s? ________ How many 1,000s? ________ How many 10,000s? ________

2. 3 7 5 2 8 . 4 5

   How many 1s? ________ How many tenths? ________ How many hundredths? ________

3. 3 4 7 5 2 0 3 4

   How many 100s? ________ How many 1,000s? ________ How many 1,000,000s? ________

Let's practice adding and subtracting. Try to complete these by memory. Knowing math facts now will help you in big ways in the future.

4. $8 + 6 =$ ☐

5. $13 - 7 =$ ☐

6. $2 + 8 =$ ☐

7. $15 + 5 =$ ☐

8. $24 - 3 =$ ☐

9. $\frac{2}{3} + 0 =$ ☐

10. $14 - 9 =$ ☐

11 $8 + 7 =$ ☐

12. $16 - 10 =$ ☐

13. $\frac{4}{5} + \frac{3}{5} =$ ☐

14. $\frac{5}{9} + \frac{2}{9} =$ ☐

15. $\frac{3}{7} + \frac{1}{7} =$ ☐

Draw the hands on the clock to show the time.

16.

11:15

17.

12:30

18.

1:45

19.

2:05

For each of the following exercises, you will pretend to buy the two items by giving the cashier \$20. Write down how much change you would receive from the cashier. These exercises will require carrying and borrowing. Be sure you use the dollar sign and keep the decimal point in the correct place.

20.

\$ 7.0 0	\$2 0.0 0
+\$1 2.0 0	−\$

\$7.00

\$12.00

21.

\$1 3.7 0	\$2 0.0 0
+\$ 5.3 0	−\$

\$13.70

\$5.30

22.

\$ 6.1 5	\$2 0.0 0
+\$ 7.8 5	−\$

\$6.15

\$7.85

Long Multiplication DAY 104

You are learning a bit of God's wisdom. Remember that God created our world and gave wisdom to the things He put here. Imagine the wisdom God used to make the cheetah able to run many times faster than a hippo. Yet the hippo is many times more suited to live in the water than a cheetah is.

What kind of wisdom did God give the cheetah?

You've already learned how to multiply some large numbers with blocks. This lesson will show you how to speed it up by doing long multiplication.

Let's multiply 12 x 3 using long multiplication. First we will arrange the exercise by putting the number with the most digits on top. In long multiplication, every number on the bottom will be multiplied with every number on the top.

Step 1. Multiply the 3 by the 2 to get 6. Write the 6 in the 1s answer place.

	1	2
x		3
		6

Step 2. Multiply the 3 by the 1 to get 3. We will then write that 3 in the 10s answer place.

We see that the answer is 36.

	1	2
x		3
	3	6

1. Now, you try the next one: 11 x 4

	1	1
x		4

Try four more long multiplications on your own.

2.

	1	1
x		5

3.

	1	2
x		4

4.

	1	3
x		2

5.

	1	4
x		2

Student Exercises

Round these numbers. Rounding is an important way to understand large numbers in God's world.

1. Round 78,123.45 to the nearest 1/10 (tenth). ________________
2. Round 78,123.45 to the nearest 1. ________________
3. Round 78,123.45 to the nearest 10. ________________
4. Round 78,123.45 to the nearest 100. ________________
5. Round 78,123.45 to the nearest 1,000. ________________
6. Round 78,123.45 to the nearest 10,000. ________________

Numbers can be split apart into additions and multiplications. Find different ways to split these numbers.

7. 10 = ____ + ____
8. 12 = ____ + ____
9. 10 = ____ × ____
10. 12 = ____ × ____
11. 20 = ____ + ____
12. 20 = ____ + ____
13. 20 = ____ × ____
14. 20 = ____ × ____

15. Suppose that 3 vans will be taking a missions team to the airport. Each van carries 11 passengers. How many total passengers will go to the airport?

x ______

16. Let's say you have a dog-breeding business, and each of your 2 female dogs has given birth to 12 puppies.

How many puppies have been born for you to sell?

x ______

17. Suppose your family is cooking for the men's breakfast at the church. There are 13 men attending, and each man will eat 3 eggs. How many eggs will you need to buy?

x ______

18. Let's try two sudoku puzzles. Here are the rules:

- Every rectangular section of 6 little boxes should include the numbers 1,2,3,4,5, and 6.
- Each horizontal line of numbers should also include the numbers 1,2,3,4,5, and 6.
- Each vertical line should also include the numbers 1,2,3,4,5, and 6.
- Each number may only be used once in each section, in each horizontal line, and in each vertical line.

Let's do part of the first puzzle together.

We will fill in the bright blue box first. We see that the bright blue box must contain a 3 because the other numbers (1,2,4,5,6) are already in the same horizontal line.

Then we can fill in the green box with a 1. That's because the rectangle of six boxes already contains a 2, 3, 4, 5, and 6.

Now, what number shall we put into the orange box? It will be a 5 or a 2, because its vertical line already contains 1, 3, 4, and 6. But we can't use a 2 because there is already a 2 in the orange box's horizontal row. So we put the 5 into the orange box. Now figure out what number to put in the purple box. Keep going on this sudoku, and then try the other one.

a.

	4	2			5
	6	5	1	2	4
4					2
	2	6	4	1	
	5	1			6
6	3		2		1

b.

	4		3	5	
3			6		2
	6	3		1	4
	2	4	5		
	3	6		2	5
2	1			6	3

Long Multiplication with Carrying

DAY 105

Today we will multiply larger numbers that require carrying. Carrying in long multiplication is similar to carrying in addition: you will be carrying a number over to the next column to the left. Let's see how carrying works in multiplication.

To help you understand carrying, let's multiply with blocks first.

3 x 17

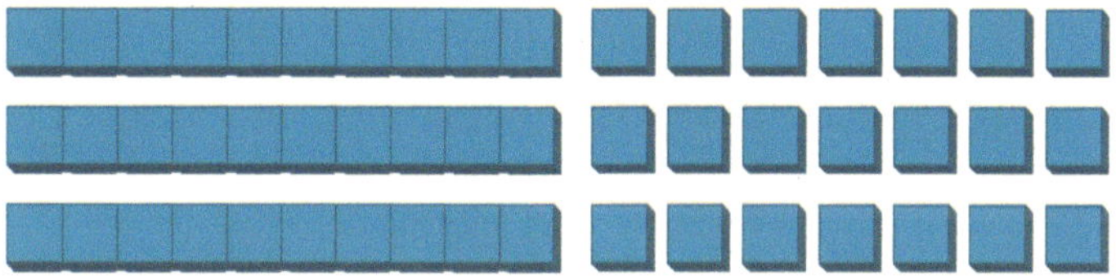

You can group all the single blocks above into groups of 10. Then put these new groups with the other chunks of 10. You will have 5 chunks of 10 and 1 single block left over. This makes an answer of 51.

Or you can look at 3 x 17 and multiply the 3 by both numbers of the 17 (the number in the 1s place and the number in the 10s place).

1. Step 1 —Multiply the 1s — 3 x 7 = ______
2. Step 2 — Multiply the 10s — 3 x 10 = ______
3. Step 3 — Add them up: ______

Now let's learn how multiply 3 x 17 using long multiplication. As before, we will put the number with the most digits on top. When multiplying these multi-digit numbers, every number on the bottom line is multiplied by every number on the top.

Step 1. Multiply 3 x 7 to get 21. Write the 1 from the 21 in the 1s answer place. What do we do with the 2 from the 21? We carry the 2 and put it above the 10s column.

Carry	2	
	1	7
x		3
		1

Step 2. Multiply 3 times the 1 from the 17. We get 3. We will add the carried 2 to the 3. So 3 + 2 = 5. Write the 5 in the 10s answer place.

The answer is 51. Now you know how to do long multiplication with carrying.

Carry	2	
	1	7
x		3
	5	1

4. Let's try another one by following the two steps. The number to carry is provided for you.

Step 1. Multiply 4 x 7. Write the 8 in the 1s answer place. Carry the 2 above the 10s column.

Step 2. Multiply 4 x 1 and add the carried 2 to that answer. Then write that answer in the 10s answer place.

Carry		2	
		1	7
	x		4

Practice more multiplication with carrying. The numbers to carry are provided.

5.

Carry		1	
		1	7
	x		2

6.

Carry		3	
		1	8
	x		4

7.

Carry		1	
		2	4
	x		3

8.

		3	2
	x		3

Often when we multiply, the answer will contain extra digits as in this example.

		100s	10s	1s
Carry			1	
			8	6
	x			3
		2	5	8

Notice that, after multiplying the last numbers (3 x 8), we took the answer (24) and added the carried 1 to get 25. We wrote the 5 in the 10s answer place, and the 2 in the 100s answer place. There is no need to carry the 2.

Student Exercises

Read the following sentences and decide how the numbers are combining: addition, subtraction, multiplication, or division. Circle the correct operation.

1. I have 10 apples and plan to eat 2 per day. (+ − × ÷)
2. In some games, points can be taken away from your score. (+ − × ÷)
3. While at the grocery store, Mom said I could choose 3 boxes of cereal to put in the cart. (+ − × ÷)
4. The 15 hours that I'm awake each day are split up into 3 parts that are almost equal: work, sleep, and play. (+ − × ÷)
5. This year for my birthday, Grandpa gave me 3 books, Dad gave me 3 books, and my cousin gave me 2 more. (+ − × ÷)
6. Each of my bookshelves holds 10 books. (+ − × ÷)
7. I took the dog for a walk three hours ago. (+ − × ÷)
8. I water my vegetable garden 2 times per day, 7 days a week. (+ − × ÷)

Add these fractions. You will only add the numerators. The denominators stay the same.

9. $\frac{3}{8} + \frac{5}{8} = \square$

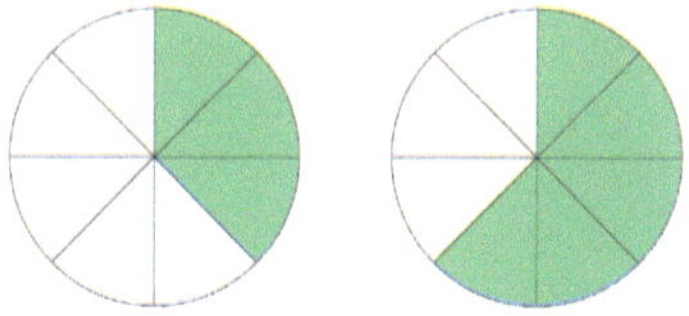

10. $\frac{7}{9} + \frac{8}{9} = \square$

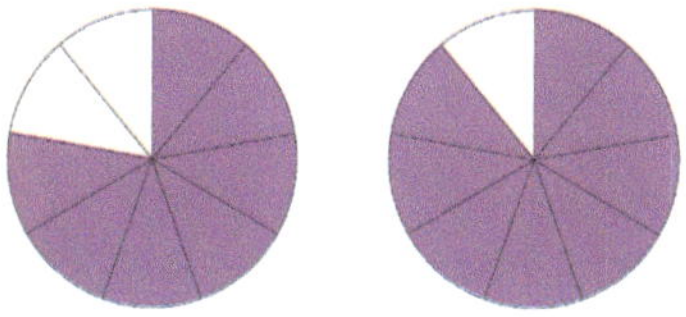

11. $\frac{7}{9} + \frac{3}{9} = \square$

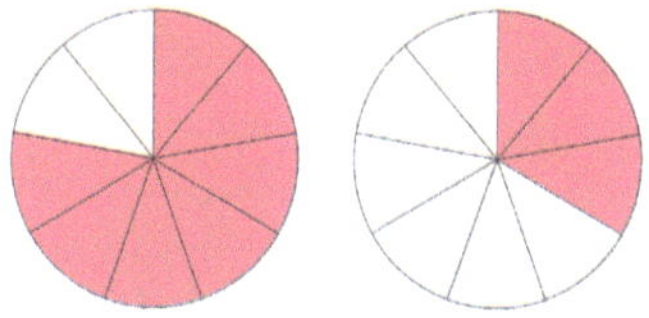

12. $\frac{8}{5} + \frac{\square}{5} = \square$

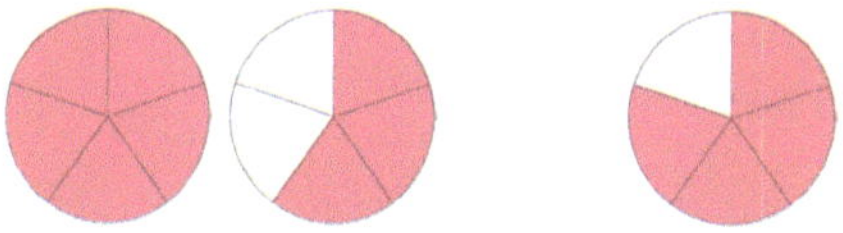

DAY 106 Practice

Here is some practice. Figure out how God made these numbers come together to make something new.

1.

	3	9
x		2

2.

	7	8
−	7	2

3.

	4 .	3	5
+	6 .	1	2

4.

	1 ,	3	4	5
+	3 ,	9	1	5

5.

	3	7
−	3	6

6.

	3	5
+	3	5

7.

	7 .	2	2
−	2 .	2	3

8.

	7	4
x		3

Which is larger or smaller? Or are they equal to each other? Write <, >, or = signs in the boxes.

9. 3 x 10 ☐ 4 x 7

10. 4 x 6 ☐ 3 x 8

11. 9 x 2 ☐ 1 x 20

12. 10 ÷ 5 ☐ 16 ÷ 4

13. 12 ÷ 2 ☐ 2 x 3

14. 15 ÷ 3 ☐ 20 ÷ 4

Everyday Math

15. Does your family like to watch birds? Let's say that, on vacation, your family saw 45 kinds of birds. Of those, 19 were new kinds that you all had never seen. How many kinds of birds had your family already seen before your vacation?

☐ − ☐ = ☐

16. Suppose that your family makes a visit to Grandma every Sunday afternoon. If you drive 34 miles roundtrip each Sunday, how many total miles would you drive in 4 weeks?

$$\begin{array}{r} 34 \\ \times \quad 4 \\ \hline \end{array}$$

17. If a road trip is 951 miles, but you make a stop for the night at 543 miles, how many miles of the trip will be left?

$$\begin{array}{r} 951 \\ -\ 543 \\ \hline \end{array}$$

18. Identify the location of each segment of the treasure map, by writing the coordinates from the picture. The location of the first piece is filled in for you.

ITEM	PLACE
X	D2

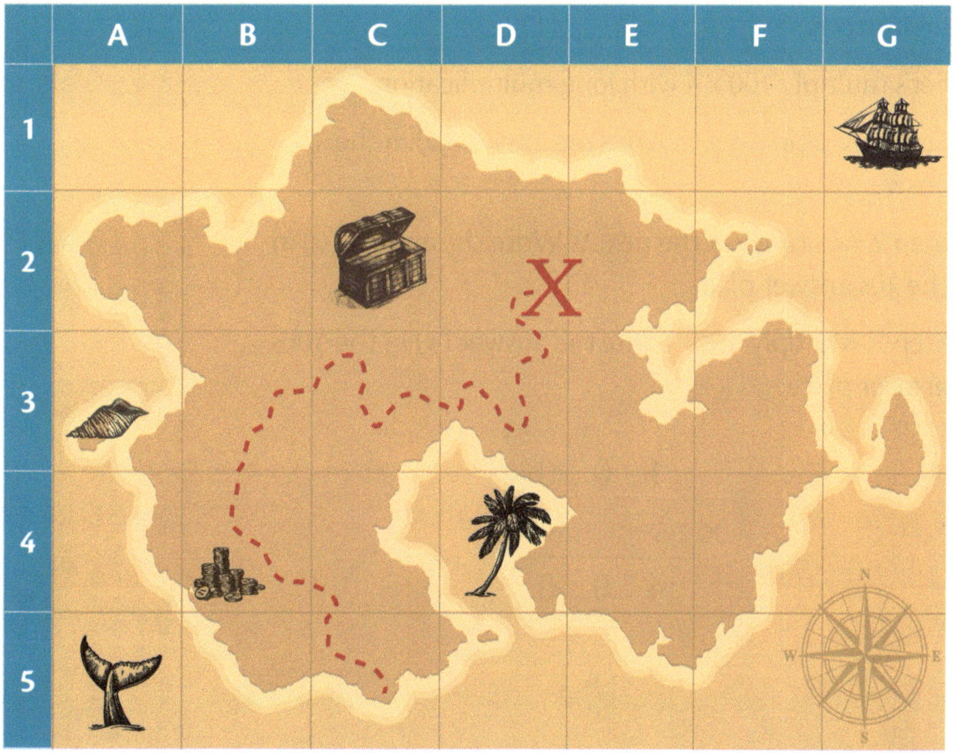

DAY 107 Longer Multiplication

God created the world with wisdom. Can you imagine the wisdom it took for God to make a dragonfly? You are learning a bit of God's wisdom in math.

Look at all these people! Let's see how many people are in this group. If you were to count them, it would take you about 7 minutes. But here, they are bunched into 4 groups of 100. Instead of counting, we could add the 4 groups together: 100 + 100 + 100 + 100. But that would still take too much time. Multiplication is a faster way to figure this out.

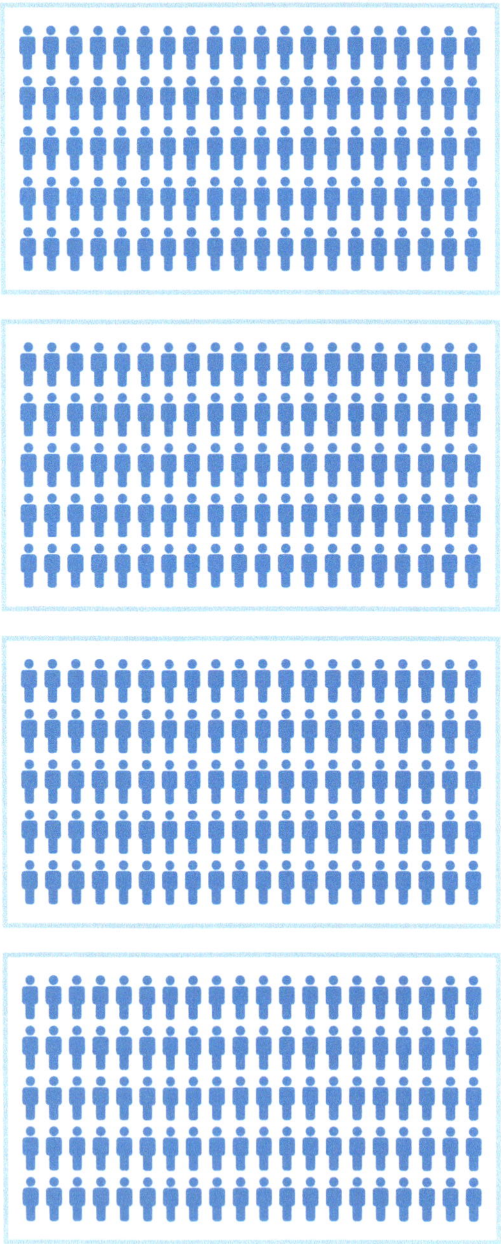

Large numbers call for long multiplication. As before, we will arrange the multiplication exercises by putting the number with the most digits on top. Then we will again multiply every number on top by every number on bottom.

Let's multiply 100 x 4 with long multiplication.

Step 1. Multiply 4 x 0. Write the answer (0) in the 1s answer place.

Step 2. Multiply 4 x the next 0. Write the answer (0) in the 10s answer place.

Step 3. Multiply 4 x 1. Write the answer (4) in the 100s answer place.

	1	0	0
x			4
	4	0	0

That was pretty easy. Now let's try a harder exercise.

Step 1. Multiply 2 x 3. Write the answer (6) in the 1s answer place.

Step 2. Multiply 2 x 2. Write the answer (4) in the 10s answer place.

Step 3. Multiply 2 x 1. Write the answer (2) in the 100s answer place.

	1	2	3
x			2
	2	4	6

Student Exercises

Practice more multiplication exercises. The last one is more challenging, but the number to carry is provided.

1.

	1	4	2
x			2

2.

	4	3	1
x			2

3.

	2	1	1
x			3

4.

Carry			1	
		2	3	5
	x			2

5.

	2	1	2
x			4

Round these numbers. Rounding is an important way to understand large numbers in God's world.

6. Round 4,345.54 to the nearest tenth ($\frac{1}{10}$). ______________

7. Round 4,345.54 to the nearest 1. ______________

8. Round 4,345.54 to the nearest 10. ______________

9. Round 4,345.54 to the nearest 100. ______________

10. Round 4,345.54 to the nearest 1,000. ______________

11. Round 4,345.54 to the nearest 10,000. ______________

Write the correct number in the box.

12. $\frac{5}{10} = \frac{50}{100} = 50\%$

13. $\frac{2}{10} = \frac{\square}{100} = 20\%$

14. $\frac{9}{10} = \frac{90}{100} = \square\ \%$

15. $\frac{\square}{10} = \frac{30}{100} = 30\%$

16. $\frac{\square}{10} = \frac{60}{100} = \square\ \%$

17. $\frac{7}{10} = \frac{\square}{100} = \square\ \%$

18. $\frac{\square}{10} = \frac{\square}{100} = 10\%$

19. $\frac{4}{10} = \frac{40}{100} = \square\ \%$

Extra Challenge

Imagine rolling a die. What would be the probability of rolling the following situations? Write the probability as a fraction with 6 in the denominator.

20. 3 ____________

21. A number bigger than 0 ____________

22. 5 or 6 ____________

23. A number that is not 4 ____________

24. 2 ____________

25. An odd number ____________

Practice DAY 108

Check out these numbers and answer the following place-value questions. Write in commas to help you read the numbers easily.

1. 4 7 8 9 7 6 4 3

   How many 10s? ________ How many 1,000s? ________ How many 10,000s? ________

2. 9 7 6 4 3 2

   How many 1s? ________ How many 100s? ________ How many 100,000s? ________

3. 5 8 7 7 5 3

   How many 100s? ________ How many 1,000s? ________ How many 1,000,000s? ________

Numbers can be split apart into additions and multiplications. Find different ways to split these numbers. Watch for whether you need to use addition or multiplication.

4. 6 = _____ + _____
5. 12 = _____ × _____
6. 15 = _____ + _____
7. 18 = _____ × _____
8. 6 = _____ + _____
9. 12 = _____ × _____
10. 15 = _____ + _____
11. 18 = _____ × _____

Speed Test

Ask your parent/teacher to time your speed on these multiplication exercises. You can practice this page every now and then to help you memorize these facts. Your goal (by the time you have completed Level 3 Math) should be to answer the exercises correctly in 90 seconds.

Average Time: 120 seconds
Target Time: 90 seconds
Best Student Time: 60 seconds
Expert/Teacher Time: 45 seconds

Your Time ________

1. 6 x 5 = ☐
2. 9 x 9 = ☐
3. 9 x 8 = ☐
4. 10 x 3 = ☐
5. 9 x 7 = ☐
6. 5 x 7 = ☐
7. 3 x 7 = ☐
8. 6 x 0 = ☐
9. 9 x 10 = ☐
10. 3 x 5 = ☐
11. 6 x 9 = ☐
12. 8 x 5 = ☐
13. 5 x 0 = ☐
14. 7 x 4 = ☐
15. 10 x 1 = ☐
16. 8 x 8 = ☐
17. 4 x 3= ☐
18. 3 x 6 = ☐
19. 6 x 6= ☐
20. 2 x 9 = ☐
21. 6 x 8 = ☐
22. 1 x 5 = ☐
23. 6 x 7 = ☐
24. 1 x 8 = ☐
25. 2 x 8 = ☐
26. 1 x 2 = ☐
27. 6 x 0 = ☐
28. 4 x 6 = ☐
29. 2 x 6= ☐
30. 5 x 5 = ☐
31. 5 x 9 = ☐
32. 3 x 9 = ☐
33. 3 x 2 = ☐
34. 5 x 2 = ☐
35. 8 x 4 = ☐
36. 8 x 7 = ☐
37. 10 x 1 = ☐
38. 5 x 4 = ☐
39. 8 x 8 = ☐
40. 4 x 9 = ☐
41. 5 x 8 = ☐
42. 4 x 4 = ☐
43. 4 x 6 = ☐
44. 7 x 7 = ☐
45. 9 x 9 = ☐

Multiplying by 10 DAY 109

> "But other [seeds] fell on good ground and yielded a crop: some a hundredfold, some sixty, some thirty." (Matthew 13:8)

This verse ties into farming. Every 1 grain (seed) of wheat sown by a farmer may yield 100 grains in the harvest. The 1 seed has been multiplied by 100 (increased by a "hundredfold"). For every 5 seeds, a hundredfold harvest would be 500 grains of wheat. God blesses us in so many ways, especially when he blesses our sowing with a generous harvest.

Today, we will increase something **tenfold** by multiplying it by 10. Imagine the number 2 as 2 seeds. Pretend to plant the 2 seeds, and show the harvest by replacing each seed with a chunk of 10. Here it is as an equation showing our harvest of 20:

2 x 10 = 20

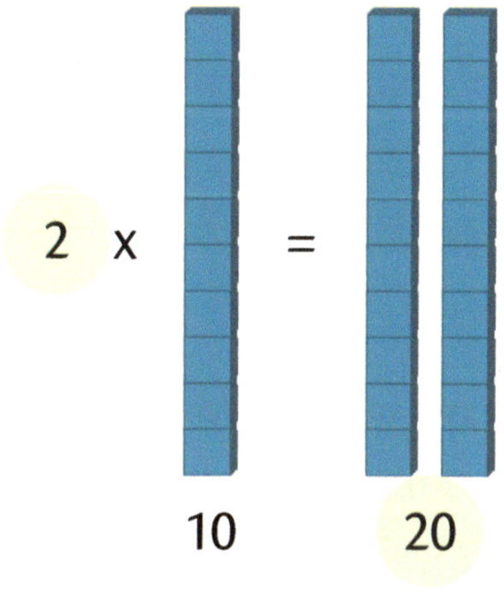

Now, what if we planted 10 seeds?

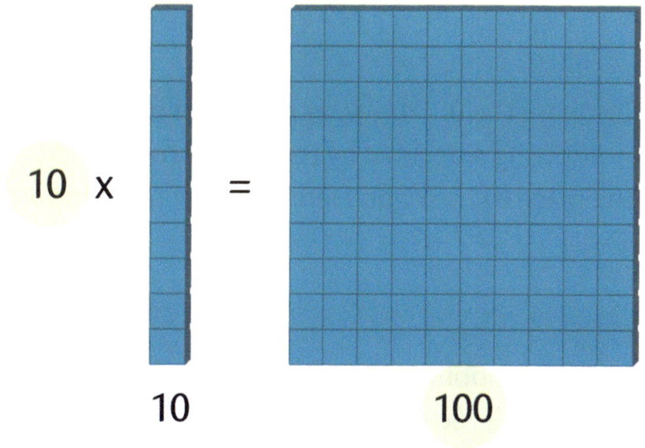

Or 30 seeds?

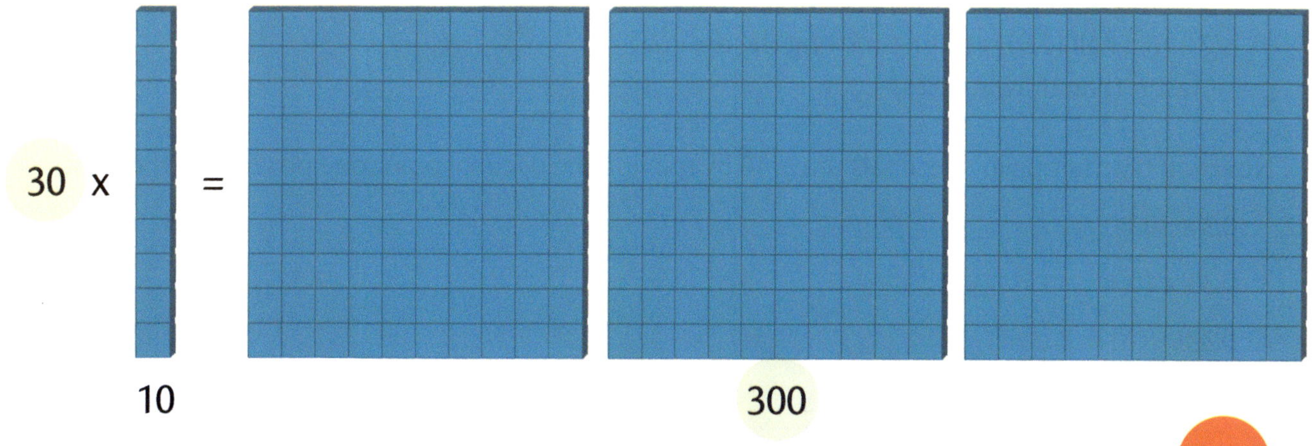

Carefully compare the 2 to the 20, the 10 to the 100, and the 30 to the 300. What's the difference? Did you noticed that when you multiply any number by 10, the number grows by one zero. 2 becomes 20. 10 becomes 100. 30 becomes 300. And, 599 multiplied by 10 would become 5990. What an easy way to multiply a whole number by 10!

Let's try a few of these yourself. Just put a 0 on the end to get the right answer. (You may need to move the comma in your answer.)

1. 5 x 10 = __________
2. 50 x10 = __________
3. 250 x 10 = __________
4. 3,550 x 10 = __________
5. 10,000,000 x 10 = __________

Now let's learn an easy way to multiply a decimal by 10.

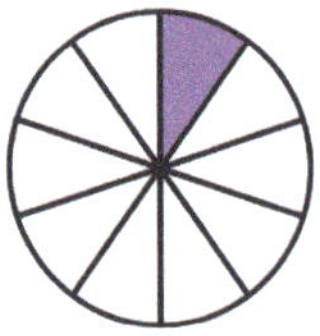

Let's take a decimal like 0.1. You should remember that 1 piece of a 10-piece pie is $\frac{1}{10}$ or 0.1. If we multiply that 1 piece of pie by 10, it equals 1 whole pie. Multiply 0.1 x 10. We write the equation this way:

0.1 x 10 = 1

Compare the 0.1 above with the 1. What happened to the decimal point? Notice that, when multiplying a decimal by ten, instead of putting a 0 on the end, you simply move the decimal point one place to the right. Here are a few more examples:

0.01 x 10 = 0.1

0.25 x 10 = 2.5

Let's think about 0.25 x 10 a little more. 0.25 as a fraction is $\frac{1}{4}$. So multiplying 10 times 0.25 makes **ten** $\frac{1}{4}$s. Look at ten $\frac{1}{4}$s using these pies. We see that ten $\frac{1}{4}$s is 2 and a half (2.5) pies. Isn't moving the decimal point faster than drawing pies?

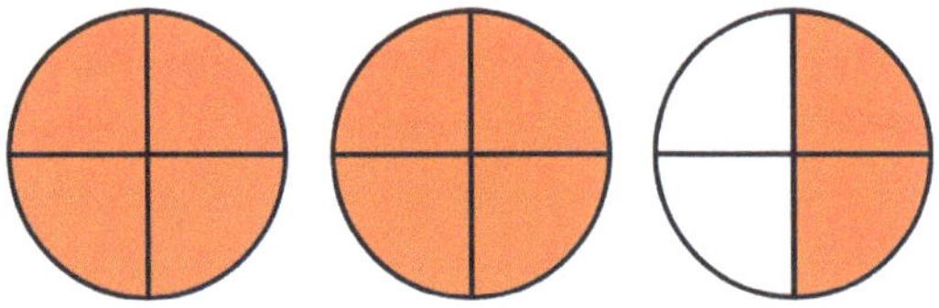

Now you try multiplying decimals times 10 by moving the decimal point one position to the right.

6. 0.3 x 10 = __________
7. 0.33 x 10 = __________
8. 0.333 x 10 = __________
9. 0.03 x 10 = __________

This next multiplication is a little different, but we still move the decimal point one place to the right, like this:

4.5 x 10 = 45.   4.5 45.

To this point, we have multiplied by 10. But what if we increase each number 100-fold, as we read in the Bible verse at the beginning of the lesson? Let's learn the easy way to multiply by 100. Instead of putting on one zero for the answer, we put on two zeroes like this:

2 x 100 = 200
15 x 100 = 1,500
345 x 100 = 34,500

Multiply by putting on more zeros.

10. 3 x 10 = 30
11. 54 x 100 = ________
12. 120 x 10 = ________
13. 543 x 100 = ________
14. 54,350 x 10 = ________
15. 1,000 x 100 = ________

For multiplying *decimals* by 100, you will need to move the decimal point two positions to the right, instead of just one.

0.01 x 100 = 1.0
0.1 x 100 = 10.

Now try these exercises. Move the decimal point to the right one or two places.

16. 0.05 X 100 = 5.
17. 0.05 X 10 = ________
18. 0.12 X 10 = ________
19. 0.12 X 100 = ________
20. 2.45 X 10 = ________
21. 2.45 X 100 = ________

An easy way to remember how many zeros to put on an answer is to count the number of zeros in the 10 or the 100 you are multiplying by. The one zero in the 10 means you put on one zero for the answer.

The number of zeros in 10 and 100 also tells you how many decimal places to move when you are multiplying decimals by 10 or 100. The one zero in 10 means you move the decimal one place for the answer.

DAY 110 Practice

Here is some practice with adding and subtracting.

1. 6 + 6 =
2. 17 – 7 =
3. 8 + 8 =
4. 5 + 5 =
5. $\frac{3}{6} - \frac{2}{6} =$
6. $\frac{3}{7} + \frac{4}{7} =$
7. 18 – 9 =
8. 7 + 7 =
9. 16 – 4 =
10. 20 – 11 =
11. 0.5 + 0.3 =
12. 0.7 – 0.2 =

Practice rounding these numbers. Rounding helps us understand numbers in God's world.

13. Round 19,928.09 to the nearest $\frac{1}{10}$. ________________
14. Round 19,928.09 to the nearest 1. ________________
15. Round 19,928.09 to the nearest 10. ________________
16. Round 19,928.09 to the nearest 100. ________________
17. Round 19,928.09 to the nearest 1,000 ________________
18. Round 19,928.09 to the nearest 10,000 ________________

When you flip a coin three times, there are 8 possible outcomes: HHH, HHT, HTT, THT, HTH, THH, TTH, TTT

Imagine flipping a coin three times. Look at the possible outcomes above to figure out the probability of each of the following things happening in those three flips. Write your answers as fractions with denominators of 8.

19. A heads on the first, flip, then a tail on the second ___________
20. Only one tails ___________
21. No tails ___________
22. Equal numbers of heads and tails ___________

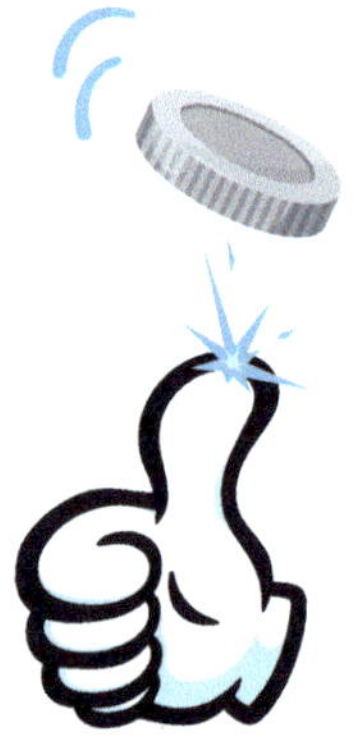

Everyday Math

23. Suppose that your family is taking a trip to see your relatives who live 143 miles away from your home. How many miles in all would you travel to go there and back again?

	1	4	3
x			2

24. Which of your relatives lives the farthest from you? How many miles (or km) in all would you travel to visit there and then get back home? Your parent/teacher may have to help you find the distance between your homes.

25. Imagine that you have started up a cookie business. You plan to make 120 cookies every week, and sell them for $1.00 each. How much money will you earn after 6 weeks?

Carry		1		
		1	2	0
	x			6

26. Now suppose that each week, you must spend $15 on cookie ingredients. How much will you spend on ingredients in 6 weeks of your cookie business?

Carry			3	
			1	5
	x			6

27. For the first six weeks of selling cookies, subtract your total expenses from the total money you took in. This is your profit.

Borrow				
	–			

DAY 111 Go Glorify God and Multiply!

This lesson integrates math into everyday life. With this activity, the student will apply God's patterns and wisdom in meaningful ways to home and community. Take a break from memory work and academic exercises to make math part of everyday life. The following are suggestions or examples. Other ideas may be added to the list..

1. A child's red wagon can carry about 200 pounds of weight. Suppose that you invited 2 of your friends to ride down a hill with you in your wagon. Your friends each weigh 60 pounds. Multiply to see how much your 2 friends weigh all together. Then add your own weight to their total. Can your wagon handle that much weight? You can check out the weight ratings for the wagons, bicycles, or anything else your family rides on.

	6	0
x		2

+			

2. Plan an imaginary party for 16 people. Include punch to drink and pizza and salad to eat. Plan for each person to drink 8 ounces of punch. How many ounces of punch should you make? __________
A gallon equals 128 ounces. How many gallons of punch will you be making? ____________

Punch

Carry			
		1	6
	x		8

Plan for each person to eat 3 pieces of pizza. How many pieces of pizza should you make?

Pizza

Carry			
		1	6
	x		3

Plan for each person to eat 4 ounces of salad. How much salad should you prepare?

Salad

Carry			
		1	6
	x		4

3. Now think of a real pizza party you could give. How many people would you like to invite? Make a list and count them up. ________

To figure out the amount of food for your pizza party, fill in the blank set of multiplication exercises.

Punch

Carry			
	X		

Pizza

Carry			
	X		

Salad

Carry			
	X		

Facts that are helpful as you prepare for the party:

- Large pizzas are usually cut into 12 pieces.
- There are 128 ounces of liquid in a gallon.
- Salads are often sold in 16-ounce bags.

4. When a mother is expecting a baby, it's a wonderful time to learn about this amazing creation of God. Let's imagine a baby inside its mother's womb when it's between five and six months old. During this month, Baby will grow about 2.5 centimeters longer. Before this month of growth, Baby's total length was about 10 times the 2.5 centimeters. How long was the baby before it grew the 2.5 centimeters? (Hint: Multiply by moving the decimal point.) ______________

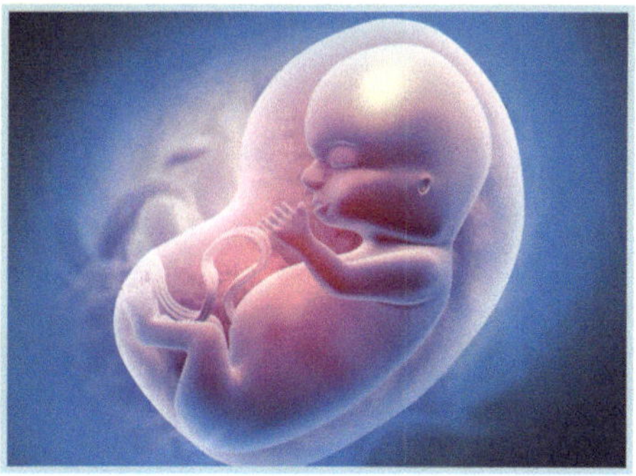

Multiply these whole numbers and decimals.

5. 2 x 10 =

6. 2 x 100 =

7. 91 x 10 =

8. 91 x 100 =

9. 6.41 x 10 =

10. 6.41 x 100 =

11. 0. 75 x 10 =

12. 0.75 x 100 =

13. Identify the location of each piece of the kitchen by finding its coordinates in the large picture. The location of the first piece is filled in for you.

WOOD	PLACE
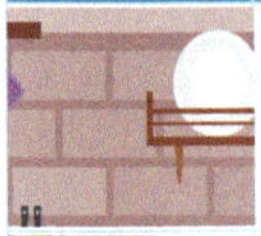	C3

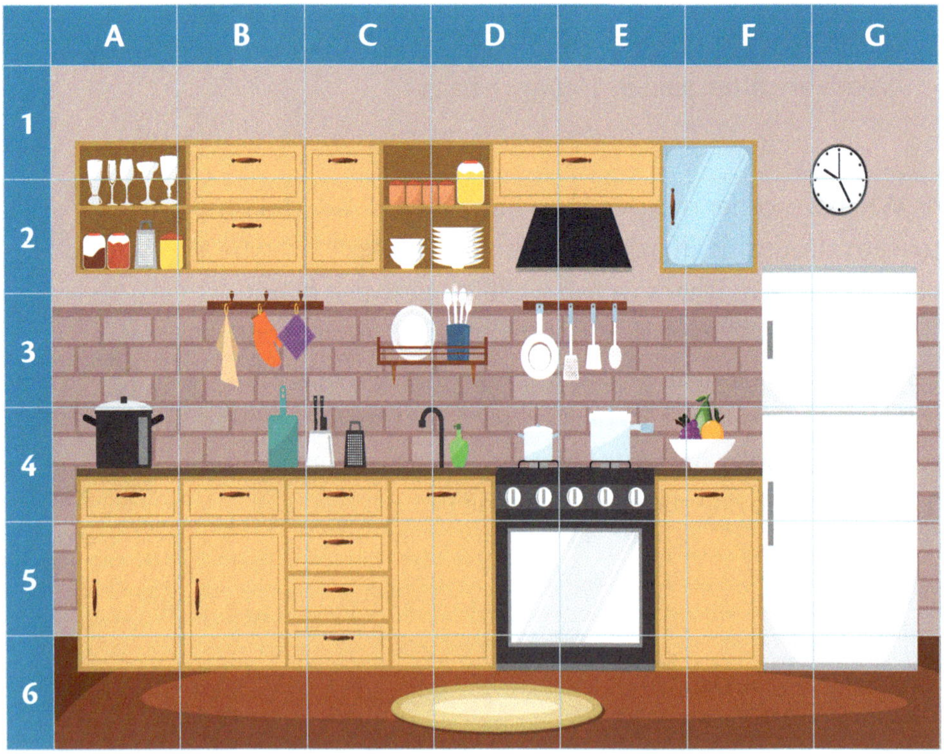

Additional Activities

For extra activities, games, and helpful videos on these math concepts, refer to **generations.org/math3**.

God made animals large and small. This small egret rides on the large elephant.

CHAPTER 10

The International System of Units

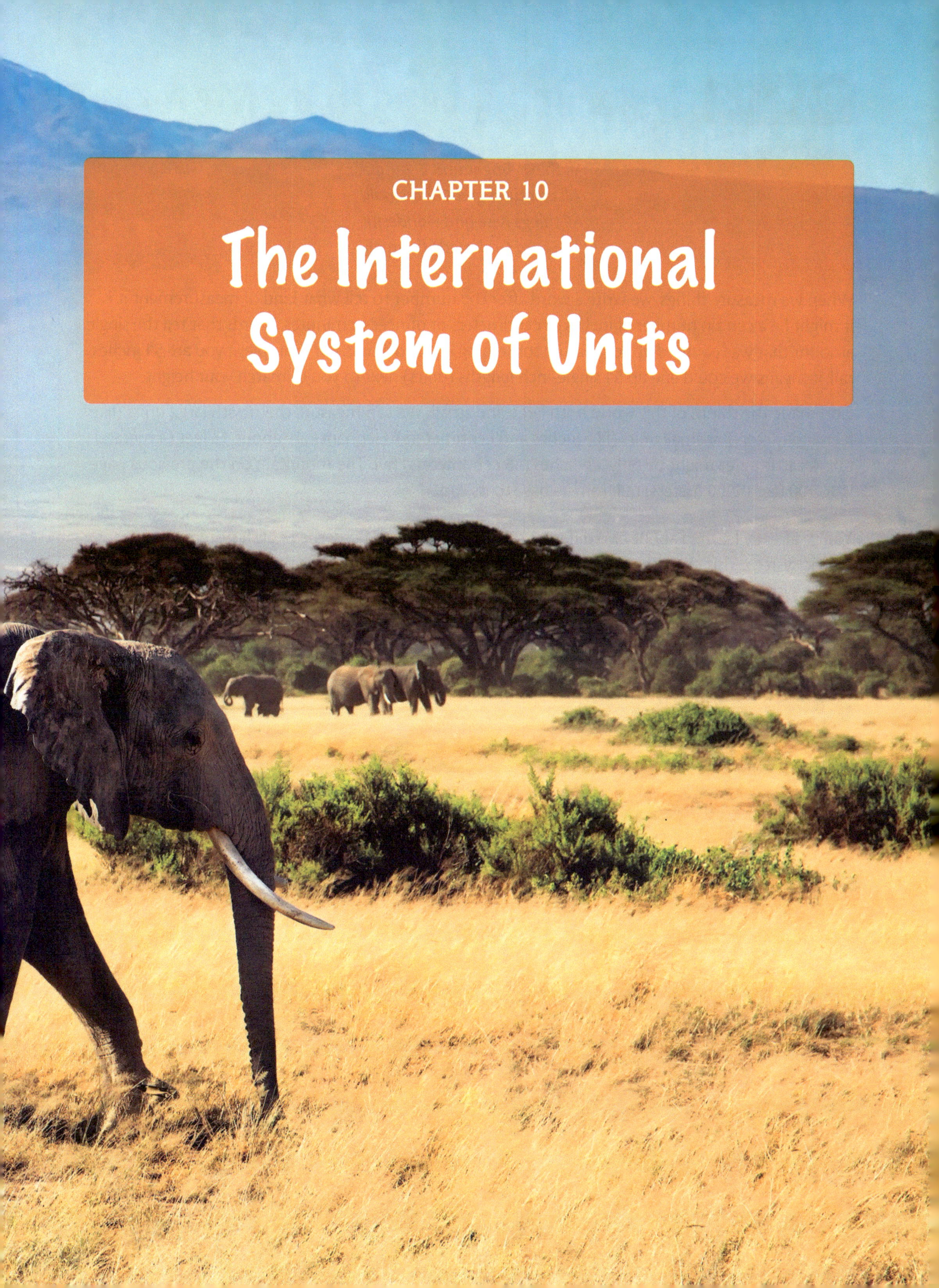

DAY 112 Measuring Length by Meters

All things bright and beautiful,
All creatures great and small,
All things wise and wonderful,
The Lord God made them all.

When we measure things, we write a word after the number to tell what kind of measurement it is. It might be a certain number of inches or centimeters or ounces or minutes. Words that tell the single amount that was counted up to measure something is called a unit. For example, if you are 51 inches tall, it means we could line up 51 single-inch lengths (units) next to you to match your height.

God made the things of this world both large and small. We can measure them with large units (like feet and meters) or small units (like inches and centimeters). Elephants are about 13 feet (4 meters) tall. A cattle egret might be only 18 inches (46 centimeters) tall. The mountain on the previous page is 16,100 feet (4900 meters) tall from its base to its top.

You've already learned about two different systems of measurement: the English system and the metric system. This chart shows the units used in each system. In this chapter, we will focus on the metric system.

UNITS OF MEASUREMENT

	LENGTH	DISTANCE	AREA	VOLUME	WEIGHT
English	Yards, Feet, Inches	Miles	Square Feet, Square Yards, Square Miles	Ounces, Quarts, Gallons	Ounces, Pounds
Metric	Millimeters, Centimeters, Meters	Kilometers	Square Centimeters, Square Meters, Square Kilometers	Milliliters, Liters	Grams, Kilograms

If you were to say, "A meerkat is 1 tall," that wouldn't mean anything to us. We would ask, "1 what?" Is the meerkat 1 foot tall, 1 meter tall, or 1 kilometer tall? Units (like feet and meters) help us communicate sizes, lengths, areas, and weights clearly.

Some countries use different languages. They also use different units to measure things. Units are like languages. Using the same system of units within a country helps everybody understand each other.

> [Jesus said,] "Therefore I say to you, do not worry about your life, what you will eat or what you will drink; nor about your body, what you will put on. Is not life more than food and the body more than clothing? Look at the birds of the air, for they neither sow nor reap nor gather into barns; yet your heavenly Father feeds them. Are you not of more value than they? Which of you by worrying can add one cubit to his stature?" Matthew 6:25-34

The wall of Jerusalem (above) is made of stone blocks about 1 cubit long. The Jewish people used the cubit as a unit of measure when Jesus walked the earth. A cubit was about 20

Below are a few metric system lengths:

- Millimeter
- Centimeter (10 Millimeters)
- Decimeter (10 Centimeters)
- Meter (10 Decimeters)
- Kilometer (1000 meters)
- 1,000 millimeters make 1 meter
- 100 centimeters make 1 meter
- 10 decimeters make 1 meter
- 1,000 meters make 1 kilometer

This ant looks like it's about 5 millimeters long. The crab (including its legs) looks like it's about 4 centimeters wide. The reason we use different-sized units is that it's easier to talk about and write numbers with less digits. It's easier to say you're taking a 3-kilometer walk than to say you're taking a 3,000,000-millimeter walk. So we measure the length of an ant using millimeters and the length of a crab using centimeters. We measure the heights of mountains using meters, and the distances between cities using kilometers.

How far can these things travel in one hour?

Snail: 1 meter

You walking: 4,000 meters (4 kilometers)

Bird flying: 30,000 meters (30 kilometers)

When we measure, it's faster to write abbreviations for our units than it is to write the whole words. Here are some common abbreviations for length in the metric system.

Millimeter = mm

Centimeter = cm

Meter = m

Kilometer = km

Let's memorize how these metric units work together:

10 mm = 1 cm

100 cm = 1 m

1,000 m = 1 km

Knowing these will help you convert (change) measurements from one metric unit to another.

In the last chapter, you learned an easy way to multiply by 10 and by 100. This will help you as we learn about the metric system. In the metric system, we can multiply by 10 or 100 (or even 1000) to get a lot of smaller units. For example, 1 centimeter x 10 = 10 millimeters.

Change these larger units into a lot of smaller units by multiplying:

1. 3 cm x 10 = _______ mm
2. 50 cm x 10 = _______ mm
3. 9.2 cm x 10 = _______ mm
4. 7 cm x ______ = _______ mm
5. 1 m x 100 = _______ cm
6. 22 m x 100 = _______ cm
7. 4.7 m x 100 = _______ cm
8. 65 m x ______ = _______ cm
9. 1 km x 1000 = _______ m
10. 39 km x 1000 = _______ m
11. 2.8 km x 1000 = _______ m
12. 6 km x ______ = _______ m

Moving the Decimal Point to Divide by 10, 100, and 1,000

You will also need to know the easy way to divide by 10, 100, and 1,000. Division is the opposite of multiplication, so when dividing, we move the decimal point to the left.

First, let's review multiplying by 10 with blocks.

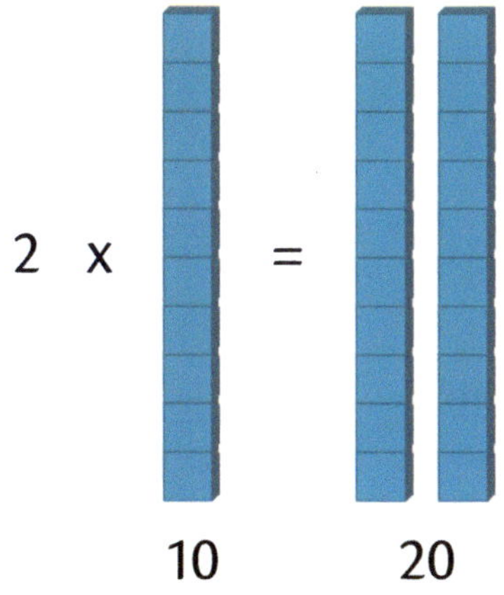

2 x 10 = 20

A zero was put on the end of 2 for the answer. This means the decimal point was moved to the right.

Now let's show how we divide by 10 with blocks.

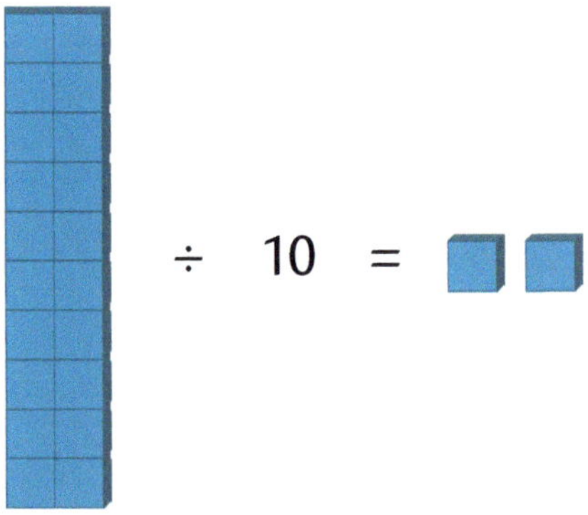

$$20 \div 10 = 2.0$$

Notice that we took the zero off the end of 20 for the answer of 2. You will also notice that the decimal point was moved one place to the left in dividing by 10.

Watch what happens to the decimal point when dividing 340 by 10. You can see that it moves **one** place to the left because the number 10 has one zero.

$$340. \div 10 = 34.0$$

Now, watch what happens when we divide a number by 100. You can see that the decimal point moves **two** positions to the left because 100 has **two** zeros.

$$340. \div 100 = 3.40$$

Now, what happens when we divide a number by 1,000? You see that the decimal point moves **three** places to the left because 1000 has **three** zeros.

$$340. \div 1000 = .340$$

Try a few of these on your own.

13. 530. ÷ 10 = __________

14. 530. ÷ 100 = __________

15. 530. ÷ 1000 = __________

16. 64. ÷ 10 = __________

17. 640. ÷ 100 = __________

18. 6400. ÷ 1000 = __________

Converting Metric Units to Different Metric Units

Let's say your little finger is 34 mm long, and you wonder how long that is in centimeters. You've memorized that there are 10 millimeters in a centimeter. So you know you have to divide the millimeters into groups of 10. Thankfully, you know the quick way to do this—you divide by moving the decimal point to the left. Since 10 has **one** zero, you will move it **one** place.

34 mm = 3.4 cm

Another way to think of this is that, because the millimeter is $\frac{1}{10}$ of a centimeter, you will divide the millimeters by 10. Because 10 has **one** zero, you will move the decimal point **one** place to the left when changing from millimeters to centimeters.

Now let's change 24 cm to meters. Remember, there are 100 cm in 1 meter. Because a centimeter is 1/100 of a meter, you will be dividing the number of centimeters by 100. And because 100 has **two** zeros, you will move the decimal point **two** places to the left when changing from centimeters to meters.

24 cm = 0.24 m

Eight-year-old children are around 128 cm tall. Let's figure out how many meters this is by moving the decimal point two places to the left.

128 cm = 1.28 m

Measure your own height using centimeters. Fill in the blank lines for your height.

__________ cm = __________ m

Try a few more of these yourself. Since you will be dividing a lot of little units into groups of a larger unit, you will move the decimal point to the left.

Sample: 99 cm = 0.99 m

19. 240 cm = ________ m

20. 140 mm = ________ cm

Sample: 1,000 m = 1 km

21. 1,250 m = ________ km

22. 955 m = ________ km

DAY 113 Practice

Here is some practice.

1.
	6	5
+	6	5

3.
	8	6
−	3	6

5.
	3.	4	5
+	5.	0	5

7.
	1,	4	5	6
+	8,	5	4	4

2.
	1	5	0
−		7	5

4.
	3	2
x		7

6.
	1	7.	3	4
−		8.	1	6

8.
	1	1	0
−		6	7

Practice counting by chunks. For each list of numbers, figure out if the list is counting up or counting down. Fill in the blanks.

9. Count by 2s: 22, 24, ______, ______, ______, ______

10. Count by 3s: 30, 33, ______, ______, ______, ______

11. Count by 4s: 40, 36, ______, ______, ______, ______

12. Count by 5s: 150, 145, ______, ______, ______, ______

13. Count by 6s: 60, 54, ______, ______, ______, ______

Use long multiplication to do these exercises.

14.
	1	2	3
x			3

15.
Carry		1	
	1	2	4
x			3

16.
Carry	1		
	1	4	3
x			3

Let's multiply and divide by 10 and by 100. When multiplying by 10, write 1 zero at the end for the answer. When multiplying by 100, write 2 zeros at the end. When dividing by 10, move the decimal point 1 place to the left. For dividing by 100, move the decimal point 2 places to the left. And for dividing by 1000, move the decimal point 3 places to the left.

Sample: 1.4 x 10 = 14

17. 35 x 10 = ________

18. 99 x 100 = ________

19. 140 x 100 = ________

Sample: 1.4 ÷ 10 = 0.14

20. 3.5 ÷ 10 = ________

21. 34.0 ÷ 100 = ________

22. 330.0 ÷ 1000 = ________

DAY 114

Multiplying, Dividing and Learning Metric Weights

> And they went to Ophir, and acquired four hundred and twenty talents of gold from there, and brought it to King Solomon. (1 Kings 9:28)

During biblical times, the talent was used as a unit of weight. In this verse, Solomon received 420 talents of gold. Each talent weighed about 76 pounds (34 kg). That's a lot of gold! This was 32,000 pounds of gold from Ophir. In 2025, this gold was worth about $1,330,000,000.

Today, we will learn about milligrams, grams, and kilograms as metric units of weight. A human hair weighs about 1 milligram. We might weigh medicine or vitamins in milligrams. A small paper clip weighs about 1 gram. We might weigh servings of food in grams.

A pineapple weighs about 1 kilogram. We might weigh cars in kilograms.

Abbreviations for weight in the metric system:

Milligram = mg

Gram = g

Kilogram = kg

Let's memorize how these units work together:

1000 mg = 1 g

1000 g = 1 kg

Nutrition Facts

Serving Size 3 oz. (85g)
Serving Per Container 2

Amount Per Serving	
Calories 200	Calories from Fat 120
	% Daily Value*
Total Fat 15g	**20 %**
Saturated Fat 5g	**28 %**
Trans Fat 3g	
Cholesterol 30mg	**10 %**
Sodium 650mg	**28 %**
Total Carbohydrate 30g	**10 %**
Dietary Fiber 0g	**0 %**
Sugars 5g	
Protein 5g	
Vitamin A 5% •	Vitamin C 2%
Calcium 15% •	Iron 5%

*Percent Daily Values are based on a 2,000 calorie diet. Your Daily Values may be higher or lower depending on your calorie needs.

	Calories	2,000	2,500
Total Fat	Less than	65g	80g
Sat Fat	Less than	20g	25g
Cholesterol	Less than	300mg	300mg
Sodium	Less than	2,400mg	2,400mg
Total Carbonhydrate		300mg	375mg
Dietary Fiber		25g	30g

Study the "Nutrition Facts" on this food label.

One serving weighs 85 grams, which is about as heavy as a box of 85 paperclips.

15 grams of the 85-gram serving are fat. That's the weight of 15 paperclips.

One serving also contains 5 grams of sugar and 5 grams of protein.

You would need 1,000 paperclips in a box to make the 1-kg weight of a pineapple.

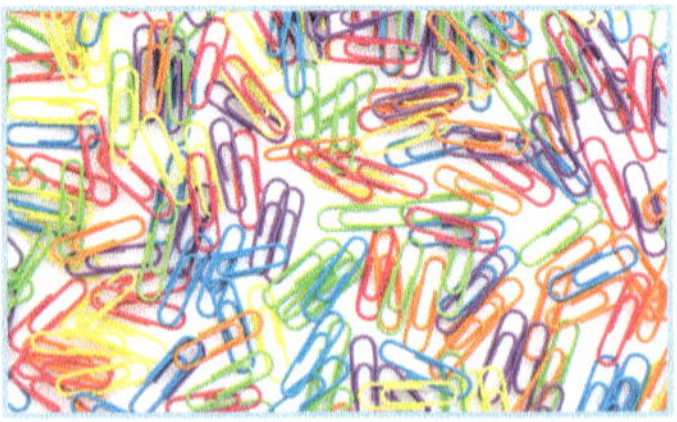

Let's convert 85 grams into kilograms. Since there are 1000 grams in a kilogram, and 1000 has three zeros, we will have to move the decimal point three places to the left. We will put a zero in front of the 85 so there will be three places for the decimal to move."

085.0 grams = 0.085 kilograms

Convert these metric weights. In each case, you will have to move the decimal point three places to the left.

Sample: 50 mg = 0.050 g

1. 120 mg = ________ g

2. 550 mg = ________ g

3. 1000 mg = ________ g

Sample: 640 g = 0.64 kg

4. 54 g = ________ kg

5. 1000 g = ________ kg

6. 2400 g = ________ kg

Student Exercises

Practice multiplying and dividing these numbers using your memory. If you get stuck, your multiplication and division table can help you.

1. $4 \times 6 =$ ☐
2. $8 \div 4 =$ ☐
3. $20 \div 10 =$ ☐
4. $4 \times 5 =$ ☐
5. $9 \times 2 =$ ☐
6. $\frac{15}{15} =$ ☐
7. $\frac{22}{2} =$ ☐
8. $5 \times 3 =$ ☐
9. $24 \div 2 =$ ☐

10. Pretend that 3 sandwiches are each cut into halves. All together there are $\frac{6}{2}$ (six-halves) of sandwich. You will pretend to split the sandwiches equally among 2 children and then among 3 children. Find the answers to the questions below by circling the amount of sandwich per child. Then write the number of pieces per child as the numerator of the fraction. This will tell you the number of halves of sandwich each child will get.

How many halves would each of 2 children get? $\frac{\square}{2}$

How many halves would each of 3 children get? $\frac{\square}{2}$

11. Pretend that you have 4 bananas, each cut into thirds. All together there are $\frac{12}{3}$ (twelve-thirds) of banana. You will pretend to split the bananas equally among 3 children, then 4 children, and then 6 children. Find the answers to the questions below by circling the amount of banana per child. Then write the number of pieces per child as the numerator of the fraction. This will tell you the number of thirds of banana each child will get.

How many thirds would each of 3 children get?

$\frac{\square}{3}$

How many thirds would each of 4 children get?

$\frac{\square}{3}$

How many thirds would each of 6 children get?

$\frac{\square}{3}$

Speed Test

Ask your parent/teacher to time your speed on these multiplication exercises. You can practice this page every now and then to help you memorize these facts. Your goal (by the time you have completed Level 3 Math) should be to answer the exercises correctly in 90 seconds.

Average Time: 120 seconds
Target Time: 90 seconds
Best Student Time: 60 seconds
Expert/Teacher Time: 45 seconds

Your Time ________

1. 6 x 5 = ☐
2. 9 x 9 = ☐
3. 9 x 8 = ☐
4. 10 x 3 = ☐
5. 9 x 7 = ☐
6. 5 x 7 = ☐
7. 3 x 7 = ☐
8. 6 x 0 = ☐
9. 9 x 10 = ☐
10. 3 x 5 = ☐
11. 6 x 9 = ☐
12. 8 x 5 = ☐
13. 5 x 0 = ☐
14. 7 x 4 = ☐
15. 10 x 1 = ☐
16. 8 x 8 = ☐
17. 4 x 3= ☐
18. 3 x 6 = ☐
19. 6 x 6= ☐
20. 2 x 9 = ☐
21. 6 x 8 = ☐
22. 1 x 5 = ☐
23. 6 x 7 = ☐
24. 1 x 8 = ☐
25. 2 x 8 = ☐
26. 1 x 2 = ☐
27. 6 x 0 = ☐
28. 4 x 6 = ☐
29. 2 x 6= ☐
30. 5 x 5 = ☐
31. 5 x 9 = ☐
32. 3 x 9 = ☐
33. 3 x 2 = ☐
34. 5 x 2 = ☐
35. 8 x 4 = ☐
36. 8 x 7 = ☐
37. 10 x 1 = ☐
38. 5 x 4 = ☐
39. 8 x 8 = ☐
40. 4 x 9 = ☐
41. 5 x 8 = ☐
42. 4 x 4 = ☐
43. 4 x 6 = ☐
44. 7 x 7 = ☐
45. 9 x 9 = ☐

DAY 115 Practice

Figure out the correct number to answer each math question. Write it in the blank. Then write an equation that includes your answer. The first one is done for you.

1. What plus 7 equals 16? 9 $9 + 7 = 16$
2. What plus 10 equals 16? ______ ______________
3. What minus 8 equals 4? ______ ______________
4. What minus 7 equals 10? ______ ______________
5. What times 2 equals 12? ______ ______________
6. What times 6 equals 18? ______ ______________
7. What divided by 2 equals 10? ______ ______________
8. What divided by 5 equals 4? ______ ______________

9. Guess the area of this rectangle to the nearest square centimeter. ______________

Measure the length and width of the rectangle and figure out its area.

______ square cm

10. Guess the area of this triangle to the nearest square centimeter. ______________

Measure and do the math.

Area of a triangle = Base x Height ÷ 2

Area = ______ x ______ ÷ 2

Area = __________ square cm

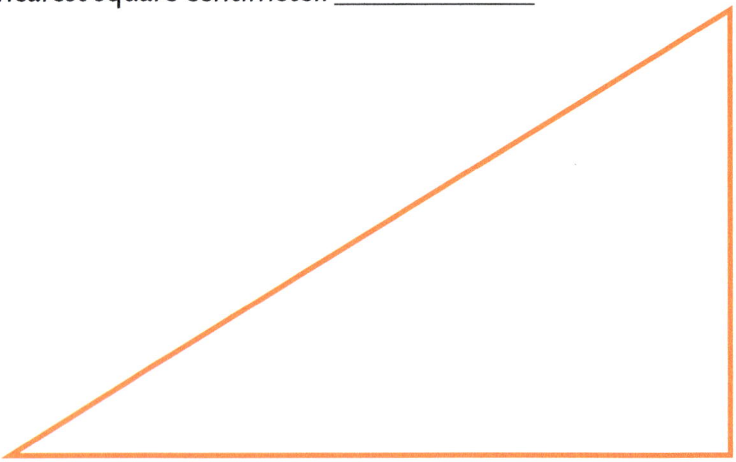

11. Guess the area of this triangle to the nearest square centimeter. ______________

Measure and do the math.

Area of a triangle = Base x Height ÷ 2

Area = _______ x ________ ÷ 2

Area = ___________ square cm

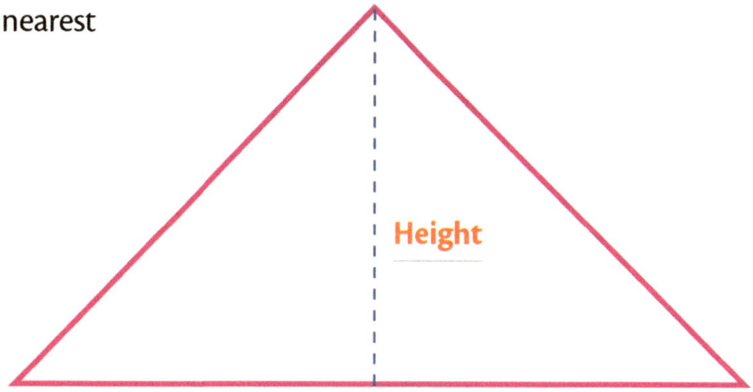

Change the fractions into mixed numbers and the mixed numbers into fractions.

12. $1\frac{1}{4} = \frac{}{}$

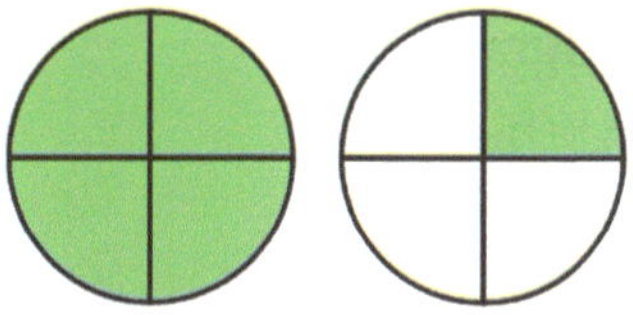

13. $3\frac{1}{3} = \frac{}{}$

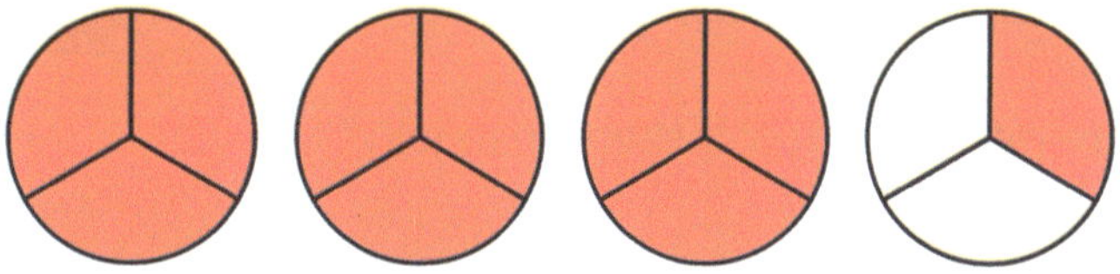

14. $2\frac{2}{5} = \frac{}{}$

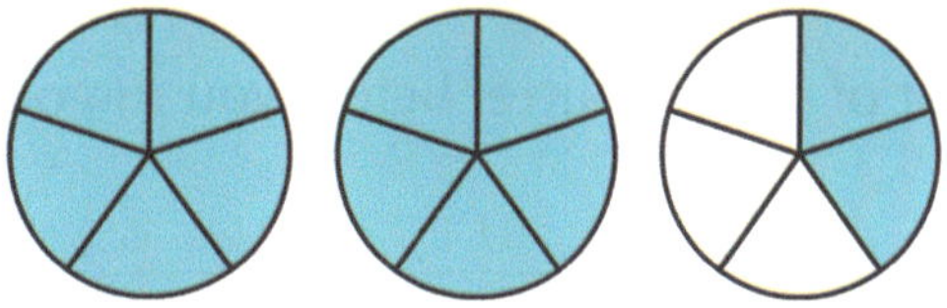

15. $\frac{6}{2} =$

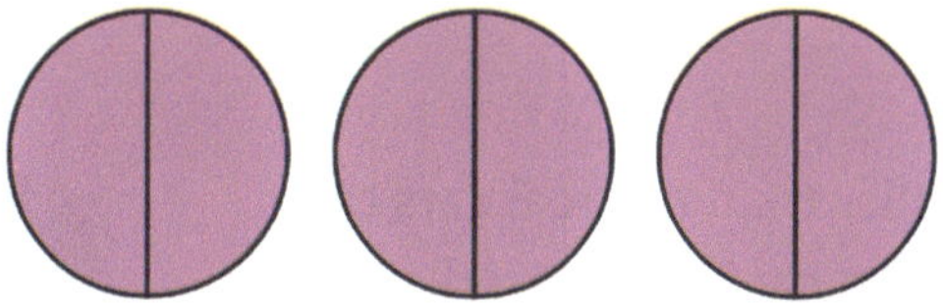

16. $\frac{8}{3} =$

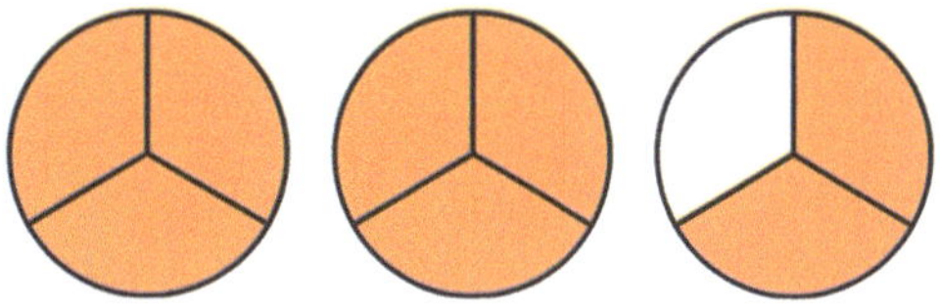

17. $\frac{9}{4} =$

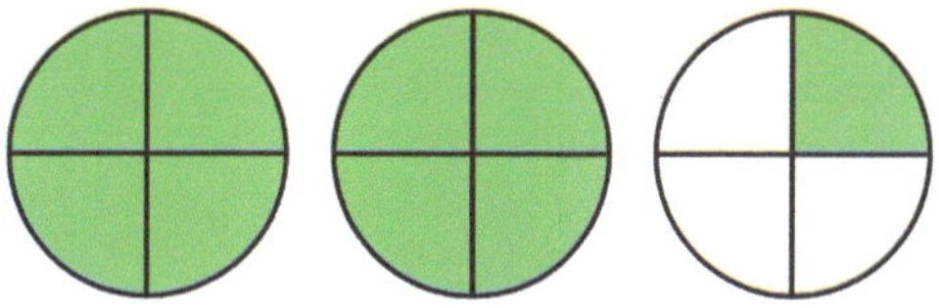

18. Carolyn knitted a blanket that is 2.4 meters long. How many centimeters long is the blanket? (Hint: Think of how many centimeters are in a meter. Then see how many zeros are in that number. Move your decimal point to the right that many places. Add a zero at the end if you need to.)

____

19. Suppose that, in your pantry, you find there is 510 grams of jelly in a large jar and 794 grams of peanut butter in another container. How many kilograms of peanut butter and jelly do you have all together? (Hint: You will need to convert grams to kilograms.)

____________________ kg

How much more peanut butter is there than jelly?

____________________ g

Do you think you will run out of peanut butter before you run out of jelly?

____

20. Model cars are a lot smaller than the real cars they look like. When a certain model car is measured, it's 24.236 inches long. About how long is that model car in feet? Round to the nearest inch first. There are 12 inches in a foot.

____

Units for Liquid Volume DAY 116

Have you memorized these metric amounts?

1 m = 1,000 mm

1 m = 100 cm

1 km = 1,000 m

1 g = 1,000 mg

1 kg = 1,000 g

Let's compare the metric system to the English system.

1 inch = 2.54 centimeters. This means an inch is about 2 ½ times longer than a centimeter.

1 foot = 30.48 centimeters

1 meter = 39.4 inches. A meter is a little over 3 feet.

1 kilometer = 0.62 miles. A kilometer is a little over 1/2 mile.

1 kilogram = 2.2 pounds. A kilogram is a little more than 2 times heavier than a pound.

1 ounce = 28.35 grams. An ounce is a lot heavier than a gram.

1 pound = 454 grams.

1 pound of apples

1 kilogram of apples

Remember, volume is a measurement of how much space is inside a three-dimensional shape. The units for liquid volume in the metric system are milliliters and liters. Let's compare the metric and English units for liquid measurements:

1 gallon = 3.79 liters (close to 4 liters)

1 ounce = 29.57 milliliters (close to 30 milliliters)

Here are some abbreviations for metric volume:

Milliliter = ml

Liter = l (lowercase L) or "L" if we don't want the lowercase L to be confused with the number one.

Since 1,000 **milli**meters are in a meter, and 1,000 **milli**grams are in a gram, you can probably tell that there are 1,000 **milli**liters in a liter. A milliliter is only one cubic centimeter. So if you were to drink water from a thimble, you would drink about 2 milliliters of water. That's not very much to drink!

Most drinks can be bought in 16-ounce, 12-ounce, or 250-ml bottles.

16 ounces = 473 ml, and 12 ounces = 355 ml

Let's convert a few of these numbers from milliliters into liters. Since there are 1,000 milliliters in a liter, we will have to move the decimal point three places to the left. Sample: 473.0 milliliters (ml) = 0.473 liters (l)

1. 250.0 milliliters (ml) = ________ liters (l)

2. 355.0 milliliters (ml) =________ liters (l)

Student Exercises

1. Suppose a friend from Japan tells you that he pumped 32 liters of gasoline into his car. There are about 4 liters in a gallon. About how many gallons did he put in his car?

32 ÷ 4 = _______ gallons

If he put 48 liters of gasoline into his car, about how many gallons would that be?

48 ÷ 4 = _______ gallons

2. If your parents filled the family car with 18 gallons of gasoline, how many liters would that be? Remember there are about 4 liters in every gallon.

18 x 4 = _______ liters

Carry		
	1	8
x		4

3. Your friend tells you he weighs 32 kilograms. There are 2.2 pounds in every kilogram, but we will round that down to 2. About how many pounds does your friend weigh?

2 x 32 kg = _______ pounds

	3	2
x		2

4. A pony weighs 280 kilograms. How many pounds does this pony weigh?

2 x 280 kg = _______ pounds

Carry			
	2	8	0
x			2

5. An eight-year-old American child is around 4 feet tall. Since there are about 30 centimeters in a foot, about how tall is an eight-year-old child in centimeters?

30 x 4 ft = ________

$$\begin{array}{rr} & 3\ 0 \\ \times & 4 \\ \hline \end{array}$$

How tall are you in centimeters?

____

Find the difference between you and the answer above by subtracting.

____

Write these fractions as decimals.

Sample: $\frac{2}{10} = 0.2$

Sample: $\frac{10}{100} = 0.1$

6. $\frac{1}{10}$ = ________

7. $\frac{4}{10}$ = ________

8. $\frac{5}{10}$ = ________

9. $\frac{9}{10}$ = ________

10. $\frac{10}{10}$ = ________

11. $\frac{40}{100}$ = ________

12. $\frac{20}{100}$ = ________

13. $\frac{50}{100}$ = ________

14. $\frac{90}{100}$ = ________

15. $\frac{100}{100}$ = ________

Finish each of these lists of shapes. Find the pattern, and then draw a picture that keeps that pattern going.

16. 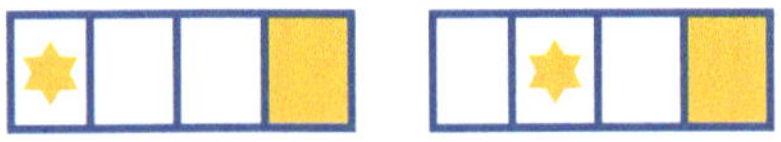

17. 

18. 

Convert the fractions into mixed numbers, and change the mixed numbers into fractions.

1. $3\frac{3}{4} = \frac{\quad}{\quad}$

2. $2\frac{1}{2} = \frac{\quad}{\quad}$

3. $1\frac{2}{3} = \frac{\quad}{\quad}$

4. $\frac{7}{5} = 1\frac{\quad}{\quad}$

5. $\frac{11}{2} = 5\frac{\quad}{\quad}$

6. $\frac{9}{3} =$

Your mother has given you a shopping list of:

12 kg apples	11 kg pears	5 kg melons
4 kg eggplant	2 kg lettuce	
3 kg bananas	2 kg peppers	

7. Using the sale prices in this advertisement, find out how much money you will spend.

Apples: 12 kg x $______ = $______

Eggplant: 4 kg x $______ = $______

Bananas: 3 kg x $______ = $______

Pears: 11 kg x $______ = $______

Lettuce: 2 kg x $______ = $______

Peppers: 2 kg x $______ = $______

Melons: 5 kg x $______ = $______

Total: ______

Find these answers.

8. 9 + 8 = ☐

9. 13 - 7 = ☐

10. 4 x 4 = ☐

11. 7 x 5 = ☐

12. 16 - 5 = ☐

13. 12 - 4 = ☐

14. 12 ÷ 2 = ☐

15. 28 ÷ 4 = ☐

16. 2 + 6 = ☐

17. 13 + 13 = ☐

18. 12 ÷ 4 = ☐

19. 20 ÷ 5 = ☐

In your mind, pretend to cut out the flat design above. Then pretend to fold it into a 3-dimensional die. Using the views of a die's surfaces (below), figure out the number of dots that belong on each blank surface of your "folded" die. Draw the dots on the flat design.

Extra Challenge

When you flip one coin 3 times in a row, there are 8 different outcomes you can get:

HHH, HHT, HTT, THT, HTH, THH, TTH, TTT

Use the list above to figure out the probability for each of the following situations. Write the probability as a fraction with an 8 in the denominator.

20. All tails

21. Two heads

22. Three heads and three tails

23. Two heads in a row

24. One heads

25. One tails

Biblical Units DAY 118

> [Bezalel] made the incense altar of acacia wood. Its length was a cubit and its width a cubit—it was square—and two cubits was its height. Its horns were of one piece with it. (Exodus 37:25)

The Bible uses a variety of units to measure length and volume. When you read the Bible, you should make sure the Word of God makes sense to you. When you read the measurements Bezalel used to make the incense altar, you should be able to picture the size of it.

1. The Incense Altar

The cubit is about 20 inches (50 cm). Half a cubit is 10 inches (25 cm).

The width of the little incense altar was 1 cubit.

How many inches (or cm) is this? __________

The length of the little incense altar was 1 cubit.

How many inches (or cm) is this? __________

The height of the little incense altar was 2 cubits.

How many inches (or cm) is this? __________

2. The Table for Showbread

> [Bezalel] made the table of acacia wood; two cubits was its length, a cubit its width, and a cubit and a half its height. (Exodus 37:10)

The width of the table for showbread was 1 cubit.

How many inches (or cm) is this? __________

The length of the table for showbread was 2 cubits.

How many inches (or cm) is this? __________

The height of the table for showbread was 1 ½ cubits.

How many inches (or cm) is this? __________

3. The Ark of the Covenant

> Then Bezalel made the ark of acacia wood; two and a half cubits was its length, a cubit and a half its width, and a cubit and a half its height. He overlaid it with pure gold inside and outside, and made a molding of gold all around it. (Exodus 37:1-2)

The width of the ark of the covenant was 1 ½ cubits.

How many inches (or cm) is this? __________

The length of the ark of the covenant was 2 ½ cubits.

How many inches (or cm) is this? __________

The height of the ark of the covenant was 1 ½ cubits.

How many inches (or cm) is this? __________

4. The Altar for Burnt Sacrifice

> [Bezalel] made the altar of burnt offering of acacia wood; five cubits was its length and five cubits its width—it was square—and its height was three cubits. (Exodus 38:1)

The width of the altar was _______ cubits.

How many inches (or cm) is this? __________

The length of the altar was _________ cubits.

How many inches (or cm) is this? __________

The height of the altar was _________ cubits.

How many inches (or cm) is this? __________

5. The Bath

> In my hearing the Lord of hosts said, "Truly, many houses shall be desolate, great and beautiful ones, without inhabitant. For ten acres of vineyard shall yield one bath . . . " (Isaiah 5:9,10)

The biblical "bath" is about 6 gallons. This passage is speaking of a huge vineyard that could only produce 6 gallons of grape juice or wine. A good vineyard should actually produce around 1,600 gallons!

> And [Huram] made the Sea of cast bronze, ten cubits from one brim to the other; it was completely round. Its height was five cubits, and a line of thirty cubits measured its circumference. . . . It contained two thousand baths. (1 Kings 7:23,26)

How big was this sea of bronze?

The diameter was __________ cubits.

How many inches (or cm is this)? ____________

The height was _________ cubits.

How many inches (or cm) is this? ___________

The circumference was _________ cubits.

How many inches (or cm) is this? ___________

How much water would this large "pond" contain? ___________ baths

How many gallons is this _________

	2	0	0	0
x				6

Student Exercises

Circle the wrong answers. Write the correct answer in the blank.

1. 0 x 6 = 6 ________
2. 3 x 8 = 11 ________
3. 4 x 12 = 48 ________
4. 12 x 7 = 82 ________
5. 3 x 9 = 11 ________
6. 11 X 9 = 99 ________
7. 5 x 8 = 40 ________
8. 9 x 7 = 69 ________

Try these long addition and subtraction exercises. Some of the numbers to carry and borrow have been included for you.

9.

Carry		1		
		3	2	1
		1	2	3
	+		4	6

11.

Carry		1		
		4	3	2
		3	9	2
	+	1	4	4

13.

Carry				
		1	1	1
		3	3	3
	+	4	5	6

10.

Borrow		5	15	
		~~6~~	~~5~~	8
	−	1	7	7

12.

Borrow			6	10
		9	~~7~~	~~0~~
	−	5	4	3

14.

Borrow				
		4	9	3
	−	3	8	8

Let's practice measuring area. First, guess the area of each of the shapes. Then measure and do the math to see how close your guess was.

15. Guess the area of this triangle to the nearest square centimeter. ____________

What is its area? Measure and do the math. ____________

Area of a triangle = base x height ÷ 2

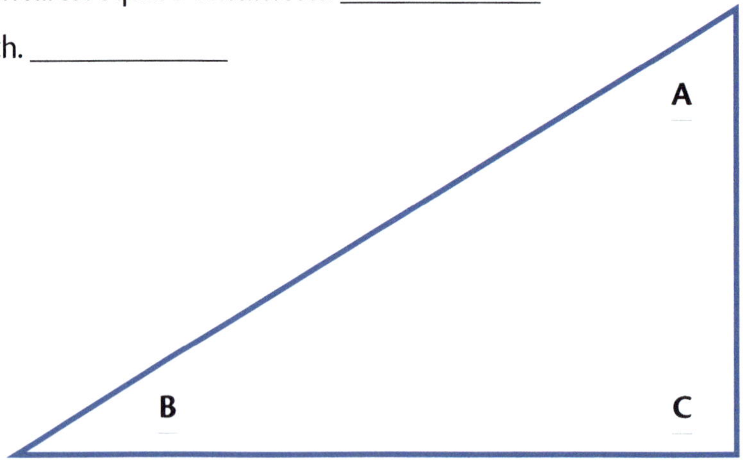

16. Guess the area of this triangle in square centimeters. ______________

What is its area? Measure and do the math.

____

Area of a triangle = base x height ÷ 2

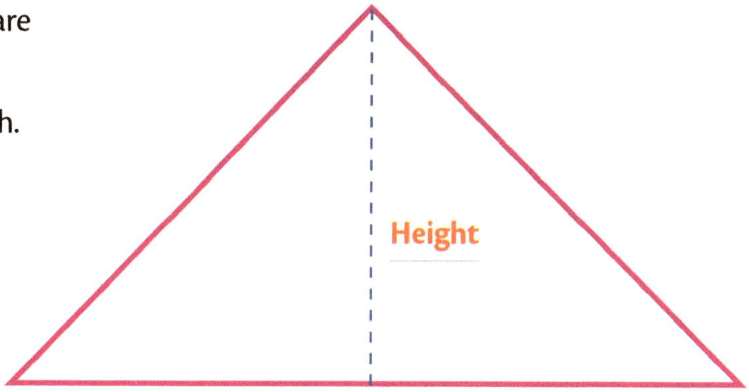

17. Guess the area of this parallelogram in square centimeters. ______________

What is its area? Measure and do the math. __________________

Area of a parallelogram = base x height

18. Guess the size of the three angles in exercise 15. A. _______ B. _______ C. _______

What is the measure of the angles? A. _______ B. _______ C. _______

Practice counting by chunks. For each list of numbers, figure out if the list is counting up or counting down. Fill in the blanks to continue the pattern.

19. Count by 12s: 12, 24, _____, _____, _____, _____

20. Count by 11s: 66, 77, _____, _____, _____, _____

21. Count by 5s: 110, 105, _____, _____, _____, _____

22. Count by 7s: 70, 63, _____, _____, _____, _____

23. Count by 9s: 0, 9, _____, _____, _____, _____

DAY 119 Practice

Let's review percents. Fill in the missing percents, decimals, and fractions. Numbers 6 and 7 are extra challenge.

Sample: $0.5 = \frac{5}{10} = \frac{50}{100} = 50\%$

1. $\square = \frac{9}{10} = \frac{90}{100} = \square$

2. $\square = \frac{\square}{10} = \frac{20}{100} = \square$

3. $0.3 = \frac{3}{10} = \square = \square$

4. $\square = \frac{4}{10} = \frac{\square}{\square} = 40\%$

5. $0.7 = \square = \square = \square$

6. $\frac{1}{4} = 0.25 = \frac{\quad}{100} = \square$

7. $1.50 = \frac{15}{10} = \frac{150}{100} = \square$

Finish each list of shapes. Find the pattern, and then draw a picture that keeps that pattern going.

8. 

9. 

10. 

11. 

12. 

13. 

14. Let's paint! Figure out the answer to each exercise below. Find that answer in the large box. Paint the answer's small square to match the colored square next to its original exercise.

15	23	43	51
37	34	27	38
39	47	50	19
21	31	30	48

27+14−7

38−16−1

15+14+14

26+18+3

14+12+4

36+12+3

17+12+9

18+14−9

25+25−19

45−17−13

34+19−5

31−18+37

37+12−10

29+16−8

31−11−1

31−15+11

Imagine rolling a die once. Figure out the probability of rolling the following results, and write each as a fraction with 6 in the denominator.

15. 5 ______

16. Not a 4 ______

17. 3 or a 4 ______

18. 8 ______

19. Not a 1 or 2 ______

20. A number bigger than 0 ______

DAY 120 Explore God's World with Math

Math is a tool we can use to learn more about God's world. Let's experiment with weighing things. This should be done outside and be supervised by a parent/teacher.

Materials:

- Various kinds of string, thread, and yarn
- A scale, covered in clear plastic
- A bucket with a handle
- A 1-quart (1 L) container
- 50 pounds (100 kg) of small rocks or gravel

Goal: To test the strength of different types of thread, yarn, and string.

1. Tie your thickest string, thread, or yarn to the pail handle with a very strong knot.
2. Weigh the bucket.
3. Use the small container to weigh a 1-pound (2 kg) amount of rocks.
4. With the bucket resting on the ground, add the rocks to the pail.
5. Carefully lift the bucket by the string a few inches off the ground.
6. If the string does not break, repeat this by adding 1 pound (2 kg) of rocks at a time until the string breaks or the bucket is full.
7. Repeat with the other strings.
8. Write your results in the table below. Be sure to first add the weight of the bucket to the rocks' weight.

TYPE OF STRING	WEIGHT AT WHICH STRING BROKE (TENSILE STRENGTH)	NOTES
1		
2		
3		
4		

Which thread or string was the strongest? ____________________

What was the weight at which the weakest string broke? ____________

How much weight could this string hold if you doubled the string? 2 x ________ = ________

How much weight could the string hold if you tripled the string? 3 x ________ = ________

Do you understand the lesson given in this portion of Scripture?

> Two are better than one, because they have a good reward for their labor. For if they fall, one will lift up his companion. But woe to him who is alone when he falls, for he has no one to help him up. Again, if two lie down together, they will keep warm; but how can one be warm alone? Though one may be overpowered by another, two can withstand him, and a threefold cord is not quickly broken. (Ecclesiastes 4:9-12)

DAY 121 Practice

Which is bigger? Fill in the blank line with either < or >.

1. kilogram (kg) ______ pound (lb)
2. inch (in) ______ centimeter (cm)
3. mile (m) ______ kilometer (km)
4. milliliter (mm) ______ liter (l)
5. gallon (g) ______ liter (l)
6. foot (ft) ______ inch (in)

Everyday Math

7. You have decided to raise two cows to sell for meat. The hay costs $200 for a ton, and each cow will eat 2 tons before it's sold. How much will you pay for hay?

Cost of Hay: 2 cows x 2 tons = 4

4 x $ ______ = $______

x			4

If you sell each cow's meat for $1,250, how much money will you earn?

Sale of Cows: 2 x $________ = $________

Profit = $________ – $________ = $________

Sale of cows				
– Cost of hay				
Profit				

8. You plan to make enough fruit salad to feed 200 people for the church fellowship meal. If each person eats 1/4 pound of fruit salad, you will need to buy 50 pounds of fruit. From the list, choose the amount of each fruit you would like to include, and write the number of pounds in the first column. Make sure they add up to 50 pounds. Multiply and write the cost of each fruit in the second column.

Oranges:	$1.00/lb	X	______ lb.	=	$______
Apples:	$2.00/lb	X	______ lb.	=	$______
Bananas:	$1.00/lb	X	______ lb.	=	$______
Pineapple:	$1.00/lb	X	______ lb.	=	$______
Kiwi Fruit:	$3.00/lb	X	______ lb.	=	$______
Grapes:	$1.00/lb	X	______ lb.	=	$______
Add it up			50 lb.		$______

9. Imagine that your family is going out to eat at a fast-food restaurant. Let's figure out how much this will cost. Fill in the table below with food they might like. Do your math on the following page Add everything up.

ORDER NUMBER	FAMILY MEMBER'S NAME	MAIN DISH OR COMBO PRICE	SIDE'S PRICE	DRINK PRICE	TOTAL FOR THIS PERSON
1					
2					
3					
4					
5					
6					
7					
8					

Add This Up: ______

Name

Carry

$.
$.
+ $.

Name

Carry

$.
$.
+ $.

Name

Carry

$.
$.
+ $.

Name

Carry

$.
$.
+ $.

Name

Carry

$.
$.
+ $.

Name

Carry

$.
$.
+ $.

Name

Carry

$.
$.
+ $.

Name

Carry

$.
$.
+ $.

10. Find the area of the shapes. Pretend each little square of the grid measures 1 foot by 1 foot. The little squares of the grid will help you find the lengths you need. In the sections that have windows and doors, find the area of each window or door. Find the area of each large section as though it had no windows and doors. Use the equations below. Write each area within its shape. Color the picture.

Area of a square = side x side

Area of a rectangle = length x width

Area of a triangle = base x height ÷ 2

Area of a circle = 3 x radius x radius

(Remember, the radius is half the length of the diameter.)

DAY 122 Go Make More Measurements

This lesson integrates math into everyday life. With this activity, the student will apply God's patterns and wisdom in meaningful ways to home and community. Take a break from memory work and academic exercises to make math part of everyday life. The following are suggestions or examples. Other ideas may be added to the list. .

Measurements are an important part of life. Let's measure large numbers today.

See what it would cost to hire someone to paint the outside of your house. The cost for a painter to paint the outside of a house in America is around $3.00 per square foot.

1. Find the surface area of your house. The areas to be painted are probably rectangles. For each rectangular section, find its area in square feet. You may have to find the area of some triangles too.

2. Add the areas of the sections together to get the total area.

3. Multiply the total square feet by $3. This is the amount it would cost for someone to paint the outside of your home.

Carry					
Sq Ft					
	x				$ 3

4. In the table below you can look for the size of your home to see how much it would cost your family to paint the house yourselves.

Cost of Paint and Supplies:

1,000 square feet $100
1,500 square feet $150
2,000 square feet $200
2,500 square feet $250
3,000 square feet $300
3,500 square feet $350
4,000 square feet $400

Cost of Professional Painting					
Cost of Paint	−				
Savings					

Subtract to see how much you would save if you painted the house yourself.

Plan to ship presents to your friends and relatives across the country next holiday season. Assume that the cost per pound will be about $4.00. In the small grids, figure out the cost to ship each package below. Use the long grid to add them up for the total cost of shipping.

Package weights:

- 3 pounds
- 5 pounds
- 1 pound
- 12 pounds
- 12 pounds
- 4 pounds

Name

5.
Carry
Weight of package
x $ 4

Name

7.
Carry
Weight of package
x $ 4

Name

9.
Carry
Weight of package
x $ 4

Name

6.
Carry
Weight of package
x $ 4

Name

8.
Carry
Weight of package
x $ 4

Name

10.
Carry
Weight of package
x $ 4

11.
Carry
$
$
$
$
$
+ $

Additional Activities

For extra activities, games, and helpful videos on these math concepts, refer to **generations.org/math3**.

What glorious
wisdom God used
to make the stars!

CHAPTER 11

Long Division

DAY 123 Introduction to Long Division

> For the LORD gives wisdom;
> From His mouth come knowledge and understanding.
> (Proverbs 2:6)

We use wisdom to learn more about the stars and about the world God made. The lessons in this chapter will require increasing your math wisdom and knowledge. Let's remember that all wisdom comes from God. You have already learned long addition, long subtraction, and long multiplication. Today you will learn long division—a way to divide numbers larger than those you see in your multiplication/division facts.

Let's first learn the names of the different numbers in a division equation:

Dividend: the number to be divided

Divisor: the number by which the dividend will be divided

Quotient: the answer

For the equation $8 \div 2 = 4$:

- 8 is the dividend.
- 2 is the divisor.
- 4 is the quotient.

Let's use $8 \div 2$ to learn how to do long division. First, we set up our dividend and divisor this way, with a bracket.

$$2\overline{)8}$$

This math expression is stated this way: "8 divided by 2."

Below you can see the names of each part of the math expression in a division bracket.

$$\begin{array}{r} \text{Quotient} \\ \text{Divisor}\overline{)\text{Dividend}} \end{array}$$

Now, let us walk through long division in four easy steps. You have already learned how to divide, multiply, and subtract. You will do these operations as steps in long division.

1. Divide

2. Multiply

3. Subtract

4. Check

Let's go through these steps for 8 ÷ 2.

Step 1. Divide. First, let's see if we can divide the dividend by the divisor. That is, can we divide 8 by 2? Yes, 8 ÷ 2 is 4. We write the number 4 above the bracket, directly over the dividend's 8.

	4
2)	8

Step 2. Multiply. We multiply the 4 by the divisor (2). 4 x 2 = 8. Now, we write this 8 under the dividend's 8.

	4
2)	8
–	8

Step 3. Subtract. We draw a bar under the bottom 8 and subtract it from the dividend's 8. Since 8 - 8 = 0, we will write 0 under the subtraction bar. So, the answer to 8 ÷ 2 is 4. The answer (quotient) is the number above the bracket.

	4
2)	8
–	8
	0

Step 4. Check. We can check our work by multiplying the quotient (4) by the divisor (2). We see that 4 x 2 = 8. Since this answer (8) is the same as our dividend, we have found the correct quotient.

Let's try another long division that's a little harder: 18 ÷ 6. This one is different because we have a two-digit dividend.

Step 1. Divide. In this case, we look at the first digit (the 1) of the dividend and ask if it can be divided by 6. No you cannot, for example, divide 1 bicycle among 6 children. So we must include the dividend's 8 with the 1 and ask if 18 can be divided by 6. Yes, 18 ÷ 6 is 3. We write the 3 in the box above the dividend's 8 because we are dividing the 18, not the 1.

		3
6)	1	8

Step 2. Multiply 3 x 6. The answer is 18. We write 18 in the box.

		3
6)	1	8
	1	8

Step 3. Subtract 18 – 18, and we write 0 in the box.

		3
6)	1	8
–	1	8
		0

Step 4. Check. Multiply the divisor (6) by the quotient (3) to see if we get the dividend (18).

$3 \times 6 = \boxed{18}$

What amazing wisdom God must have used to create the peacock with such beautiful patterns in its feathers!

Student Exercises

Practice of a few of these on your own:

1.

4)	1	6
–		

2.

5)	2	5
–		

3.

6)	3	6
–		

Decide which number is larger, and write the correct symbol in the circle.

4. 4.5 ◯ 5.4

5. $\frac{1}{2}$ ◯ $\frac{1}{3}$

6. 1.42 ◯ 1.63

7. $\frac{5}{8}$ ◯ $\frac{2}{8}$

Do these additions and subtractions by memory. If one is taking you too long, take a guess. Then go back after you finish the others, and take the time to figure out the hard one. Change the answer if you need to, but maybe you had it right!

8. $3 + 3 =$ ☐

9. $16 - 8 =$ ☐

10. $10 + 10 =$ ☐

11. $12 + 12 =$ ☐

12. $14 - 9 =$ ☐

13. $8 + 7 =$ ☐

14. $16 - 10 =$ ☐

15. $24 - 3 =$ ☐

16. $\frac{5}{9} + \frac{2}{9} =$ ☐

17. $\frac{3}{7} + \frac{1}{7} =$ ☐

18. $\frac{5}{9} - \frac{2}{9} =$ ☐

19. $\frac{3}{7} - \frac{1}{7} =$ ☐

DAY 124 Practice

Practice multiplying and dividing these numbers. If there are any you don't know, take a guess and check the multiplication table to see if you're right.

1. $9 \times 4 =$ ☐
2. $45 \div 9 =$ ☐
3. $9 \times 9 =$ ☐
4. $\frac{40}{8} =$ ☐
5. $7 \times 7 =$ ☐
6. $54 \div 9 =$ ☐
7. $9 \times 3 =$ ☐
8. $64 \div 8 =$ ☐
9. $8 \times 6 =$ ☐
10. $\frac{49}{7} =$ ☐

Practice of a few of these on your own:

11. $4\overline{)24}$
12. $5\overline{)30}$
13. $6\overline{)42}$

14.

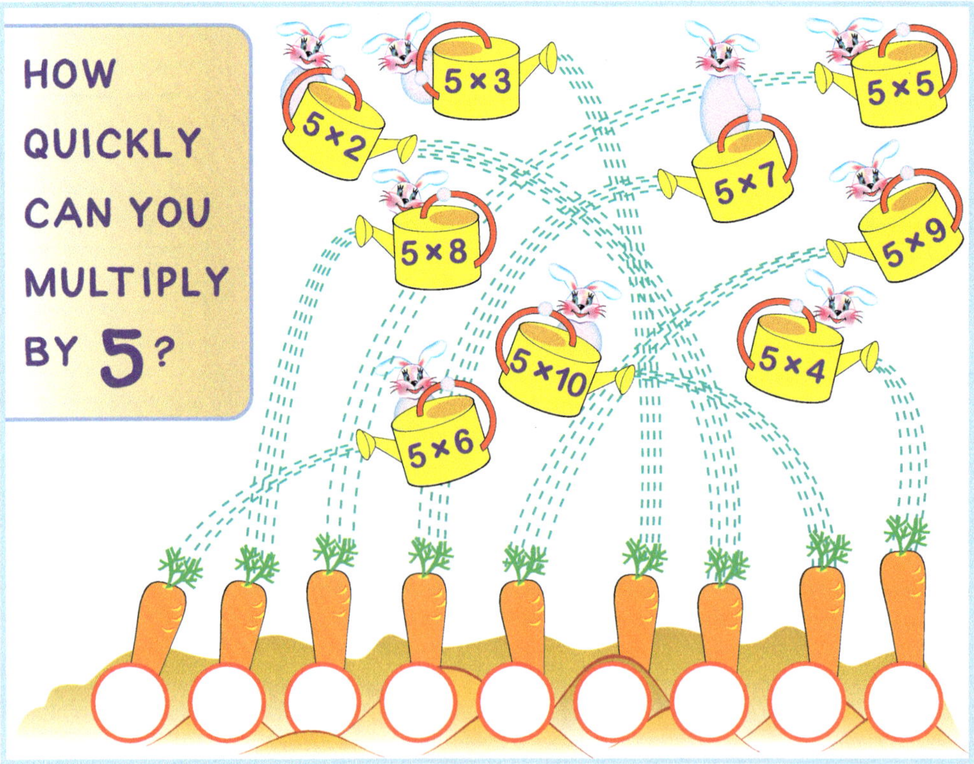

More Long Division

DAY 125

Today we are going to divide larger numbers. Pretend that your parents gave you 246 marbles to divide equally with your brother. How many marbles would each of you get? First, let's split 246 in half using blocks. Split the 2 chunks of 100 equally into 2 parts. Split the chunks of 10 equally. Split the single blocks equally.

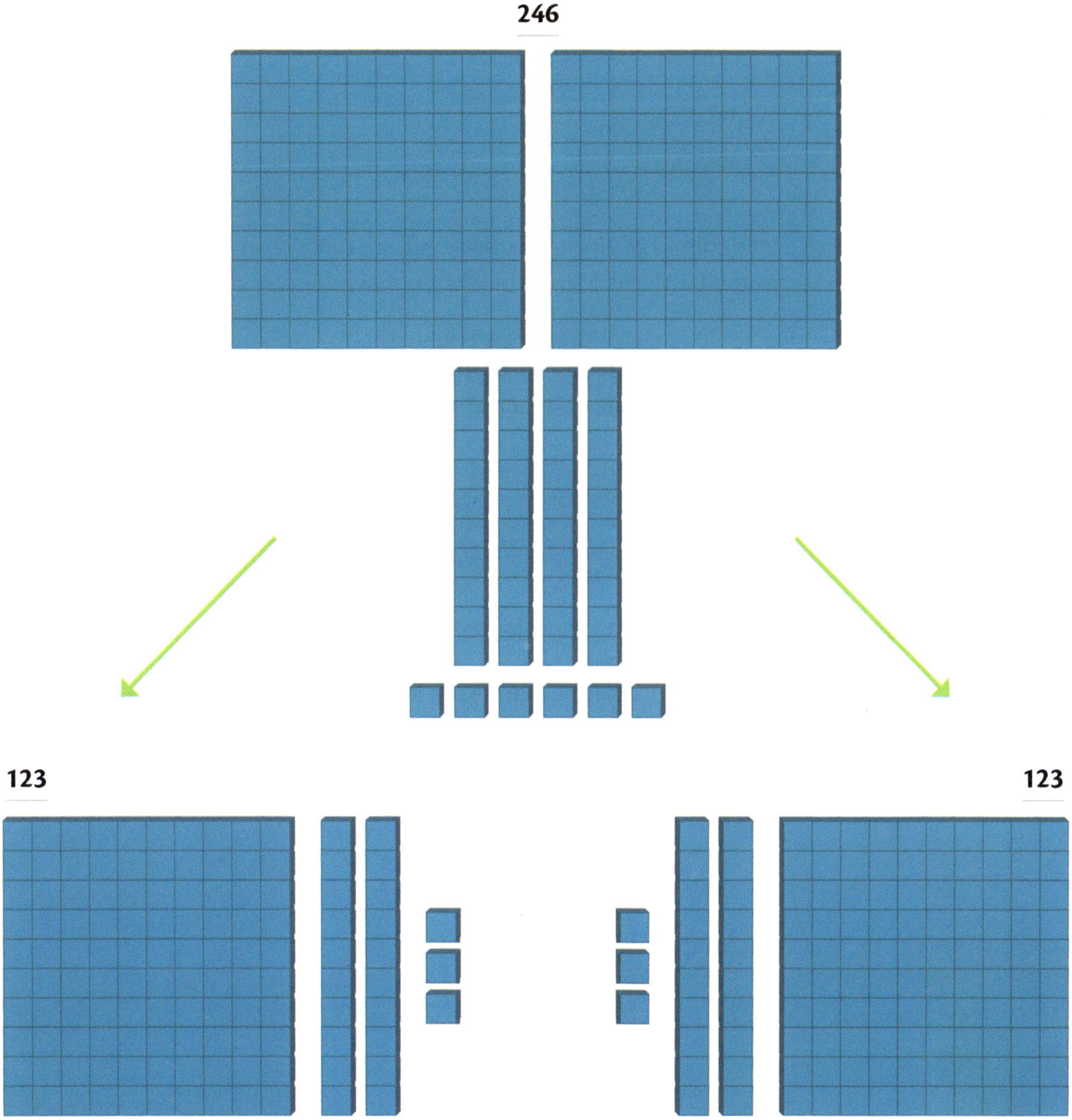

How many marbles will each of you get? _______

It wasn't so hard to split 246 using blocks. Now you will learn how to divide large numbers into equal parts using long division. Remember, long division can divide a larger group into a certain number of equal-sized smaller groups. Let's start with this one:

$$42 \div 3$$

In this case, we are dividing 42 blocks into 3 equal groups. You can see that we have shown 42 blocks as 4 chunks of 10 and 2 single blocks.

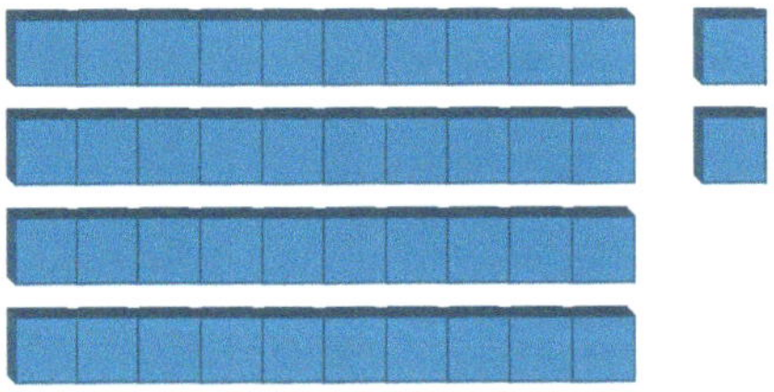

We know we can divide 3 of the chunks of 10 into 3 groups, but we must change the extra chunk of 10 into single blocks to be grouped with the 2 single blocks. Soon we will also divide the 12 single blocks into 3 equal groups.

Break this group of 42 blocks into 3 equal groups. (You will need to include a 10-block and some single blocks in each group). Draw a circle around each group.

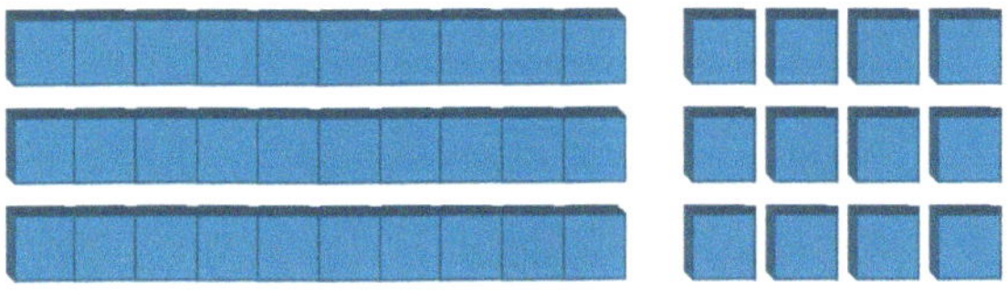

Each group contains 14 blocks, so 42 ÷ 3 = 14. Now let's divide 42 into 3 parts using long division.

Step 1. Divide. To divide 4 by 3, ask yourself how many 3s fit into 4. One 3 fits into 4. You may write the 1 above the 4 because the 1 three fits into that 4.

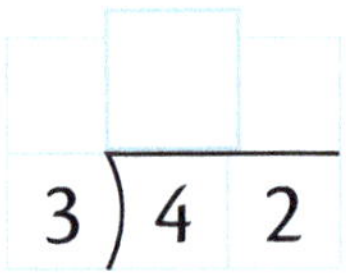

Step 2. Multiply 1 x 3, and place the answer (3) under the 4.

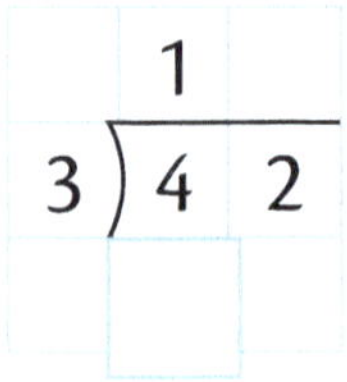

Step 3. Subtract 4 – 3. We will write the answer (1) under the 3. Now don't forget about the 2 in the dividend. We'll bring the 2 down next to the 1. This is because we are subtracting 2 – 0, which equals 2.

	1	
3)	4	2
–	3	
	1	2

We're not finished yet. We still need to divide the 12 by 3. Now 12 is the new dividend, and 3 is still the divisor.

Step 1 again. Divide again. How many 3s fit into 12. The answer is 4. We'll write a 4 next to the 1 in the quotient.

	1	4
3)	4	2
–	3	
	1	2

Step 2 again. Multiply again. Multiply 3 x 4. You may write the answer (12) under the 12 from the subtraction.

Step 3 again. Subtract again. Subtract 12 – 12. Write the answer (0) at the bottom.

	1	4
3)	4	2
–	3	
	1	2
–		

Step 4. Check. The answer to 42 ÷ 3 = 14. Check the answer by multiplying the quotient by the divisor.

14 x 3 = ______

Carry			
		1	4
	x		3

As you can see, we had to repeat the steps in this longer long division. If you always follow the steps and be careful with your math facts, you will get correct answers.

Now try another one. The steps are provided, but you can write all the numbers in the boxes.

Step 1. Divide. Ask yourself how many 3s fit into the 3 in the dividend. One 3 fits into 3. Write the 1 in the box above the dividend.

	☐	
3)	3	6

Step 2. Multiply 1 x 3, and write the answer (3) in the box under the dividend's 3.

	1	
3)	3	6
	☐	

Step 3. Subtract 3 – 3, and write the answer (0) underneath. At the same time, bring the 6 down, because you are also subtracting 6 - 0 = 6. You're not finished yet. You have a new dividend.

	1	
3)	3	6
–	3	
	☐	☐

Step 1 again. Divide again. Ask yourself how many 3s are in 6? The answer is 2. You will write a 2 next to the 1 in the quotient.

	1	☐
3)	3	6
–	3	
	0	6

Step 2 again. Multiply again. Multiply 2 x 3. Write the answer (6) under the 6 from the subtraction.

Step 3 again. Subtract again. Subtract 6 – 6. Write the answer (0) at the bottom.

	1	2
3)	3	6
–	3	
	0	6
–	☐	☐
	☐	☐

Step 4. Check the answer (12) by multiplying it by the divisor.

12 x 3 = ______

Carry			
		1	2
	x		3

Student Exercises

Finish these long division exercises on your own.

1.

	1	
2)	2	4
–	2	
–		

2.

	1	
4)	4	4
–	4	
–		

3.

	1	
5)	5	5
–	5	
–		

Figure out the correct number to answer each math question. Write it in the blank. Then write an equation that includes your answer. The first one has been done for you.

1. What plus 4 equals 13? 9 9 + 4 = 13
2. What plus 10 equals 13? ______ ______________
3. What minus 8 equals 8? ______ ______________
4. What minus 7 equals 8? ______ ______________
5. What times 2 equals 10? ______ ______________
6. What times 4 equals 8? ______ ______________
7. What divided by 2 equals 10? ______ ______________
8. What divided by 5 equals 4? ______ ______________

Below are shapes that have been folded and have creases. There is a smaller orange shape inside. For each one, figure out how many small orange shapes fit in its large shape. Write the number next to the shape.

9. 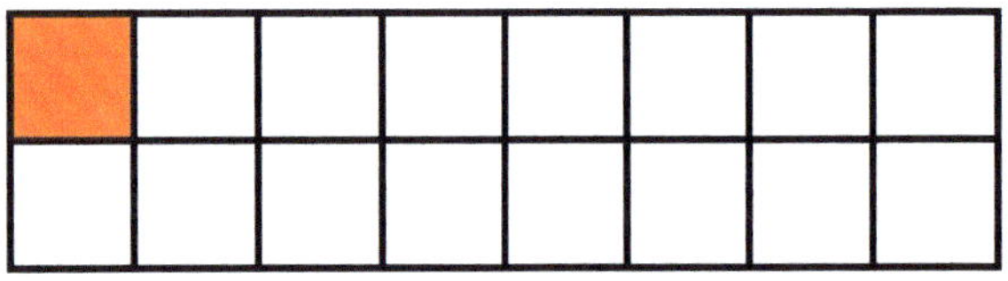

10. 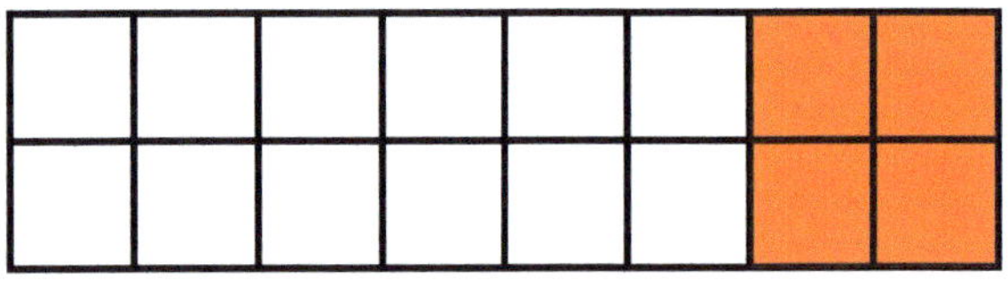

11. 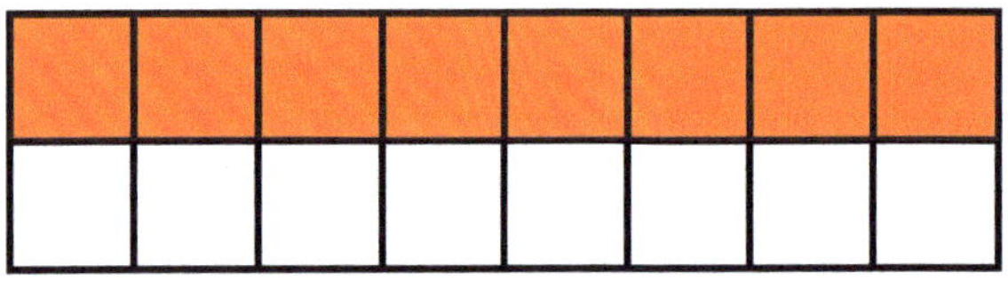

12. 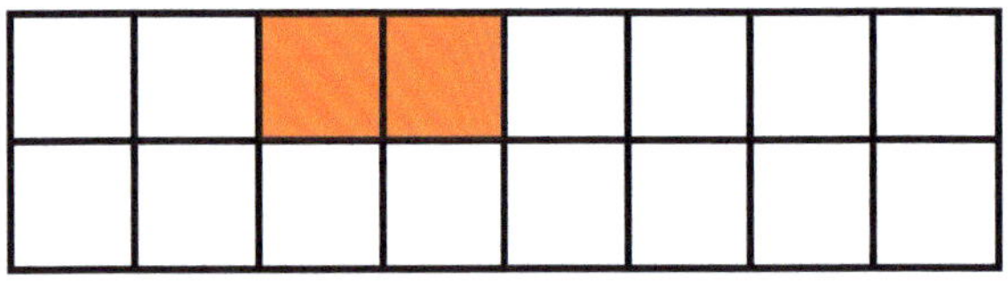

13. 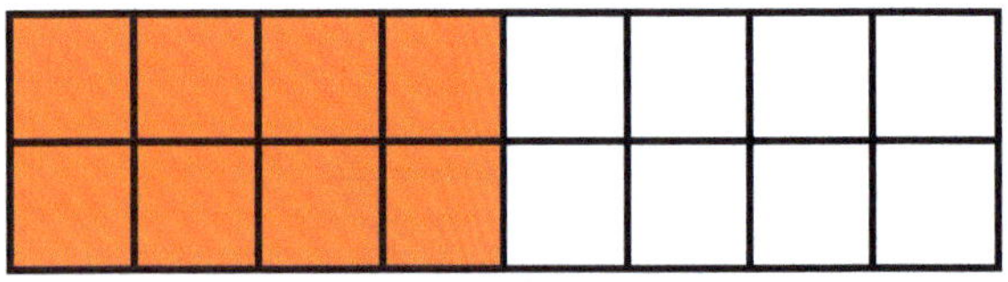

14. 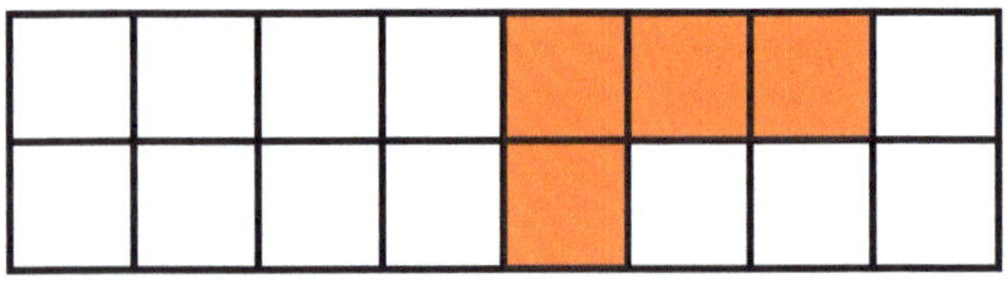

Practice DAY 126

Practice rounding these numbers.

1. Round 3,748,583 to the nearest 100. ____________
2. Round 3,748,583 to the nearest 10,000 ____________
3. Round 3,748,583 to the nearest 100,000 ____________
4. Round 56,132 to the nearest 1,000 ____________
5. Round 56,132 to the nearest 10,000 ____________
6. Round 56,132 to the nearest 100,000 ____________

Practice these long division and multiplication exercises. Parts of the answers and some carries have been provided for you.

7.

	1	
6)	6	6
–	6	
–		

8.

	1	
7)	8	4
–	7	
–		

9.

	1	
8)	9	6
–	8	
–		

10.

Carry			1	
		1	1	2
	x			5

11.

Carry		1		
		3	2	1
	x			5

12.

Carry		1	2	
		1	3	4
	x			5

Everyday Math

Read the following sentences and decide how the numbers in the story are combining—addition, subtraction, multiplication, or division. Circle the correct operation.

13. We divided up the cards among 5 players. (+, –, ÷, x)

14. Our family will be supporting another orphan this year, which will increase our giving by $600. (+, –, ÷, x)

15. You filled six cartons with 5 pounds of apples each. (+, –, ÷, x)

16. You will give away 7 balls from your collection of 15 balls. (+, –, ÷, x)

17. On our family vacation, we will drive 300 miles the first day, 400 miles the second day, and 500 miles the third day. (+, –, ÷, x)

18. There are 30 players for dodgeball, and we will separate them into two equal groups. (+, –, ÷, x)

What amazing wisdom God used to create this frog and all its skills!

19. Here is a floor plan (map of the floor) for a small house with three rooms. Let's figure out the area of each room and the area of the whole house.

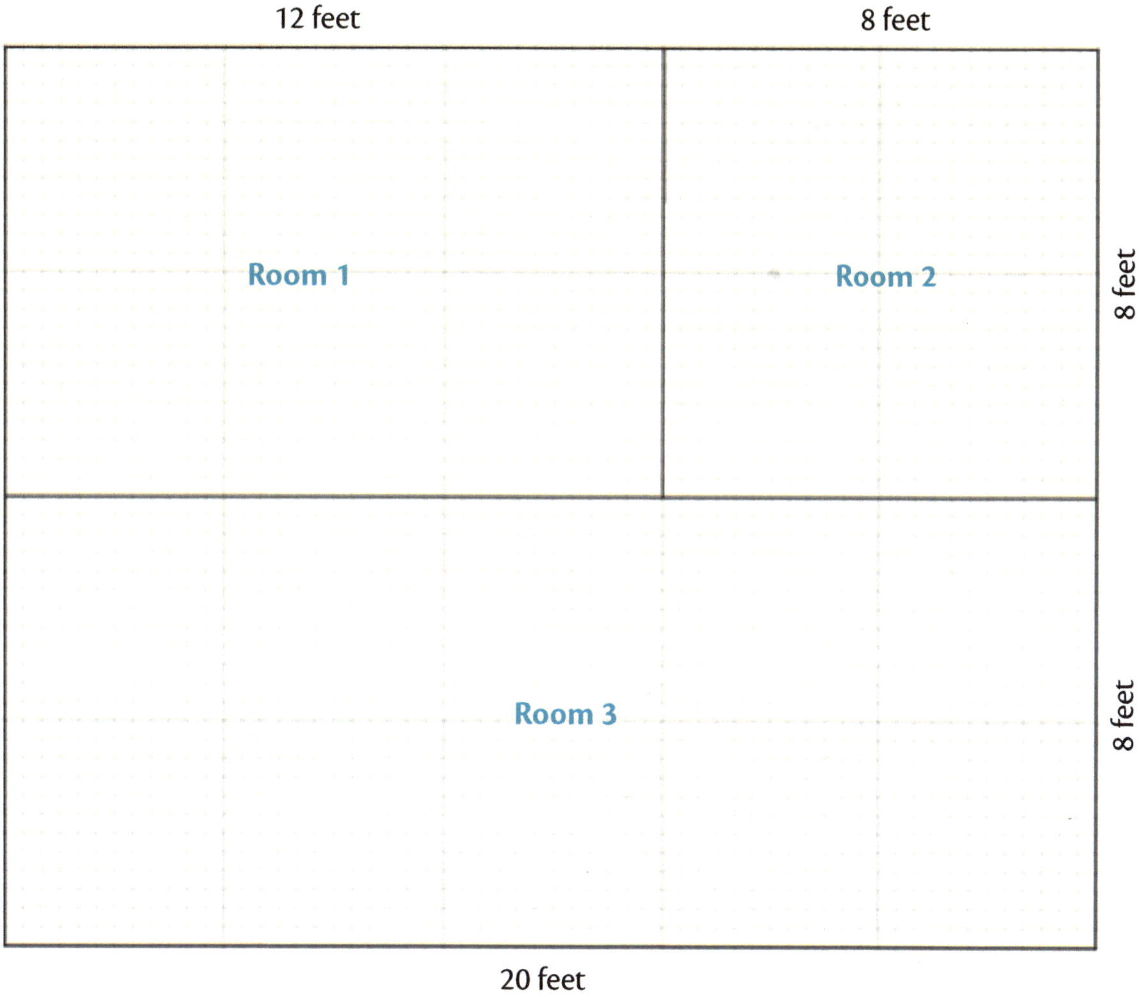

Room 1 = ________ X ________ = ________ square feet

Room 2 = ________ X ________ = ________ square feet

Room 3 = ________ X ________ = ________ square feet

Add them up! Total area of house = ____________ square feet

20. Here is a floor plan of a cabin with four rooms. Let's figure out the area of each room and the area of the whole cabin.

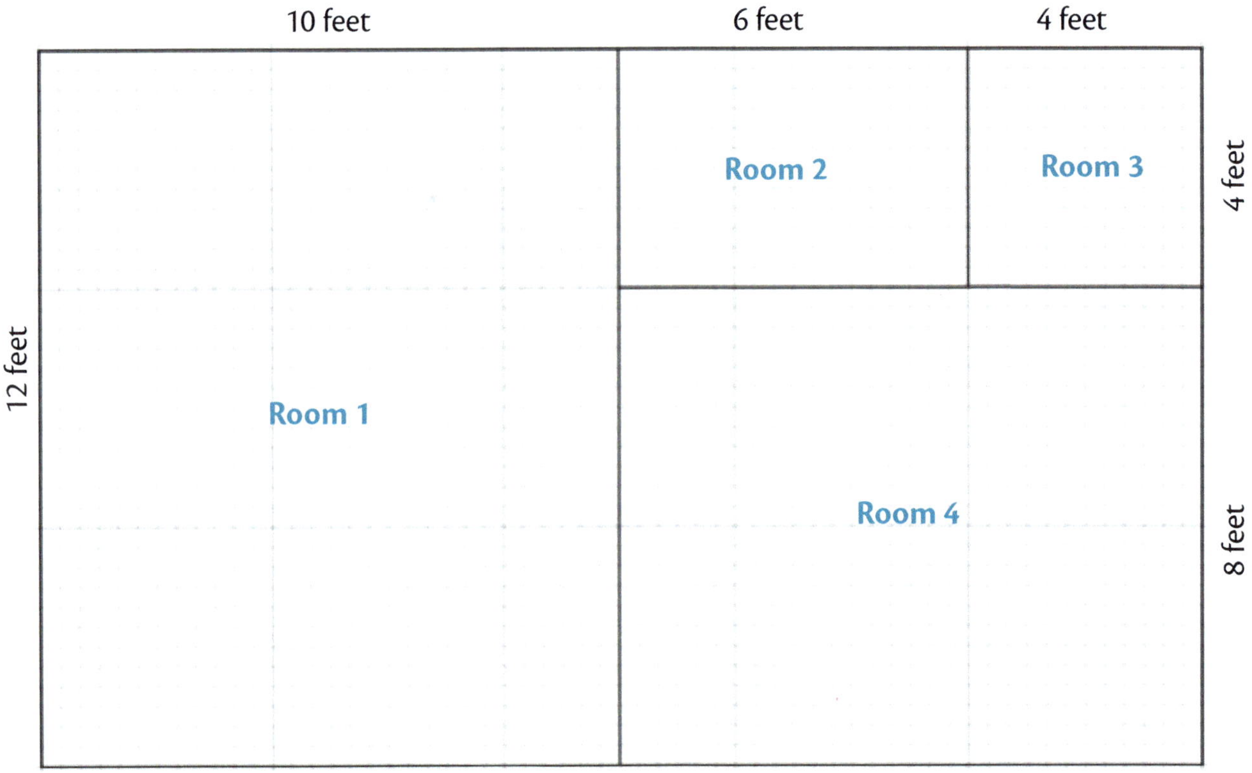

Room 1 = ________ X ________ = ________ square feet

Room 2 = ________ X ________ = ________ square feet

Room 3 = ________ X ________ = ________ square feet

Room 4 = ________ X ________ = ________ square feet

Add them up! Total area of the cabin = ____________ square feet

Longer Long Division, Comparisons, and Counting by Chunks

DAY 127

> "He is your praise, and He is your God, who has done for you these great and awesome things which your eyes have seen." (Deuteronomy 10:21)

Let's praise God for showing us how to do long division.

Let's divide some three-digit dividends.

Let's divide 125 into 5 parts.

5)	1	2	5

Step 1. Divide. Does 5 fit into 1? No. Does 5 fit into 12? Yes. How many times does 5 fit into 12? The answer is 2. Write the answer (2) above the 2 of the 12 since we are dividing the 12.

		☐	
5)	1	2	5

Step 2. Multiply 2 x 5, and write the answer (10) under the dividend's 12.

		2	
5)	1	2	5
–	☐	☐	

Step 3. Subtract 12 – 10 and write the answer (2) under the 10. Bring the 5 down to the right of the 2 because 5 – 0 = 5.

		2	
5)	1	2	5
–	1	0	
		☐	☐

Step 1 again. Divide again. How many 5s fit into the 25? The answer is 5. You will write a 5 next to the quotient's 2 at the top.

		2	☐
5)	1	2	5
–	1	0	
		2	5

Step 2 again. Multiply again. Multiply 5 x 5. Write the answer (25) under the 25 from the subtraction.

Step 3 again. Subtract again. Subtract 25 – 25, and write the answer (0) at the bottom.

		2	5
5)	1	2	5
–	1	0	
		2	5
–			

Step 4. Check. Check your answer (25) by multiplying 25 x 5.

Carry			
		2	5
x			5

God used His wonderful wisdom to create orcas. What skills has God given them?

Student Exercises

Practice these on your own.

1.

4)	6	8
–		
–		

2.

5)	1	4	5
–			
–			

3.

3)	1	0	8
–			
–			

Decide which number is larger and write the correct symbol in the circle.

4. $\frac{3}{4}$ ◯ $\frac{1}{4}$

5. 0 ◯ 4

6. $\frac{5}{7}$ ◯ $\frac{5}{6}$

7. 1.2 ◯ 12

Practice counting by chunks. Figure out if each list is counting up or counting down. Fill in the blanks to continue the pattern.

8. Count by 5s: 50, 45, ______, ______, ______, ______

9. Count by 6s: 60, 54, ______, ______, ______, ______

10. Count by 7s: 70, 63, ______, ______, ______, ______

Using all the tiles on the right (each one only once), fill in the equations on the left.

11. ☐ × ☐ = 60

12. 8 ☐ 10 = ☐

13. 10 × ☐ = 40

14. ☐ × 10 ☐ 20

15. ☐ × ☐ = 100

16. 10 ☐ 7 = ☐

17. ☐ × 10 ☐ 50

DAY 128 Practice

Figure out the correct number to answer each math question. Write it in the blank. Then write an equation that includes your answer. The first one is done for you.

1. What plus 8 equals 14? 6 $6 + 8 = 14$
2. What plus 3 equals 16? ______ ________________
3. What minus 7 equals 10? ______ ________________
4. What minus 6 equals 9? ______ ________________
5. What times 5 equals 20? ______ ________________
6. What times 3 equals 12? ______ ________________
7. What divided by 1 equals 10? ______ ________________
8. What divided by 3 equals 5? ______ ________________

Add these fractions. Fill in the pies for the answer.

9. $\frac{4}{5} + \frac{2}{5} = \text{—}$

10. 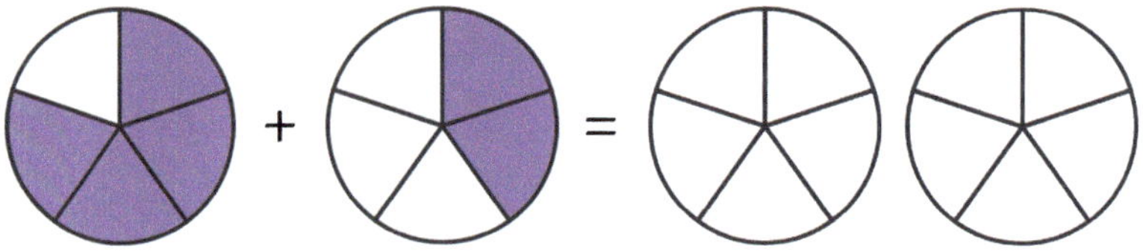

11. $\frac{4}{9} + \frac{7}{9} = \text{—}$

12. 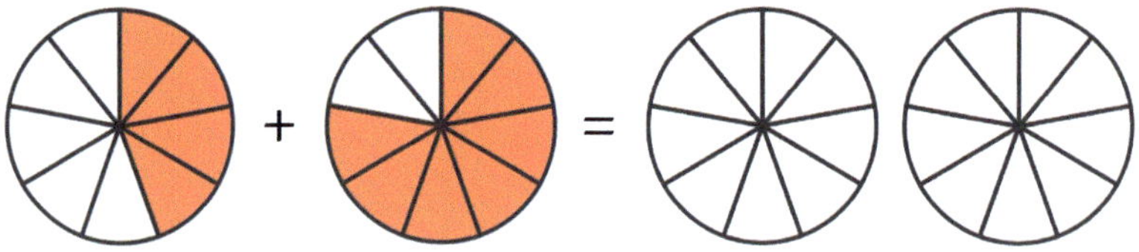

13. $\frac{7}{6} + \frac{5}{6} = \text{—}$

14. 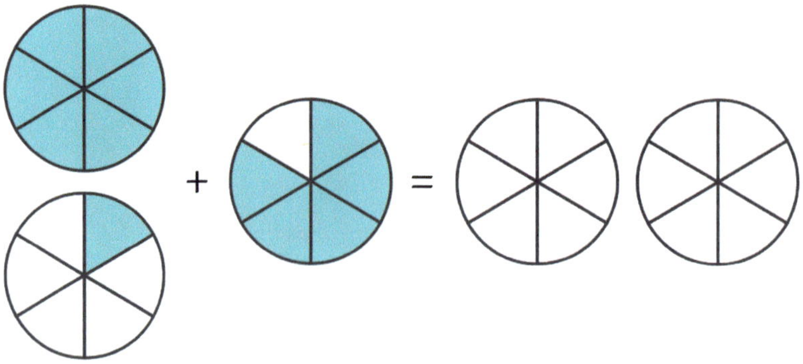

15. Suppose you've saved $69 to buy gifts for 3 loved ones. You wish to divide the money up equally in 3 parts. How much will there be to spend for each gift?

$$3\overline{)69}$$

16. If your mom gave you 55 animal crackers to split up evenly between 5 children, how many crackers will each child receive?

$$5\overline{)55}$$

Solve these multiplication problems.

17.

Carry			
	5	2	1
x			6

19.

Carry			
	4	1	8
x			6

18.

Carry			
	3	5	9
x			2

20.

Carry		
	7	7
x		3

Write in the correct number.

21. 99 centimeters = 0.99 meters

360 cm = ________ m

22. 1,000 meters = 1 kilometer

1,550 m = _______ km

23. 12 inches = 1 foot

36 in = ________ ft

24. 1 foot = 12 inches

4 ft = ________ in

DAY 129 Longer Division Using Boxes

Let's look at long division another way—with boxes. This will help you understand what's happening when we divide large numbers into equal groups. Let's start with a story.

Your grandfather has given you and two cousins his collection of marbles. He wants you to divide up the 75 marbles among the 3 of you. That's 75 ÷ 3. Let's use boxes for a fun way to figure it out.

Step 1 — We write the total number of marbles in the box as shown. (We call this the dividend). These are the marbles we are dividing up among the 3 children. We write the number of children outside of the box. (We call this the divisor.)

Think of 75 as 70 + 5. First we will divide up the 70 marbles. That's 7 tens. Can we divide 7 tens into 3 equal groups? No. But how many tens can we give to each of the 3 children? Can we give 30 marbles to each of the 3 children? Of course not! That would be 90 marbles in all, and there are only 75 marbles. Can we give 20 marbles (2 sets of 10) to each child? Yes! If we give 20 marbles to each of the 3 children, that would be 60 marbles in all. We will write 20 marbles above the box. Each child gets 20 marbles so far.

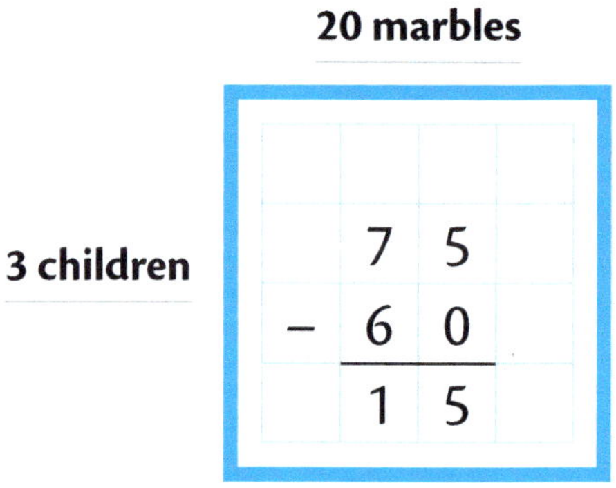

Next, we will multiply 3 x 20 to show a total of 60 marbles. We subtract this 60 marbles from the 75 marbles.

Step 2 — That leaves 15 marbles to divide up among the 3 children. Now we make a second box. Once more we write the number of children outside the box as the divisor. We write 15 inside the second box as the dividend. Now, we divide 15 by 3 (the 3 children), and we see that each child gets 5 more marbles. We write 5 above the second box. Now multiply 3 x 5 to get 15 marbles.

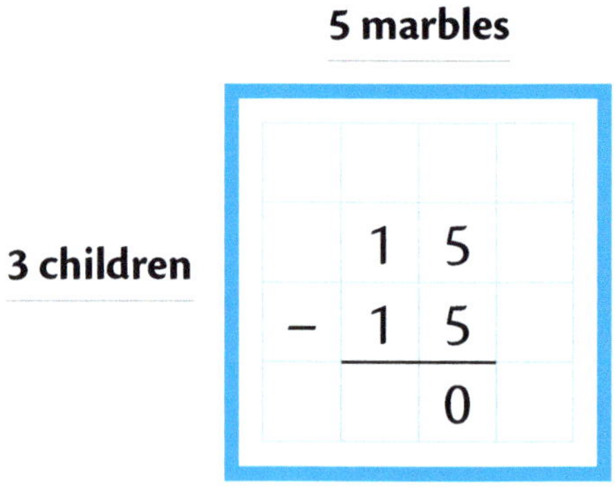

Subtract this 15 from 15 to get zero. Now there are 0 marbles left to split up among the children.

Step 3 — We place the boxes together and add up the two amounts of marbles. This shows how many marbles each child will get.

20 marbles (3 children)

	7	5
−	6	0
	1	5

5 marbles (3 children)

	1	5
−	1	5
		0

20 + 5 = 25 marbles/child

1. Suppose your uncle has given you 92 baseball cards to divide up among yourself and 3 cousins. How many baseball cards does each child get? Continue working on the partially-finished box division below.

20 cards (4 children)

	9	2
−	8	0

−		

2. Perhaps your mother has laid out 95 grapes to share among 5 children. How many grapes will each child get? Complete the box division below.

10 grapes (5 children)

	9	5
−	5	0

−		

Let's review the fast way to **multiply** numbers by 10 and 100.

- When you multiply a number by 10, you will either:
  - put one zero on the end (Example: 10 x 32 = 320)
  - or move the decimal point one place to the right. (Example: 10 X 1.5 = 15)
- When you multiply a number by 100, you will either:
  - put two zeros on the end (Example: 100 X 32 = 3200)
  - or move the decimal point two places to the right. (Example: 100 X 0.12 = 12)

Let's review the fast way to **divide** numbers by 10 and 100. (Remember, division is the opposite of multiplication.)

- When you divide a number by 10, you will either:
  - remove one zero from the end (Example: 320 ÷ 10 = 32)
  - or move the decimal point one place to the left. (Examples: 32. ÷ 10 = 3.2; 2.0 ÷ 10 = 0.2)
- When you divide a number by 100, you will either:
  - remove two zeros from the end (Example: 3200 ÷ 100 = 32)
  - or move the decimal point two places to the left. (Example: 32.0 ÷ 100 = 0.32)

Look at where these mountain goats live! What wisdom and skills did God give these animals? Can you imagine the great wisdom God has to create animals that can safely live in such places?

Student Exercises

Try a few of these on your own:

1. 560 ÷ 10 = ☐
2. 6.5 ÷ 10 = ☐
3. 54. ÷ 10 = ☐

Here is some practice with adding and subtracting. Use your wisdom of long addition and long subtraction to figure out how God made these numbers come together to make something new.

4. $\begin{array}{r} 15 \\ +\ 18 \\ \hline \end{array}$

6. $\begin{array}{r} 4.5 \\ +\ 5.6 \\ \hline \end{array}$

8. $\begin{array}{r} 35 \\ -\ 16 \\ \hline \end{array}$

10. $\begin{array}{r} 2.4 \\ -\ 2.3 \\ \hline \end{array}$

5. $\begin{array}{r} 78 \\ -\ 42 \\ \hline \end{array}$

7. $\begin{array}{r} 146 \\ +\ 287 \\ \hline \end{array}$

9. $\begin{array}{r} 37 \\ +\ 21 \\ \hline \end{array}$

11. $\begin{array}{r} 253 \\ -\ 160 \\ \hline \end{array}$

Practice multiplying and dividing these numbers by using your memory or by moving decimal points.

12. $6 \times 3 =$ ☐

13. $15 \div 15 =$ ☐

14. $4 \times 4 =$ ☐

15. $\frac{8}{2} =$ ☐

16. $23.4 \div 10 =$ ☐

17. $10 \div 2 =$ ☐

18. $12 \times 2 =$ ☐

19. $12 \div 2 =$ ☐

20. $23.4 \times 10 =$ ☐

21. $\frac{3}{3} =$ ☐

DAY 130 Practice

For each list of numbers, figure out if the list is counting up or counting down. Then, fill in the blanks to continue the pattern.

1. Count by 2s: 36, 34, ____, ____, ____, ____
2. Count by 0.1s (tenths): 0.3, 0.4, ____, ____, ____, ____
3. Count by 5s: 75, 70, ____, ____, ____, ____
4. Count by 5s: 45, 50, ____, ____, ____, ____
5. Count by 10s: 450, 460, ____, ____, ____, ____
6. Count by 10s: 320, 310, ____, ____, ____, ____

Decide which number is larger and write the correct symbol in the circle.

7. 3.42 ◯ 3.12

8. $\frac{3}{5}$ ◯ $\frac{3}{1}$

9. 123 ◯ 132

10. $\frac{3}{2}$ ◯ $\frac{4}{2}$

Everyday Math

11. Let's say your neighbors have asked you to take care of their dogs for 3 weeks. They will pay you $8 per day for the work. How much money will you earn in the 3 weeks?

How many days are in 3 weeks?
3 x 7 = ________

How much will you be paid?
$8 x _____ = ________

x ______

12. Suppose that your neighbor pays you to wash her car and vacuum its interior. This chart shows four months of your earnings. You plan to give a tithe of this money in Sunday's church offering. This means you must figure out what $\frac{1}{10}$ of the money is by dividing it by 10. You know how to divide by 10 quickly—by moving the decimal point one place to the left. Do the divisions for each month and add them up!

January	\$40	÷	10	=	
February	\$60	÷	10	=	
March	\$80	÷	10	=	
April	\$120	÷	10	=	
TOTAL					

13. Complete each vertical and horizontal equation with its missing number.

14. How much money do you find here? __________

Speed Test

Ask your parent/teacher to time your speed on these multiplication exercises. You can practice this page every now and then to help you memorize these facts. Your goal (by the time you have completed Level 3 Math) should be to answer the exercises correctly in 90 seconds.

Average Time: 120 seconds

Target Time: 90 seconds

Best Student Time: 60 seconds

Expert/Teacher Time: 45 seconds

Your Time ________

1. 6 x 5 =
2. 9 x 9 =
3. 9 x 8 =
4. 10 x 3 =
5. 9 x 7 =
6. 5 x 7 =
7. 3 x 7 =
8. 6 x 0 =
9. 9 x 10 =
10. 3 x 5 =
11. 6 x 9 =
12. 8 x 5 =
13. 5 x 0 =
14. 7 x 4 =
15. 10 x 1 =
16. 8 x 8 =
17. 4 x 3 =
18. 3 x 6 =
19. 6 x 6 =
20. 2 x 9 =
21. 6 x 8 =
22. 1 x 5 =
23. 6 x 7 =
24. 1 x 8 =
25. 2 x 8 =
26. 1 x 2 =
27. 6 x 0 =
28. 4 x 6 =
29. 2 x 6 =
30. 5 x 5 =
31. 5 x 9 =
32. 3 x 9 =
33. 3 x 2 =
34. 5 x 2 =
35. 8 x 4 =
36. 8 x 7 =
37. 10 x 1 =
38. 5 x 4 =
39. 8 x 8 =
40. 4 x 9 =
41. 5 x 8 =
42. 4 x 4 =
43. 4 x 6 =
44. 7 x 7 =
45. 9 x 9 =

DAY 131 Practice

Figure out the correct number to answer each math question. Write it in the blank. Then write an equation that includes your answer. The first one is done for you.

1. What plus 6 equals 15? 9 $9 + 6 = 15$
2. What plus 8 equals 13? ______ ____________________
3. What minus 5 equals 11? ______ ____________________
4. What minus 2 equals 5? ______ ____________________
5. What times 6 equals 18? ______ ____________________
6. What times 9 equals 9? ______ ____________________
7. What divided by 4 equals 4? ______ ____________________
8. What divided by 6 equals 3? ______ ____________________

Fill in the blanks with the right word: **dividend**, **divisor**, or **quotient**.

9. ____________________ ÷ ____________________ = ____________________

10. 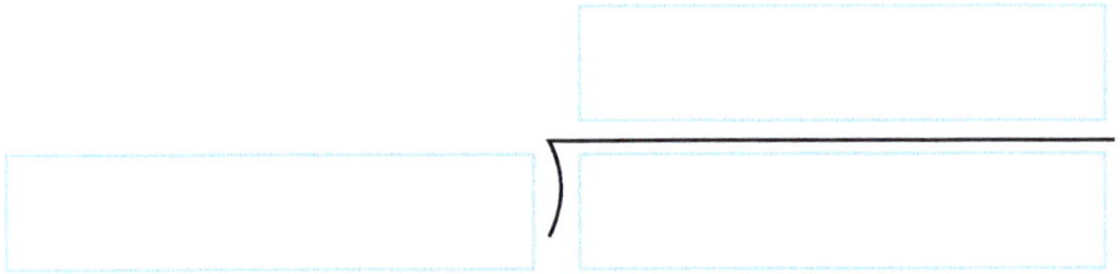

Should we desire to read a fraction of the Bible or the whole thing?

Add and subtract these fractions. Circle the answers that are equal to 0 or 1.

11. $\frac{6}{7} + \frac{5}{7} = \square$

12. $\frac{3}{9} + \frac{6}{9} = \square$

13. $\frac{13}{17} + \frac{2}{17} = \square$

14. $\frac{6}{7} - \frac{2}{7} = \square$

15. $\frac{14}{19} - \frac{5}{19} = \square$

16. $\frac{5}{9} - \frac{5}{9} = \square$

Try these long additions and subtractions.

17.

Carry			
	1	4	5
	5	6	9
+	2	4	2

19.

Carry			
	5	2	1
	3	7	9
+	2	9	9

21.

Borrow			
	1	8	7
−		9	9

18.

Carry			
	1	2	3
	4	3	1
+	5	6	7

20.

Borrow			
	4	5	2
−	3	3	3

22.

Borrow			
	4	8	7
−	2	9	8

23. Kelly baked 42 brownies to sell. She will take home the brownies that don't sell. If Kelly sells 37 brownies, how many will she take home?

24. The area inside a museum's rectangular picture frame is 48 square feet. Two of the sides of the frame are each 6 feet long. How long are each of the other sides?

$\square \div \square = \square$

25. Henry and Jacob had a pile of snow that they completely used up to make snowballs. Henry used $\frac{3}{4}$ of the snow pile and Jacob used $\frac{1}{4}$. How much more of the snow pile did Henry use than Jacob?

$\square - \square = \square$

Multiplication & Division in Real Life!

DAY 132

This lesson integrates math into everyday life. With this activity, the student will apply God's patterns and wisdom in meaningful ways to home and community. Take a break from memory work and academic exercises to make math part of everyday life. The following are suggestions or examples. Other ideas may be added to the list.

1. Double a recipe to make twice as much food. Find a recipe for a meal or dessert, and double the amount of each ingredient.

If the recipe calls for $\frac{1}{2}$ teaspoon of something, just add $\frac{1}{2} + \frac{1}{2}$. That makes 1 whole teaspoon.

If the recipe calls for ¼ teaspoon, add $\frac{1}{4} + \frac{1}{4}$. That makes $\frac{2}{4}$ ($\frac{1}{2}$).

If the recipe calls for 3 cups, multiply 3 cups x 2. That makes 6 cups.

2. Plan a pizza night for the family. Order some pizzas and divide the total number of pieces evenly among all who are eating. If the number of people is less than 10, you can use long division to see how many pieces each person may have. (There might be some extra pieces if they don't divide evenly.) If the number of people is exactly 10, you can move the decimal point (on the total number of pieces) to the left one place. If the number of people is more than 10, you will have to use a calculator or ask for help. Someday you will learn how to do long division with large divisors. Here's the equation to use if the number of people is less than 10:

Number of pieces of pizza ÷ number of people eating = amount of pizza per person

3. Get a notebook and plan how much you will spend on presents for an upcoming holiday. Decide how much total money you are going to spend (dividend). Divide that by the number of presents you would like to buy (divisor). Your quotient is the amount you can spend on each present.

4. Plan a road trip that will last a few days. Take the number of total miles (dividend) and divide them by the number of days you plan to travel (divisor). That will give you the number of miles you can aim to travel each day of your trip (quotient).

5. Divide up the yard work. Choose a chore like weeding rows in the vegetable garden or clipping the lawn edges after Dad mows. Measure the distance to be covered (in feet or meters) and divide it between yourself and any sisters or brothers able to participate.

6. Help the Apostles get to the Jerusalem Council (Acts 15). Draw a line following these coordinates from the apostles position all the way to Jerusalem.

H5, H4, H3, H2, G2, F2, F3, F4, F5, F6, E6, D6, C6, B6, B5, B4, A4, A3, A2

Additional Activities

For extra activities, games, and helpful videos on these math concepts, refer to **generations.org/math3**.

Making watches requires a lot of carefulness and skill.

CHAPTER 12

Getting It Right; Doing It Faster

DAY 133 Adding Decimals . . . Carefully

> He who has a slack hand becomes poor,
> But the hand of the diligent makes rich.
> He who gathers in summer is a wise son;
> He who sleeps in the harvest is a son who causes shame. (Proverbs 10:4-5)

> The lazy man does not roast what he took in hunting. (Proverbs 12:27)

> As vinegar to the teeth and smoke to the eyes,So is the lazy man to those who send him. (Proverbs 10:26)

God tells us to be diligent about the things He wants us to do. Getting things done without putting them off is a sign of diligence. Doing things fast is important, but doing them carefully is even more important. A lazy person is slow at working because he is afraid of hard work. He is not dependable. He makes a lot of mistakes because he doesn't try very hard. A lazy person is not careful to cut the meat off the animal he took in the hunt. That's just hard work!

> The way of the lazy man is like a hedge of thorns,
> But the way of the upright is a highway.
> (Proverbs 15:19)

Sometimes math is hard because we don't understand it. But sometimes math is hard because we are kind of lazy. The way of a lazy man is like a hedge of thorns blocking his progress. His laziness keeps him from working hard to accomplish something.

Let's work on accuracy today. **Accuracy** means doing something without making mistakes. How do we avoid making mistakes with math? By being careful! We need to be extra careful with numbers that have decimals or a lot of digits.

Compare these two adding exercises. Which one is accurate?

$$\begin{array}{r} 23.244 \\ +\ \ 2.341 \\ \hline 255.85 \end{array} \qquad \begin{array}{r} 23.244 \\ +\ \ 2.341 \\ \hline 25.585 \end{array}$$

In the first example, the answer's decimal point was put in the wrong place. This may seem like only a small mistake, but the answer is very wrong. This answer is much too large. One quick way to test this answer is to round the numbers we're adding. Then we add the rounded numbers together. These numbers round to 23 and 2, and they add up to 25. We know that 25 is not an exact answer, but it's a good guess. Now you can see how much larger 255.85 is than 25.585. And you know that 25.585 has the decimal point in the correct place. This is the accurate answer.

Approximate: A number is **approximate** when we say it's amount is around a certain number, but we do not say its exact amount. 25.285 is **approximately** 25.

Let's work on accuracy for these exercises. Be sure to put the answer's decimal point in the right place! For each of the first three exercises, round the numbers in your head to the nearest whole number. Next, add the rounded numbers to get an approximate answer. Then go back and complete all six exercises for their exact answers.

1.

$$\begin{array}{r} 8.45 \\ +\ 7.34 \\ \hline \end{array}$$

Approximate Answer

Exact Answer

2.

$$\begin{array}{r} 12.34 \\ +\ 22.22 \\ \hline \end{array}$$

Approximate Answer

Exact Answer

3.

$$\begin{array}{r} 30.45 \\ +\ 40.42 \\ \hline \end{array}$$

Approximate Answer

Exact Answer

4.

$$\begin{array}{r} 123.456 \\ +\ 111.111 \\ \hline \end{array}$$

5.

$$\begin{array}{r} 654.321 \\ +\ 222.222 \\ \hline \end{array}$$

6.

$$\begin{array}{r} 345{,}567.876 \\ +\ 111{,}111.012 \\ \hline \end{array}$$

Each lettered triangle, A - F, is missing a colored section. Choose from the colored shapes, 1 - 6, at the bottom to complete each triangle so it matches one of the "trees" at the top.

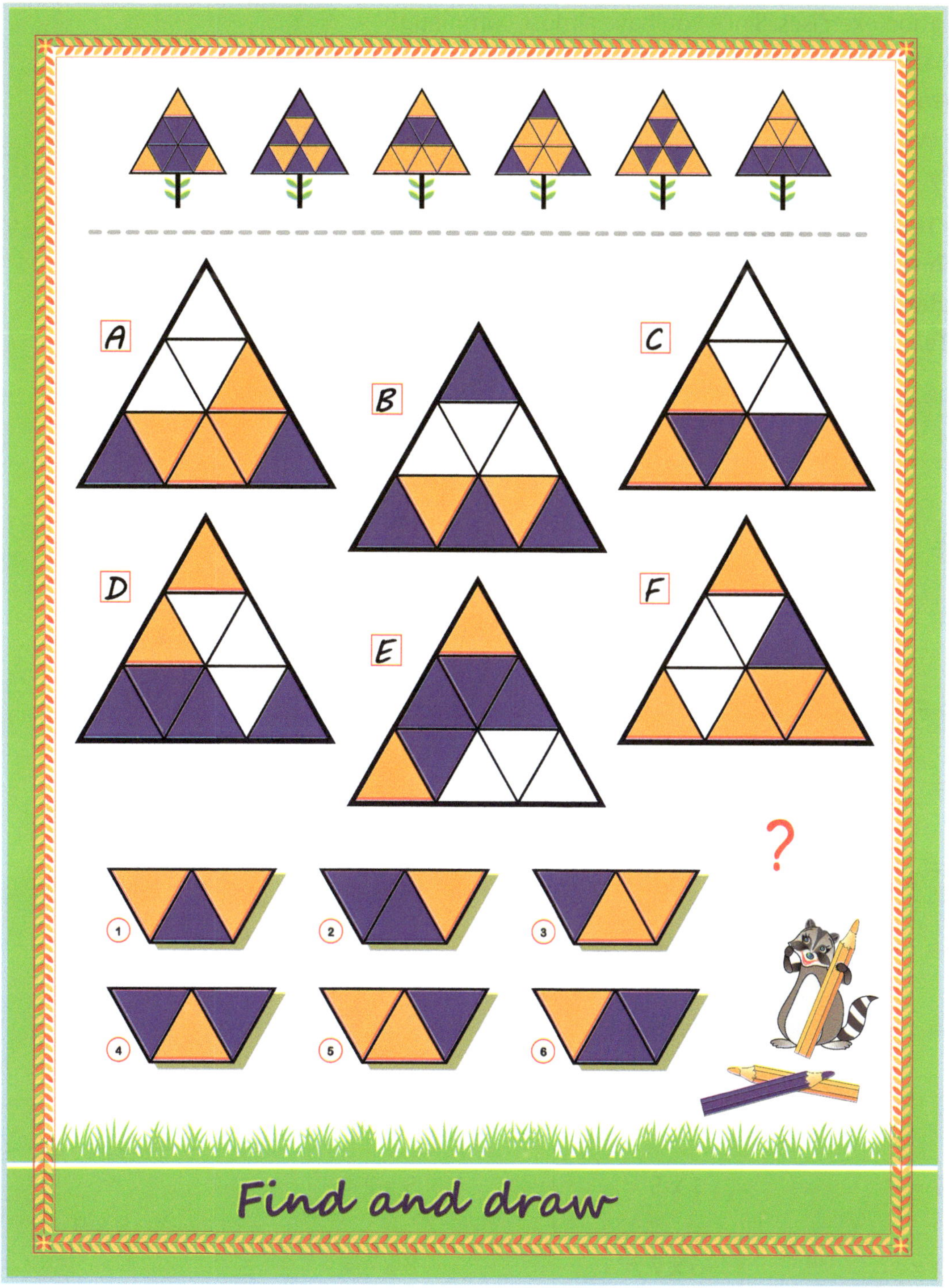

Speed Test

Your parent/teacher should time your speed on these multiplication exercises. Since you have almost completed Level 3 Math, your time should be 90 seconds. But this time, when you are finished, add a 5-second penalty to your score for every mistake you made. Do the drill fast, but do it accurately.

Average Time: 120 seconds
Target Time: 90 seconds
Best Student Time: 60 seconds
Expert/Teacher Time: 45 seconds

Your Time ________

1. 6 x 5 =
2. 9 x 9 =
3. 9 x 8 =
4. 10 x 3 =
5. 9 x 7 =
6. 5 x 7 =
7. 3 x 7 =
8. 6 x 0 =
9. 9 x 10 =
10. 3 x 5 =
11. 6 x 9 =
12. 8 x 5 =
13. 5 x 0 =
14. 7 x 4 =
15. 10 x 1 =
16. 8 x 8 =
17. 4 x 3 =
18. 3 x 6 =
19. 6 x 6 =
20. 2 x 9 =
21. 6 x 8 =
22. 1 x 5 =
23. 6 x 7 =
24. 1 x 8 =
25. 2 x 8 =
26. 1 x 2 =
27. 6 x 0 =
28. 4 x 6 =
29. 2 x 6 =
30. 5 x 5 =
31. 5 x 9 =
32. 3 x 9 =
33. 3 x 2 =
34. 5 x 2 =
35. 8 x 4 =
36. 8 x 7 =
37. 10 x 1 =
38. 5 x 4 =
39. 8 x 8 =
40. 4 x 9 =
41. 5 x 8 =
42. 4 x 4 =
43. 4 x 6 =
44. 7 x 7 =
45. 9 x 9 =

Practice DAY 134

Which number is bigger? Write the correct symbol in the circle.

1.	4.5 ◯ 4.56	3.	$\frac{5}{6}$ ◯ $\frac{5}{10}$
2.	$\frac{3}{4}$ ◯ $\frac{1}{4}$	4.	0.25 ◯ 0.35

Practice adding and subtracting by memory.

5. $\frac{4}{5} + \frac{2}{5} =$ ☐

6. $\frac{3}{7} + \frac{1}{7} =$ ☐

7. $0.6 + 0.2 =$ ☐

8. $\frac{6}{7} + \frac{3}{7} =$ ☐

Round these numbers to the nearest 10 (for 2-digit numbers) or nearest 100 (for 3-digit numbers). Add the rounded numbers in your head for an approximate answer for each exercise. The first one is shown as an example. (The symbol ≈ means *is approximately equal to.*)

81 + 13 ≈ 80 + 10 = 90

9. 48 + 28 ≈ ________ + ________ = ________

10. 212 + 100 ≈ ________ + ________ = ________

11. 345 + 265 ≈ ________ + ________ = ________

12. 567 + 343 ≈ ________ + ________ = ________

Practice these long divisions.

13. $3\overline{)39}$

14. $3\overline{)66}$

15. $3\overline{)99}$

For each exercise, find the area of its colored rectangles. Add the areas to find the total.

16.

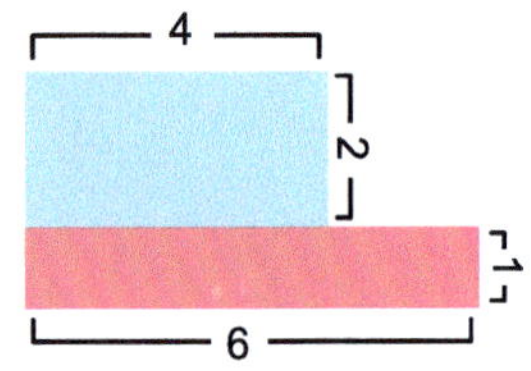

(___x___) + (___x___)

___ + ___

Area = ________

17.

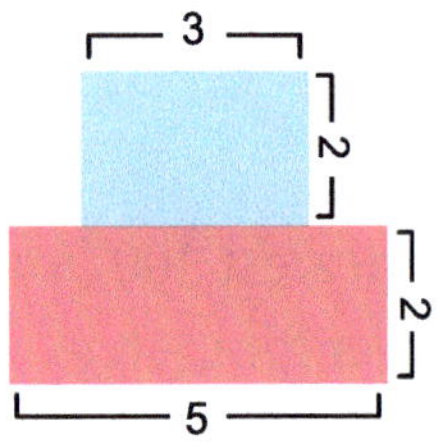

(___x___) + (___x___)

___ + ___

Area = ________

18.

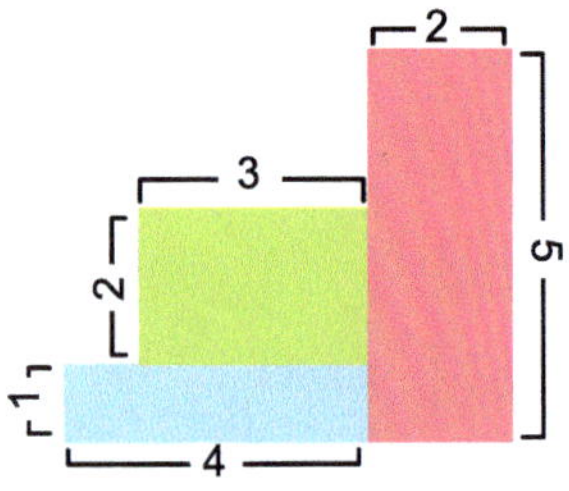

(__x__) + (__x__) + (__x__)

___ + ___ + ___

Area = ________

19.

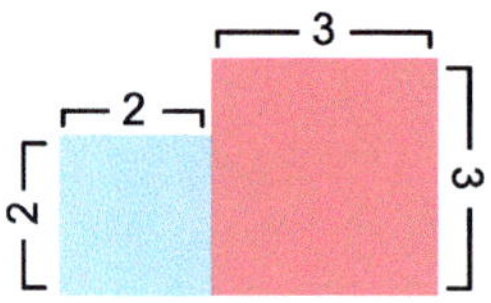

(___x___) + (___x___)

___ + ___

Area = ________

20.

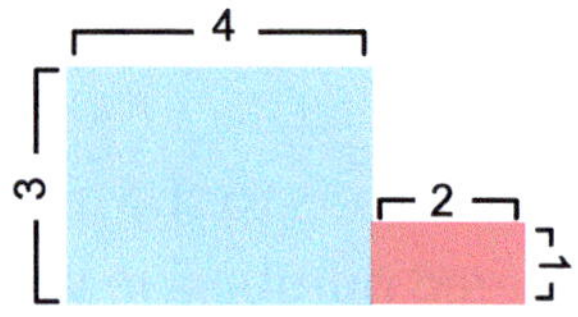

(___x___) + (___x___)

___ + ___

Area = ________

21.

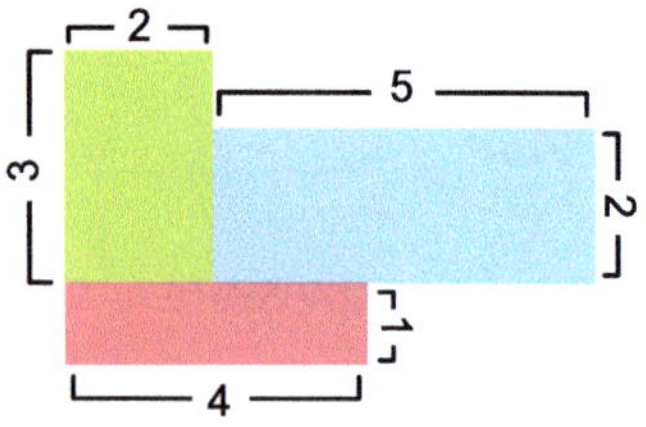

(__x__) + (__x__) + (__x__)

___ + ___ + ___

Area = ________

Subtracting Decimals . . . Carefully DAY 135

Don't lose your footing. Keep trusting in Christ for help.

You're near the end of your math book!

> But as for me, my feet had almost stumbled;
> My steps had nearly slipped.
> (Psalm 73:2)

Today you will carefully subtract numbers with a lot of digits.

First, let's find an approximate answer to this subtraction.

$$\begin{array}{r} 30.487 \\ -\ 20.356 \\ \hline \end{array}$$

We will round the first number to 30 and the second number to 20. Then, we'll subtract 30 - 20 for an approximate answer of 10. Now let's fill in the exact answer, making sure the decimal point is in the right place.

$$\begin{array}{r} 30.487 \\ +\ 20.356 \\ \hline 10.131 \end{array}$$

Since 10.131 is very close to our estimate of 10, it appears we have the right answer. Our decimal point is in the right place.

Both addition and subtraction exercises can be harder when you have to carry or borrow several times. Let's try a harder one.

You have already learned how to double borrow when two digits in a row are too small to subtract from. For this exercise, the first three digits are too small to subtract from. Study this subtraction example to see how borrowing was needed three times in a row.

Borrow			8	11		10	15
		4	~~9~~	~~2~~	.	~~1~~	~~5~~
	−	3	8	5	.	5	6
		1	0	6	.	5	9

Let's try checking our answer by approximating. We decide to round to the nearest 100 because that's something we can subtract in our head. We round 492 to 500, and we round 385 to 400. Then we subtract 400 - 300 and get 100. That's close to the 106.59 we got by long subtraction, so our answer is probably right.

Now let's try a long addition exercise with five digits. Study the example below to see how it's possible to carry several times in a row.

Carry		1	1	1	1	
		5	6	5.	9	8
	+	3	9	4.	5	7
		9	6	0.	5	5

Let's check our answer by approximating. We round the top number to 600 and round the bottom number to 400. Then we add 600 + 400 and get 1000. That's close to the 960.55 we got by long addition.

Student Exercises

Add and subtract these exercises carefully! Be sure to put the decimal points in the right places. The numbers to borrow and carry have been provided for you. Check your answers by rounding to get approximate answers.

1.

Carry		1	1	1		
		3	3	9.	9	4
	+	3	9	4.	5	5

Approximate Answer

2.

Carry			1	1	1	
		1	4	4.	6	1
	+	4	3	7.	5	9

Approximate Answer

3.

Borrow			2	13	15	
		8	~~3~~	~~4~~.	~~5~~	3
	−	3	2	5.	6	1

Approximate Answer

Everyday Math

Let's subtract decimals with everyday math.

Suppose that, at 3:00 pm on July 7th, the temperature in Denver, Colorado was 91.7°F. The next morning, at 5:00 am, the temperature was 49.8°F. Subtract to find the difference between these temperatures. Subtract to find the Celsius difference too.

Fahrenheit

4.

Borrow			
		9	1.7
	−	4	9.8

Celsius

5.

Borrow			
		3	3.2
	−		9.9

The same day, the temperature at 3:00 pm in Honolulu, Hawaii was 85.8°F (29.9°C). The next morning at 5:00 am, the temperature in Honolulu was 77.9°F (25.5°C). Find the differences.

Fahrenheit

6.

Borrow			
		8	5.8
	−	7	7.9

Celcius

7.

Borrow			
		2	9.9
	−	2	5.5

What do you notice about the temperature swings in Denver compared with Honolulu's temperature swings? What do you think makes them different?

Optional Investigation: What are the summer and winter temperature swings where you live?

8. Find the answers to the equations. Color the areas according to the colors that match your answers.

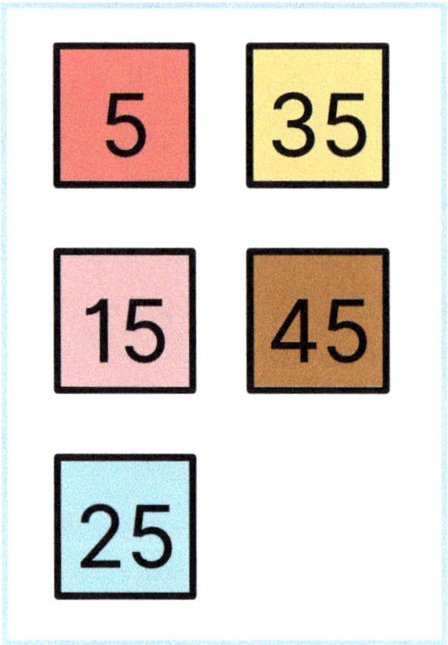

Read the temperatures. Write the correct number of degrees on the line below each thermometer.

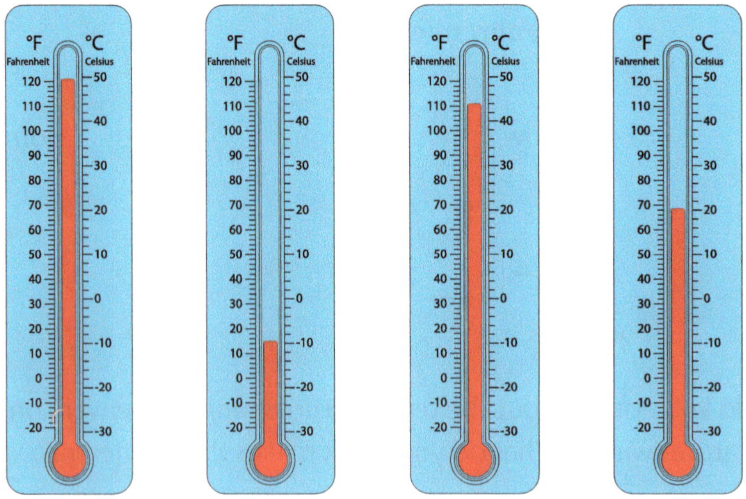

9. _____ °F 10. _____ °C 11. _____ °F 12. _____ °C

Practice DAY 136

Guess the line lengths and angle measurements below. Then measure them to see how close your guesses were. Use centimeters for the lines.

1

2

3

		YOUR GUESS	YOUR MEASUREMENT
1.	Line #1		
2.	Line #2		
3.	Line #3		

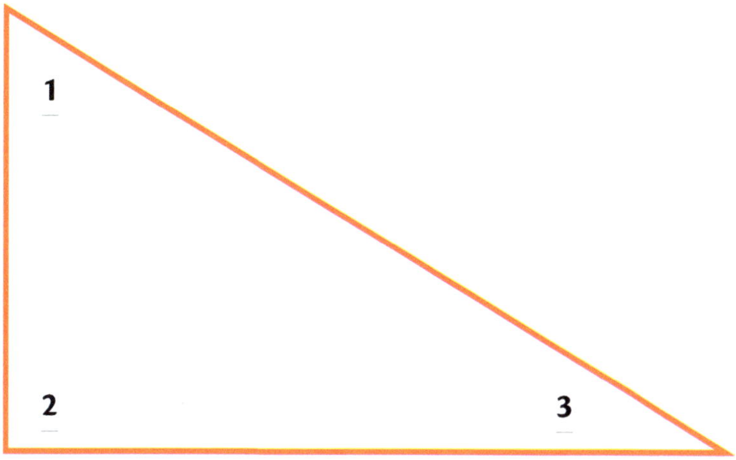

		YOUR GUESS	YOUR MEASUREMENT
4.	Angle #1		
5.	Angle #2		
6.	Angle #3		

Can you find different ways to split these numbers apart?

7. 6 = ____ + ____

8. 6 = ____ + ____

9. 12 = ____ x ____

10. 12 = ____ x ____

11. 15 = ____ + ____

12. 15 = ____ + ____

13. 18 = ____ x ____

14. 18 = ____ x ____

Everyday Math

15. Suppose you have a new picture to hang on a wall. You would like the picture to be centered (with the same amount of wall space on both sides). The wall is 72.75 inches wide and the picture is 36.75 inches wide. How much wall space should be on each side of the picture?

(Hint: Subtract the picture's width from the wall's width. Then use a calculator to divide the answer by 2.)

$$\begin{array}{r} \_\_.\_\_ \\ - \_\_.\_\_ \\ \hline \end{array}$$

____ ÷ ____ = ____

16. Suppose you started a business selling stuffed animals, yoyos, and shoes on the internet. You bought the products at low prices and sold them at higher prices. Let's fill out this table to see what your profit was after a few sales.

For each category below, subtract your cost per item from its selling price. Then multiply to find out your profit. The stuffed-animal category is filled in for you. Do the same for the 6 yo-yos and shoes.

Which of the three products gives you the most profit? ______________________

Which product is the most popular? ______________________

PRODUCT	COST PER ITEM	SELLING PRICE PER ITEM	PROFIT PER ITEM	NUMBER OF ITEMS SOLD	TOTAL PROFIT
Stuffed Animals	$5	$8	$3	10	$30
Yoyos	$10	$20		6	
Shoes	$15	$25		5	

17. Add the total profits for each category to find your overall profit:

$ __________ + $ __________ + $ __________ = $ __________

Figure out the answer to each exercise below. Then find its answer in the large box. Color that answer's surrounding square to match the colored pattern next to its original exercise.

12	7	23	15
14	21	33	27
6	34	11	30
18	5	17	25

13+5−4

17−8+21

14+3+8

26−8+3

6+3+3

10+12+5

15−8−2

25+9−11

23+3−11

19−7−5

25−14+22

31−12−2

27+4−13

21+7+6

16+7−12

18−3−9

Refresher on Calculators DAY 137

The calculator is a helpful tool and a faster way to add, subtract, multiply, and divide. God gave scientists the wisdom to invent calculators so people could do math quickly. Find a calculator and try calculating:

5 x 19

Push the following buttons in this order:

5
x
1
9
=

1. What answer is shown on the screen? ________

Now try calculating: 189 ÷ 3

Push the following buttons in this order:

1
8
9
÷
3
=

2. What is the answer? _______

Warning! It's easy to make mistakes on calculators too. You can push the wrong numbers or put the decimal point in the wrong place. It's still a good habit to round and get an approximate answer in your head before you use the calculator to get the exact answer.

Let's approximate the answer for this addition by rounding to the 10s place.

43 + 52 + 38 +14

43 rounds to: 40

52 rounds to: 50

38 rounds to: 40

14 rounds to: 10

Add these rounded numbers in your head quickly. You should get 140.

Now use a calculator to add. (Your parent/teacher can provide you with a calculator or let you use one on a phone or the internet.)

$$43 + 52 + 38 + 14 = 147$$

You can see that the number 147 is pretty close to our guess of 140.

Estimate and add: 18 + 28 + 34 + 42.

3. 18 rounded to nearest ten: ________
4. 28 rounded to nearest ten: ________
5. 34 rounded to nearest ten: ________
6. 42 rounded to nearest ten: ________
7. Estimated total: ________
8. Calculator total: ________

Estimate and add: 20.22 + 13.45 + 14.87 + 34.87.

9. 20.22 rounded to nearest whole number: ________
10. 13.45 rounded to nearest whole number: ________
11. 14.87 rounded to nearest whole number: ________
12. 34.87 rounded to nearest whole number: ________
13. Estimated total: ________
14. Calculator total: ________

Student Exercises

Can you beat the calculator?

Time yourself while adding the following numbers in your head or by hand as quickly as you can. Write the answers. Then time yourself as you add the same numbers with a calculator. Be sure to write the answers again. Compare the times, and cross check your answers. Did you beat the calculator?

Add in your head or by hand:

1. $\begin{array}{r} 8 \\ +\ 9 \\ \hline \end{array}$

2. $\begin{array}{r} 78 \\ +\ 19 \\ \hline \end{array}$

3. $\begin{array}{r} 234 \\ +\ 287 \\ \hline \end{array}$

4. $\begin{array}{r} 24 \\ 18 \\ +\ 32 \\ \hline \end{array}$

Your Time ________

Add with a calculator:

$\begin{array}{r} 8 \\ +\ 9 \\ \hline \end{array}$

$\begin{array}{r} 78 \\ +\ 19 \\ \hline \end{array}$

$\begin{array}{r} 234 \\ +\ 287 \\ \hline \end{array}$

$\begin{array}{r} 24 \\ 18 \\ +\ 32 \\ \hline \end{array}$

Your Time ________

Calculators can be slower than your head when you are doing simple adding and subtracting. But now let's use a calculator to try some exercises that would take much longer by hand. Fill in the answers.

5. 34,873 + 99,537 = __________

6. 924.847 + 487.384 = __________

7. 87,653 – 9,456 = __________

8. 9 X 949 = __________

9. 497 ÷ 7 = __________

Everyday Math

10. Suppose you are helping set up a special event at your church. For lunch, you will be using round tables that can each fit 6 people comfortably. If 15 tables are set, how many people can attend?

11. Suppose that 28 people are attending a pro-life prayer meeting. The pastor asks everyone to gather into groups of 4 to pray. How many groups are formed?

12. During a volleyball game, 12 players are on the court—6 from each team. God gave each player (and you) 12 pairs of ribs to protect the heart and lungs. Let's count the pairs of ribs that make up the rib cages of 12 people.

Count by 12s (12 times): _____, _____, _____, _____, _____,

_____, _____, _____, _____, _____, _____, _____

Now multiply this number by 2 to find out how many individual ribs are inside the 12 people.

_______ x 2 = _______

DAY 138 Practice

Finish each list of shapes. Find the pattern, and then draw a picture that keeps that pattern going.

1. 

2. 

3. 

4. 

Do these multiplication problems.

5.

	1	1	1
x			3

6.

	2	2	2
x			3

7.

	3	3	3
x			3

8.

Carry		1		
		2	4	1
	x			3

Multiply and divide by 10.

9. 4 x 10 = _______

10. 56 x 10 = _______

11. 23 x 10 = _______

12. 177 x 10 = _______

13. 20 ÷ 10 = _______

14. 350 ÷ 10 = _______

15. 1500 ÷ 10 = _______

16. 236 ÷ 10 = _______

Change mixed numbers into fractions and fractions into mixed numbers. Pie drawings are included for the first exercise. Draw and color in your own pies for the rest of the exercises.

17.	$2\frac{3}{4} =$	$\frac{\quad}{4}$	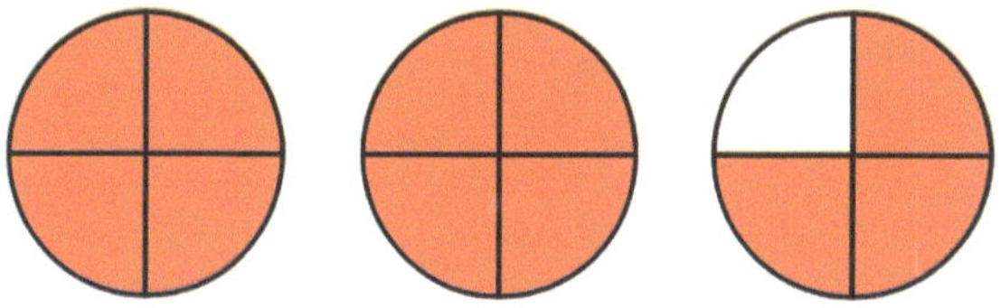
18.	$1\frac{1}{2} =$		
19.	$3\frac{2}{3} =$		
20.	$\frac{7}{3} =$		
21.	$\frac{12}{5} =$		
22.	$\frac{9}{2} =$		

Decode this message. Find each letter using the coordinates on the attached chart. The first letter is provided for you.

	A	B	C	D
1	W	L	R	T
2	O	X	M	C
3	D	H	Q	J
4	B	P	U	O
5	Y	V	I	S
6	Z	G	K	A

A1	C5	D1	B3
W			

B6	A2	A3

	A	B	C	D
1	U	T	L	R
2	M	H	N	D
3	E	O	B	Y
4	Q	A	G	Z
5	W	I	V	P
6	X	N	K	S

B4	C1	C1

B1	B2	B5	B6	C4	D6

	A	B	C	D
1	E	I	U	T
2	F	S	F	C
3	H	J	P	D
4	K	G	C	R
5	O	V	M	Z
6	L	Q	B	A

D6	D4	A1

C3	A5	B2	B2	B1	C6	A6	A1

	A	B	C	D
1	M	A	Z	6
2	D	H	X	D
3	1	Y	9	V
4	R	E	K	W
5	Y	V	L	F
6	T	:	2	J

A1	B1	A6	A6	B2	B4	D4

A3	C3	B6	C6	D1

Using Math to Keep Track of Things DAY 139

> Be diligent to know the state of your flocks,
> And attend to your herds;
> For riches are not forever,
> Nor does a crown endure to all generations.
> When the hay is removed, and the tender grass shows itself,
> And the herbs of the mountains are gathered in,
> The lambs will provide your clothing,
> And the goats the price of a field;
> You shall have enough goats 'milk for your food,
> For the food of your household,
> And the nourishment of your maidservants.
> (Proverbs 27:23-27)

These verses show that God wants us to keep track of the money and other things He's given us. Ranchers and farmers shouldn't let their cattle and chickens wander off. We shouldn't leave our bicycles out in the rain to get rusted. It's not a good idea to leave money lying around where thieves would be tempted to steal it.

We should be good stewards of the things God has put under our care. That means we need to take care of things and not waste money. Of course, sharing with others is not wasting.

God has given people the ability to make and use spreadsheets to keep track of things. This is a very useful math tool. There are many practical ways we can use spreadsheets. Today, let's learn how to add and subtract using a spreadsheet.

Here is a simple spreadsheet. Each little box is called a cell. Each cell's location is named by coordinates. You can see that the number 30 has been put in the cell with the coordinates A1. The number 42 has been put into cell B3.

	A	B	C	D	E
1	30				
2					
3		42			
4					
5					
6					

Spreadsheets are very flexible. You can type numbers wherever you like by moving the cursor to any cell on the spreadsheet. On the previous page, you can see that the cursor is located on cell A1.

Almost every computer has a spreadsheet application. You can also find spreadsheets to use online.

Let's put the numbers 34 and 28 into the B column of this spreadsheet. We'll put these numbers into cells B3 and B4.

	A	B
1		
2		
3		34
4		28
5		

Now, let's say we would like to add 34 and 28 together and put the answer in Cell B5. We would go to cell B5 and type this formula:

=B3 + B4

Then we would press "enter" or "return."

	A	B
1		
2		
3		34
4		28
5	**Sum Total**	62

The answer is 62. Notice that the words "Sum Total" have been put into cell A5, right next to B5. This helps us remember what the number in B5 stands for. It stands for the addition of the two numbers above it. After we press enter, we cannot see the formula we typed into cell B5. We see the answer instead of the formula. But if we click on that cell, it will show up.

Subtracting is just as easy. Let's find the answer to 34–28? We go to cell B5 and type this formula:

=B3 – B4

The answer that shows up is 6.

	A	B
1		
2		
3		34
4		28
5	**Difference**	6

Now, let's type the numbers 3, 5, 7, 9, 11, and 13 into the B column. We put these numbers into cells B3, B4, B5, B6, B7, and B8. We would like to add these numbers and put the answer in B9. Let's see how a spreadsheet helps us quickly add all the numbers from B3 down to B8.

	A	B
1		
2		
3		3
4		5
5		7
6		9
7		11
8		13

In cell B9, we type the formula:

=sum(b3:b8)

Typing "b3:b8" is like saying "everything from B3 through B8."

1		
2		
3		3
4		5
5		7
6		9
7		11
8		13
9	**Sum Total**	48

The spreadsheet will automatically put the answer (48) in cell B9.

Notice that we added the words "Sum Total" in A9 right next to B9. This is so we can remember that the number in B9 is the addition of all the numbers above it. You can't see the formula that we typed into cell B9. All you see is its answer. But remember, you can click on it and it will show up.

A spreadsheet is a great way to add up a lot of numbers. If we put 100 numbers in B1 to B100, we could add them all up by typing the formula below. Then boom, we have the answer!

=sum(b1:b100)

Note: Some spreadsheets might use slightly different formulas. You can read the manual for each spreadsheet to make sure you have the right formulas for what you want to do.

How do we use math to look after our "herds?" Let's pretend that your cat had five kittens, and you are selling them to your neighbors for pets. You sold the kittens for $15, $25, $22, $32, and $10. These numbers have already been put on this spreadsheet. What formula would you type into the blue box to find the total amount you made from selling the kittens? Fill out the formula below:

	A	B
1		
2		
3		15
4		25
5		22
6		32
7		10
8	**Sum Total**	

=sum(   )

1. Then ask your parent/teacher to help you find a spreadsheet on a computer or online. Type in the numbers and let the spreadsheet figure out the total: $ ____________

Use the color guides below to make a picture. For each coordinate pair listed in the color guides, find that square on the graph and fill it in with the guide's color.

A1, B1, C1, D1, E1, F1, G1, H1, I1, J1, K1, A2, C2, D2, E2, F2, G2, H2, I2, K2, A3, D3, E3, F3, G3, H3, K3, A4, E4, F4, G4, K4, A5, K5, A6, K6, A7, K7, A8, B8, J8, K8, A9, B9, C9, I9, J9, K9, A10, B10, C10, D10, H10, I10, J10, K10, A11, B11, C11, D11, E11, F11, G11, H11, I11, J11, K11

B3, C3, I3, J3, B4, D4, H4, J4, B5, C5, D5, E5, F5, G5, H5, I5, J5, B6, C6, D6, E6, F6, G6, H6, I6, J6, C7, E7, F7, G7, I7, D8, E8, F8, G8, H8, E9, F9, G9

C4, I4, B7, J7, C8, I8, D9, H9, E10, G10

B2, J2, D7, H7, F10

	A	B	C	D	E	F	G	H	I	J	K
1											
2											
3											
4											
5											
6											
7											
8											
9											
10											
11											

Practice DAY 140

Numbers can be split apart with addition and multiplication. Can you find different ways to split these numbers apart? Fill in the blanks.

1. 6 = ___ + ___
2. 6 = ___ + ___
3. 18 = ___ x ___
4. 18 = ___ x ___
5. 16 = ___ + ___
6. 16 = ___ + ___
7. 24 = ___ x ___
8. 24 = ___ x ___

> Even to your old age, I am He,
> And even to gray hairs I will carry you!
> I have made, and I will bear;
> Even I will carry, and will deliver you.
> (Isaiah 46:4)

Thank You, God, that You will carry me through math and all of life!

Practice these multiplication exercises.

9.

	1	3	3
x			2

10.

	3	2	1
x			4

11.

Carry				
		1	1	2
	x			5

12.

Carry				
		1	1	3
	x			7

Practice long addition. Notice that, when a column adds up to a number in the 20s, you will need to carry a 2.

13.

Carry		1	2	
		3	3	8
		2	4	5
	+	4	5	7

14.

Carry		2	2	
		4	3	9
		1	9	7
	+	3	7	6

15.

Carry				
		1	5	1
		3	6	4
	+	3	4	5

Let's see if you know the spreadsheet formulas for adding up the numbers in columns B, C, and D. What formula would you write in the blue cells? The first one is provided for you.

B11=sum(b1:b10)

16. C11=______________

17. D11=______________

	A	B	C	D
1		6	10	120
2		8	15	140
3		10	20	160
4		12	25	180
5		14	30	200
6		16	35	220
7		18	40	240
8		20	45	260
9		22	50	280
10		24	55	300
11	**Sum Total**			

Comparing Fractions

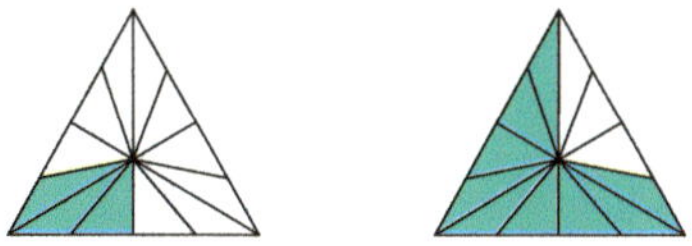

$\frac{3}{12}$ < $\frac{9}{12}$

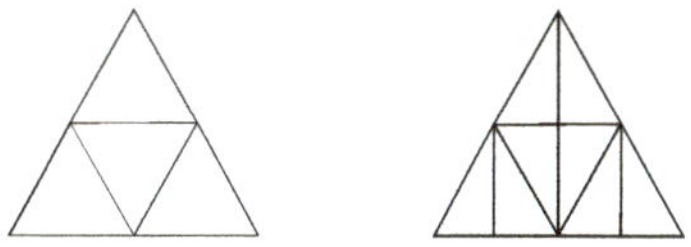

18. $\frac{3}{4}$ ◯ $\frac{5}{8}$

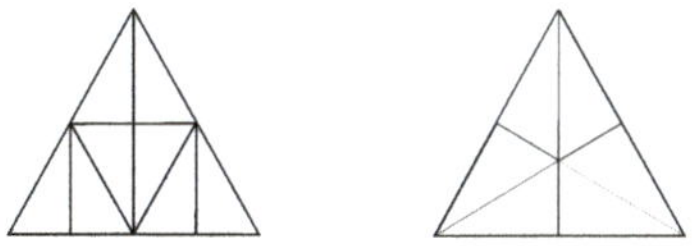

19. $\frac{3}{8}$ ◯ $\frac{3}{6}$

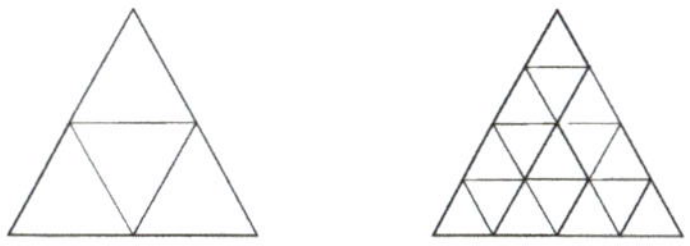

20. $\frac{2}{4}$ ◯ $\frac{12}{16}$

+ ADD or SUBTRACT? −

21. 78 ☐ 10 = 68
22. 55 ☐ 5 = 50
23. 54 ☐ 5 = 59
24. 60 ☐ 20 = 80
25. 97 ☐ 7 = 90
26. 31 ☐ 8 = 39
27. 82 ☐ 8 = 90
28. 45 ☐ 5 = 50
29. 74 ☐ 10 = 64
30. 21 ☐ 9 = 30

DAY 141 Using Math to Study God's World

We learn math because God gave it to us. He gave us math as a useful tool for many things. Today let's use math to help us with science—studying God's creation.

You've already learned about volume. You've also learned about weight. We use science to compare one thing about God's creation with another. We can compare the volume of something to its weight. Some large volumes have small weights, and some small volumes have large weights.

This bag of rocks weighs 3 kilograms. But the bag of cotton balls weighs only about 30 grams. The rocks weigh about 100 times more than the bag of cotton balls. But both bags have the same volume.

Compare a rock to a bag of cotton balls or whipped cream. A small rock can be heavy, and a big glob of whipped cream or a bag of cotton balls can be light. We use the word *density* to describe one difference between a rock and whipped cream. We can think of density as how packed something is. The whipped cream has a low density. It has been whipped and is not very packed. A rock has a high density. Its minerals are packed closely together.

A certain volume of whipped cream or feathers would weigh less than the same volume of rocks. Scientists have a math formula for talking about density. **Density** is the weight of something divided by its volume. (Scientists usually use the word *mass* instead of weight, but we are going to stick with the word *weight*.)

$$\text{Density} = \text{Weight} \div \text{Volume}$$

Activity

Now, let's try using math with a science-discovery lesson in the kitchen. We will find the density of different substances.

Supplies you will need:

1. One cup of five of the following:
    - Cookie dough
    - Ice Cream
    - Mud
    - Sand
    - Flour
    - Vegetable Oil
    - Sugar
    - Milk
    - Whipped Cream
    - Cotton Balls
    - Rocks
    - Ice
    - Bananas
    - Popped Popcorn
2. A 1-cup measuring cup
3. A calculator
4. A kitchen scale that weighs in grams

Let's explore the density of five different substances. Do the steps below for each substance.

Step 1 — Weigh the empty measuring cup.

Weight of Cup = __________ grams

Step 2 — Fill the cup to the top with the substance. Do not pack it down. (To weigh ice, first weigh the empty cup. Then fill it with water and freeze it before weighing it again.)

Step 3 — Weigh the cup with the substance in it.

Weight of cup + substance = _______________ grams

Subtract the weight of the empty cup from the weight of the full cup.

Full cup – Empty cup = ____________ grams

This is the weight of the substance only. Record this weight on the chart below.

Step 4 — The volume of one cup is 237 ml. Using a calculator, divide the weight of each substance by 237 to find its density.

Weight ÷ 237 = Density

SUBSTANCE	WEIGHT	DENSITY

Step 5. Study your chart. Which substance has the highest density? _____________________

Which substance has the lowest density? ________________________________

Practice DAY 142

Fill in the blanks for these equations using your memory. Watch for the operation signs:

1. 4 x _____ = 32
2. 24 ÷ _____ = 12
3. 40 ÷ _____ = 10
4. _____ x 5 = 30
5. 10 + _____ = 22
6. 6 + _____ = 18
7. _____ − 4 = 14
8. _____ ÷ 2 = 3
9. _____ + 12 = 26
10. _____ ÷ 4 = 4

Try these long division exercises. The first one has been started for you.

11.

	1	
4)	6	4
−	4	
	2	4
−		

12.

3)	3	3
−		
−		

13.

4)	4	4
−		
−		

14.

4)	8	4
−		
−		

Everyday Math

15. New hymnals cost $30 each. Your church would like to buy 100 new hymnals. How much will the church need to spend?

16. Pretend that your family starts a window washing business, and that your dad will pay you $8 for every window you wash. How many windows will you need to wash to buy a $50 gift for your mom? How much money will you have left over?

________ ÷ ________ = ________

____

17. Lets pretend you ran a lemonade stand at the fair. You made $26.50 on the first day it opened. If you paid your mom back $7 for ingredients, how much money did you have left as profit that day?

–			

18. Suppose that your family car needs 4 new tires. If the best tires for the price cost $165 each, how much will your family pay for all 4 tires?

	1	6	5
x			4

19. **Extra challenge**: Imagine that a lot of people at your church signed up to provide 23 dinners for a family whose mother is in the hospital. Divide to see how many weeks the family will receive meals.

____

After the last full week of dinners, how many more days of meals remain? ________

Charting Progress DAY 143

A trend is the general way something is going. For example, even if you get weak after having the flu, your trend is to get stronger as you are growing up. Today, we will be charting trends using a line graph. Line graphs are graphs where the coordinates are connected with a line.

Trend graphs can help you see how things change or how you are progressing. Let's pretend that you and a friend are getting ready to compete in a 100 meter race. You both work out every day except Sunday. On Saturdays, you time yourselves running the 100 meter race. Here is the data you have collected.

100 METER RUNNING TIME (IN SECONDS)

	YOU	YOUR FRIEND
Week 1	27	28
Week 2	25	26
Week 3	24	25
Week 4	21	22
Week 5	20	20
Week 6	19	19
Week 7	18	19
Week 8	17	18

Trend graphs really do help you see your progress. Let's put these running times on a trend graph using dots.

Across the bottom of the graph are the weeks of your timed runs. (The horizontal line of a graph is usually called the *x-axis*.)

Up the side of the graph is shown the possible number of seconds for the running times from 0 to 30. (The vertical line of a graph is usually called the *y-axis*.)

The blue dots mark the coordinates for your time each week. Drawing the dots on a graph is usually called *plotting the points*. Now draw a line to connect the blue dots in the chart. Next, plot the points of your friend's coordinates with red dots. Connect the red dots.

Look at the trend graph. You can see that your best improvement occurred between week 3 and week 4. That's the steepest part of the chart. When was your friend's best improvement?

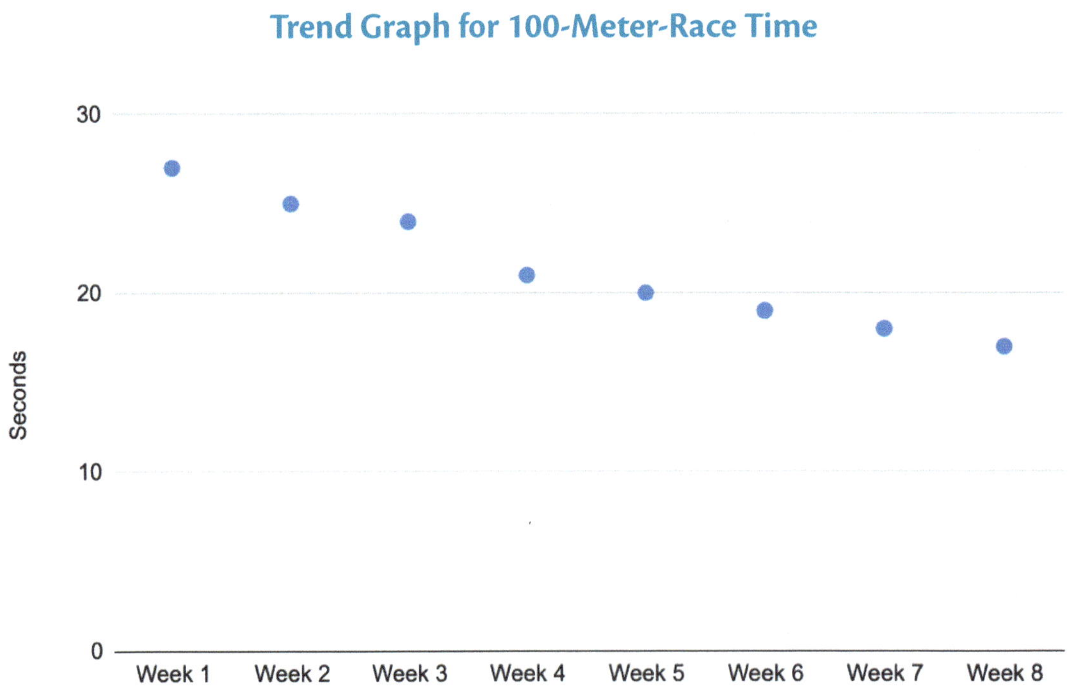

Student Exercises

Complete these addition and subtraction exercises. You will need to borrow and carry for each of them.

1.

Carry				
		5	3	5
	+	3	7	6

2.

Carry				
		6	4	9
	+	3	9	3

3.

Borrow				
		6	4	2
	–	3	4	7

4.

Borrow				
		6	2	4
	–	5	5	4

5. Let's make a trend graph showing the weight of boys and girls as they grow. We'll get our information from the table on the next page. The table shows the approximate weight of boys and girls at different ages. The information for boys has already been shown on the line graph as blue dots at the correct coordinates. Go ahead and connect the blue dots, making a line. Then, use a pink marker to plot the points for the girls' weight coordinates. Connect the pink dots.

TABLE OF APPROXIMATE WEIGHT OF BOYS AND GIRLS FROM BIRTH TO AGE 8

	BOY (POUNDS)	GIRL (POUNDS)
Birth	8	8
Year 1	21	19
Year 2	28	27
Year 3	32	31
Year 4	36	35
Year 5	41	40
Year 6	46	45
Year 7	51	50
Year 8	57	57

Trend Graph for Approximate Weight of Boys and Girls

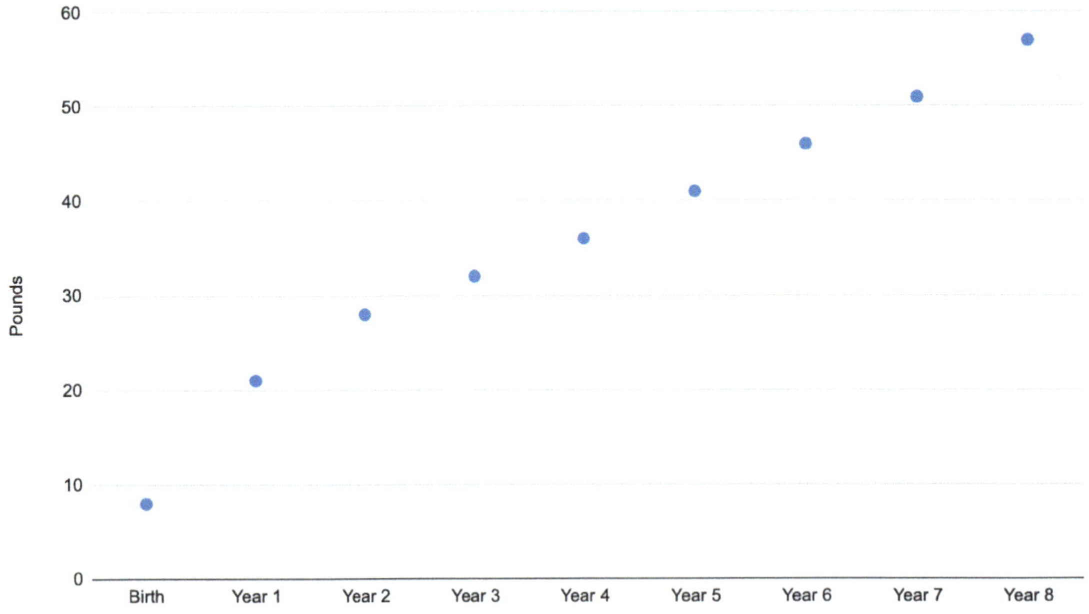

Between what two ages does a child gain the most weight? Hint: Look for the steepest part of the trend line. ____________ and ____________

At what ages are boys usually heavier than girls? ____________________

At what ages do they weigh about the same? ____________________

Fill in the correct fraction as shown by the purple squares.

6.

7.

Activity

Help the koala get to the eucalytus. To make the path, figure out the answer to each equation. Then find that number in the grid and color its square yellow.

1. 11+4= ◯
2. 12+7= ◯
3. 18+6= ◯
4. 19+9= ◯
5. 21+9= ◯
6. 25+8= ◯
7. 26+5= ◯
8. 28+7= ◯
9. 33+7= ◯
10. 34+9= ◯
11. 35+3= ◯
12. 37+7= ◯
13. 41+9= ◯
14. 43+6= ◯
15. 46+9= ◯
16. 48+5= ◯
17. 54+9= ◯
18. 51+8= ◯
19. 61+5= ◯
20. 64+5= ◯

			18	20	17	30	33	31	12	89	73
		→	15	19	24	28	75	35	52	72	67
			25	42	22	38	43	40	50	93	57
20	27	57	36	26	50	44	66	69	→		
10	25	16	19	32	49	32	59	60			
89	12	67	42	19	55	53	63	68			

DAY 144 Practice

This lesson integrates math into everyday life. With this activity, the student will apply God's patterns and wisdom in meaningful ways to home and community. Take a break from memory work and academic exercises to make math part of everyday life. The following are suggestions or examples. Other ideas may be added to the list. .

1. Use math to conduct your own scientific discoveries. Let's see how much dirt your family gets on the floor in one day. Sweep up as much of the house as you can with a broom and dustpan. Then wait a day before sweeping again. On that day:

Weigh a piece of paper. _______________

Sweep up the one day's amount of dirt and put it on the paper. Weigh the paper with its dirt. ____________________

Subtract out the weight of the paper.

Weight of dirt only: __________

Now that you have this information, how much dirt would you expect to sweep up in a week? (Multiply the weight of one day's dirt by 7.) Weight of dirt after a week _______

How much dirt would you expect to sweep up in 4 weeks? (Multiply the weight of one week's dirt by 4.) Weight of dirt after 4 weeks: ________

How often do you think you should sweep the floors in your house? ________________

2. Do you ever try to be more efficient when you clean your room? To be efficient means to do the most amount of work with the least amount of effort. It means doing something well, but doing it without wasting time or money.

Let's find out how efficiently your family car runs. Cars that are not efficient are expensive to drive. They use more gas. Efficient cars can drive more miles for every gallon of gas (or more kilometers per liter). Cars that are less efficient get fewer miles per gallon (or fewer km per liter). They will require more gas to go the same distance.

Step 1 — You can find out how many miles your car has been driven on one gallon of gas. This is how we measure its efficiency. First, fill the gas tank all the way up. Before you drive away from the gas station, have your parent help you write down the car's mileage. Mileage: __________

Step 2 — The next time your car needs gas:

Look at the pump to see how much gas it took to fill the tank all the way up again. This is how much gas your car has used since its last fill up.

Amount of gas to fill the tank again ______________________________

Write down the car's mileage again. Mileage: __________

Step 3 — Subtract to find the number of miles your car has been driven since the last fill up.

Second mileage – First mileage = Number of miles driven on the amount of gas

________________ – ________________ = ___________________

Step 4 — Divide the number of miles driven by the amount of gas the car used. You may use a calculator.

Miles driven ÷ Gallons of gas added to tank = Efficiency (miles driven on one gallon of gas)

Your car's efficiency = ____________ miles per gallon

Note: You can do the same study using kilometers per liter.

Extra Challenge

Now, let's look at how many gallons of gas your car uses to go 200 miles and 400 miles. Use the formula below and a calculator to find the answers.

200 miles ÷ Your car's miles per gallon = _________ gallons

400 miles ÷ Your car's miles per gallon = _________ gallons

Suppose that your car is usually driven about 200 miles per month. Let's see how much you would pay for gas in a month. Look at a gas station's sign to see how much gas costs per gallon. Use this formula:

Gallons used (from above) x Price Per Gallon (from the sign) = $ _____________

3. Do a real-life trend graph for some exercises you are working on. Maybe make a table for the number of pushups or situps you do for a week. Then make a graph to plot the points. Connect the points with a line.

You are discovering more of God's world. God's wisdom is sometimes hidden in His world. But God wants us to discover it and be amazed. That's what Proverbs 25:2 tells us:

> *It is* the glory of God to conceal a matter,
> But the glory of kings *is* to search out a matter.

Answer Key

Chapter 1—Day 1

1. completed
2. 8÷1=1=8
3. 7x2=9=14
4. 6÷2=12=3
5. 10-1=11=9
6. 12-4=16=8
7. 4x2=6=8
8. 10÷1=11=10
9. 4x2=8
10. 8x2=16
11. 4x4=16
12. 6x6=36
13. 5x5=25
14. 6x4=24
15. 3x7=21
16. 1x10=10

Chapter 1—Day 2

1. 20
2. 30
3. 100
4. 100
5. 300
6. 500
7. 1,234,567
8. 10,000
9. 100,000
10. 1,000,000
    Million (1,000,000): 1
    Hundred Thousand (100,000): 2
    Ten-Thousand (10,000): 3
    Thousand (1,000): 4
    Hundred (100): 5
    Ten (10): 6
    One (1): 7

Student Exercises

		1,000,000	100,000	10,000	1,000	100	10	1
1.	1,234,567	1	2	3	4	5	6	7
2.	32,431			3	2	4	3	1
3.	145,943		1	4	5	9	4	3
4.	95,703			9	5	7	0	3
5.	999,999		9	9	9	9	9	9
6.	1,999,991	1	9	9	9	9	9	1
7.	458,392		4	5	8	3	9	2
8.	1,010,010	1	0	1	0	0	1	0
9.	9,999,999	9	9	9	9	9	9	9
10.	8,275,384	8	2	7	5	3	8	4

11. 32
12. 77
13. 77
14. 89
15. 46
16. 89
17. 66
18. 89
19. completed
20. orange square, orange triangle, blue circle
21. yellow star, red star
22. yellow square, pink square, green star, pink square, yellow star, green circle
23. pink circle, pink star, pink diamond, blue diamond, blue circle, blue star
24. 1, 2, 3, 4
25. 1, 2, 3
26. 1, 2

Chapter 1—Day 3

1. 5+9=14
2. 23+46=69
3. 321+542=863
4. 111+181=292
5. 4+8=12
6. 421+511=932
7. 217+121= 338

8. 16-8=8
9. 45-21=24
10. 57-14=43
11. 397-101=296
12. 17-5=12
13. 45-14=31
14. 57-17=40
15. 852-342= 50
16. 9
17. 6
18. 14
19. 17
20. 4
21. 2
22. 6
23. 4

Chapter 1—Day 4

1. completed
2. 12
3. 17
4. 18
5. 33
6. 24
7. 18
8. 28
9. 27
10. 15
11. 6÷3=2
12. 9÷3=3
13. 20÷2=10
14. 16÷4=4
15. 30÷10=3
16. 15÷3=5
17. 15÷5=3
18. 18÷6=3

Chapter 1—Day 5

1. 35+22+14=71
2. 17+23+57=97
3. 68+21+35=124

Student Exercises

1. 51+35+99=185
2. 25+37+41=103
3. 82+33+26=141
4. 13+67+19=99
5. 14+26+15=55
6. 24+41+3=68
7. 72+52+48=172
8. 65+36+11=112
9. empty
10. orange triangle
11. red star, yellow star
12. empty
13. green star
14. empty
15. 2
16. 1, 2

Chapter 1—Day 6

1. 26,783
   10s: 8
   1,000s: 6
   10,000s: 2
2. 3,152,345
   1s: 5
   100s: 3
   100,000s: 1
3. 708,310,743
   100s: 7
   1,000s: 0
   1,000,000s: 8
4. 35,781
   10s: 8
   1,000s: 5
   10,000s: 3
5. 56,921
   10s: 2
   100s: 9
   10,000s: 5
6. 125,690
   1s: 0
   1,000s: 5
   10,000s: 2
7. 347,552
   10s: 5
   10,000s: 4
   1,000,000s: none
8. 1,067,254
   1s: 4
   100s: 2
   10,000s: 6
9. 3×4=12
10. 6×1= 6
11. 2×5=10
12. 7×3=21
13. 7×8=56

14. 5×6=30
15. 8×5=40
16. 8×9=72

Chapter 1—Day 7

1. 60
2. 100
3. 40
4. 100
5. 500
6. 1,000
7. 600
8. 1,000
9. 70,000
10. 100,000

Extra Challenge

11. 950,720
12. 950,700
13. 951,000
14. 950,000
15. 1,000,000
16. 1,000,000

Student Exercises

1. ≤
2. ≥
3. ≤
4. ≤
5. ≤
6. ≤
7. ≥
8. ≤
9. ≤
10. ≥
11. ≥
12. ≥

Chapter 1—Day 8

1. 3 × 2 = 6
2. 6×2=12
3. 4×4=16
4. 5×3=15
5. 2×4=8
6. 6×3=18
7. 1×11=11
8. 3×1=3
9. 4×3=12
10. 6×8=48
11. 5×7=35
12. 6×6=36
13. 10×9=90
14. 7×5=35
15. 8×8=64
16. 4×9=36
17. 6×4=24

Everyday Math

18. Answers will vary
19. 1x7=7, 3x7=21
20. 2x8=16
21. Answers will vary
22. 3x2=6
23. Answers will vary, 60 degrees
24. Answers will vary, 130 degrees
25. Answers will vary. 3 in
26. Answers will vary. 2+1+2+1=6
27. Answers will vary. 2+2+2+2=8

Chapter 1—Day 9

1. 524+398=922
2. 658+285=943
3. 453+547=1,000

Student Exercises

1. 5+5=10
2. 7+7=14
3. 9+9=18
4. 11+11=22
5. 6+6=12
6. 8+8=16
7. 10+10=20
8. 12+12=24
9. 24–12=12
10. 20–10=10
11. 16 –8=8
12. 12–6=6
13. 22–11=11
14. 18–9=9
15. 14–7=7
16. 9
17. 5
18. 15
19. 12
20. 5
21. 1
22. 12
23. 12

Chapter 1—Day 10—Practice

1. 17+23+57=97
2. 28+61+55=144
3. 35+24+14=73
4. 72+47+12=131
5. 68+21+35=124
6. 79+13+32=124
7. 329,752,038
   10s: 3
   1,000s: 2
   10,000s: 5
8. 2,684
   1s: 4
   100s: 6
   1,000s: 2
9. 2,973,453
   100s: 4
   1,000s: 3
   1,000,000s: 2
10. Answers will vary, 40 degrees
11. Answers will vary, 80 degrees
12. Answers will vary, 110 degrees
13. Answers will vary, 3 in
14. Answers will vary, 3x2=6

Chapter 1—Day 11

Student Exercises

1. 365,400
2. 367,000
3. 400,000
4. 1,346,000
5. 1,300,000
6. 1,000,000
7. 7+14=21
8. 5+10=15
9. 3+6=9
10. 1+2=3
11. 6+12=18
12. 4+8=12
13. 2+4=6
14. 0+0=0
15. 20-9=11
16. 18-9=9
17. 16-9=7
18. 14-9=5
19. 19-9=10
20. 17-9=8
21. 15-9=6
22. 13-9=4

Chapter 1—Day 12—Practice

1. 6÷2=3
2. 10÷2=5
3. 6÷3=2
4. 12÷3=4
5. 8÷2=4
6. 12÷2=6
7. 9÷3=3
8. 15÷3=5
9. 16÷4=4
10. 42÷6=7
11. 50÷10=5
12. 24÷6=4
13. 24÷8=3
14. 42÷7=6
15. 50÷5=10
16. 24÷4=6
17. 346,859
   10s: 5
   1,000s: 6
   10,000s: 4
18. 28,091,645
   1s: 5
   100s: 6
   100,000s: none

Chapter 1—Day 13

Pg 34:
643-237=406
425-247=178

Student Exercises

1. 67-19=48
2. 945-656=289
3. 523-358=165
4. 3,748,600
5. 3,750,000
6. 3,700,000
7. 56,000
8. 100,000
9. 0
10. 7x4=28
11. 6x2=12
12. 3x9=27
13. 5x5=25

Everyday Math

14. 45+52+48=145
15. 68+77+89=234
16. 20+35+46+52=153

Chapter 1— Day 14—Practice

Student Exercises

1. eagle
2. hummingbird
3. stars
4. woman
5. fish
6. gecko
7. 265
8. 322
9. 605
10. 220
11. 99
12. 514
13. 581
14. 486
15. 322

Everyday Math

16. 4 more miles
16-12=4
17. 18 total pieces of candy
6x3=18
18. The perimeter is 6 ft
2+1+2+1=6

Chapter 1—Day 15

Answers will vary.

Chapter 2—Day 16

Student Exercises

Pg. 46: Answers will vary, check work.

1. 20, 22, 24, 26, 28, 30
2. 30, 33, 36, 39, 42, 45
3. 50, 55, 60, 65, 70, 75
4. 40, 48, 56, 64, 72, 80
5. 100, 110, 120, 130, 140, 150

	Number	1,000,000	100,000	10,000	1,000	100	10	1
6.	123,456		1	2	3	4	5	6
7.	7,654,321	7	6	5	4	3	2	1
8.	234,987		2	3	4	9	8	7
9.	99,999			9	9	9	9	9
10.	100,004		1	0	0	0	0	4
11.	404,040		4	0	4	0	4	0

12. 3x5=15
13. 6x3=18
14. 4x7=28
15. 9x2=18
16. 4x8=32
17. 5x7=35
18. 3x8=24
19. 7x4=28

Chapter 2—Day 17—Practice

1. 3×4=12
2. 6×1=6
3. 5×4=20
4. 1×5=5
5. 2×5=10
6. 7×3=21
7. 1×2=12
8. 2×10=20
9. 7×8=56
10. 5×6=30
11. 7×6=42
12. 10×7=70
13. 8×5=40
14. 8×9=72
15. 3×9=27
16. 6×6=36
17. 7,638,051
    10s: 5
    1,000s: 8
    10,000s: 3
18. 904,735,234
    100s: 2
    1,000s: 5
    1,000,000s: 4
19. 4,729
    1s: 9
    100s: 7
    100,000s: 0
20. 567,208
    1s: 8
    10s: 0
    10,000s: 6

Chapter 2—Day 18—Activity

Answers will vary.

Chapter 2—Day 19—Practice

1. Answers will vary
2. 120 degrees

3. Answers will vary
4. 175 degrees
5. Answers will vary, 4 inches
6. Answers will vary, 5.5 inches
7. 3x2=6
8. 5x1=5
9. 4x3=12
10. 5x2=10
11. 3x2=6
12. 3,152,345
    10s: 4
    1,000s: 2
    10,000s: 5
13. 708,310,743
    1s: 3
    100s: 7
    100,000s: 3
14. 68,425
    100s: 4
    1,000s: 8
    1,000,000s: 0
15. 324+146=470
16. 452-134=318
17. 535+166=701
18. 631-253=378
19. 616+187=803
20. 154-35=119

Chapter 2—Day 20

1. pink diamond
2. 2
3.
4. 
5. gray circle
6. orange half circle
7. 6, 9, 12, 15, 18
8. 10, 15, 20, 25, 30

Student Exercises

1. 20, 18, 16, 14, 12, 10
2. 30, 27, 24, 21, 18, 15
3. 50, 45, 40, 35, 30, 25
4. 72, 64, 56, 48, 40, 32
5. 100, 90, 80, 70, 60, 50
6. 8
7. 8
8. 1
9. 17
10. 9
11. 3
12. 12
13. 18

Chapter 2—Day 21—Practice

1. 132+547=679
2. 317+158=475
3. 257+434=691
4. 763-354+409
5. 532-159=373
6. 484+446=930
7. 394+199=593
8. 154+54=208
9. 175-86=89
10. 852-146=706
11. 400+599=999
12. 835+67=902
13. 249+31=280
14. 426-31=395
15. 645-256=389

Everyday Math

16. 212+72=284
17. 35+46+39+28=148, 165-148=17
18. 26+37+55=118
19. 106-79=27
20. 135-86=49, 49-22=27
21. 16
22. 8
23. 4
24. 4
25. 4
26. 2

Chapter 2—Day 22

1. 
2. 
3. 
4. 
5.

Student Exercises

Positions: 6
1. 1:40
2. 1:50
3. 2:00

Positions: 4
4. 1:45

5. 2:00
Positions: 12
6. 1:25
7. 1:30
8. 1:35
9. 1:40
10. 1:45
11. 1:50
12. 1:55
13. 2:00
14. 8÷4=2
15. 16÷4=4
16. 5÷5=1
17. 15÷5=3
18. 12÷4=3
19. 20÷4=5
20. 10÷5=2
21. 20÷5=4
22. 36÷6=6
23. 36÷1=36
24. 45÷5=9
25. 32÷4=8
26. 36÷4=9
27. 36÷9=4
28. 45÷9=5
29. 32÷8=4

5	3	4	6	7	8	9	1	2
6	7	2	1	9	5	3	4	8
1	9	8	3	4	2	5	6	7
8	5	9	7	6	1	4	2	3
4	2	6	8	5	3	7	9	1
7	1	3	9	2	4	8	5	6
9	6	1	5	3	7	2	8	4
2	8	7	4	1	9	6	3	5
3	4	5	2	8	6	1	7	9

Chapter 2—Day 23—Practice

1. 835+227=1,062
2. 647+153=800
3. 280+530=810
4. 912+189=1,101
5. 824-539=285
6. 431-57=374
7. 783-325=458
8. 576-172=404
9. 8x4=32
10. 12x2=24
11. 4x4=16
12. 2x9=18
13. 5x5=25
14. 0x3=0
15. 5x3=15
16. 8x6=48
17. 108-35=73
18. 48-26=22
19. 33+30+31=94
20. 115-40=75

Chapter 2—Day 24

Student Exercises

1. 9
2. 3
3. 16
4. 15
5. 5
6. 2
7. 20
8. 20
9. Answers will vary, 30 degrees
10. Answers will vary, 90 degrees
11. Answers will vary, 120 degrees
12. Answers will vary, 13 cm
13. Answers will vary, 15 cm
14. 5x7=45 cm
15. 2x2=4 in
16. 5x12=60

Chapter 2—Day 25

1. 15, 18, 21, 24, 27, 30, 33, 36
2. 12, 18, 24, 30, 36, 42, 48, 54
3. 0, 9, 18, 27, 36, 45, 54, 63
4. 6+8=14
5. 8+6=14
6. 7+7=14
7. 8+8=16
8. 8+3=11
9. 3+8=11
10. 15,334+22,749=38,083
11.14-7=7
12. 16-7=9
13. 18-7=11
14. 19-7=12
15. 20-7=13
16. 32-12=23
17. 34,565-13,431=21,134

Chapter 2—Day 26

Answers will vary.

Chapter 3—Day 27

1. 9+4=13
2. 13+7=20
3. 8x4=32
4. 10x5=50
5. 6-2=4
6. 11-4=7
7. 10÷2=5
8. 12÷4=3
9. 8+6=14
10. 12+3=15
11. 16÷4=4
12. 25÷5=5

1. 212
2. 312

Activity

1. 510
2. 12
3. 46
4. Answers will vary
5. Answers will vary
6. Answers will vary, eg. 44
7. Answers will vary, eg. 63
8. Answers will vary, eg. 24

Student Exercises

1-6. Answers will vary, check work.

Chapter 3—Day 28—Practice

1. 5+3=8
2. 13-5=8
3. 5x4=20
4. 56÷8=7
5. 12+7=19
6. 16-4=12
7. 6x7=42
8. 12÷4=3
9. 4+13=17
10. 10-3=7
11. 10x10=100
12. 50÷5=10
13. $\frac{2}{5}$
14. $\frac{8}{12}$
15. $\frac{5}{9}$
16. $1\frac{3}{4}$
17. $1\frac{7}{12}$
18. $2\frac{1}{12}$
19. ≥
20. ≤
21. ≥
22. ≤
23. ≤
24. ≥
25. ≥
26. ≤
27. ≥
28. ≥

Chapter 3—Day 29—Practice

1. 7+4=11
2. 12-5=7
3. 5x6=30
4. 42÷7=6
5. 15+2=17
6. 10-3=7
7. 7x8=56
8. 16÷4=4
9. 6+12=18
10. 9-4=5
11. 7x10=70
12. 30÷10=3
13. 50, 52, 54, 56, 58, 60
14. 40, 44, 48, 52, 56, 60
15. 0, 6, 12, 18, 24, 30
16. 0, 8, 16, 24, 32, 40
17. 40, 48, 56, 64, 72, 80
18. 
19. 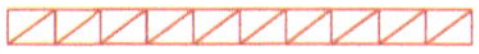
20. 
21. 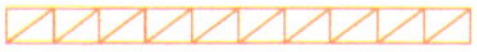
22. 
23. 4,248,900
24. 4,248,900
25. 4,248,900
26. 4,248,900
27. 7,900,000
28. 8,000,000

Chapter 3—Day 30

1. 1
2. 3
3. 4

Student Exercises

1. 305+210+134=649

2. 111+460+328=899
3. 430+156+193=779
4. 444+333+222+111=1,110
5. 1590-390=1,201
6. 1459-258=1,201
7. 4132-126=4,006
8. 2999-1998=1,001
9. 7131+386=7,517
10. 3095+477=3,572
11. 5964-956=5,008
12. 8663-427=8,236
13. 212 , 52 , 5÷2
14. 412 , 92 , 9÷2
15. 312 , 72 , 7÷2
16. 712 , 152 , 15÷2
17. 113 , 43 , 4÷3
18. 213 , 73 , 7÷3
19. 912 , 192 , 19÷2

Chapter 3—Day 31

1. 4+4=8
2. 3+12=15
3. 7x4=28
4. 5x3=15
5. 9-2=7
6. 8-5=3
7. 12÷3=4
8. 81÷9=9
9. 9+6=15
10. 12+6=18
11. 70÷7=10
12. 45÷9=5

Pg. 95. $\frac{5}{6} - \frac{2}{5} = \frac{3}{5}$

1. $\frac{9}{5}$
2. $\frac{7}{10}$
3. $\frac{3}{3}$
4. $\frac{4}{4}$
5. $\frac{1}{2}$
6. $\frac{2}{7}$
7. 1
8. 0
9. $\frac{0}{10}$, 0
10. $\frac{9}{9}$, 1
11. $\frac{0}{5}$, 0
12. $\frac{10}{10}$, 1

13-18. Circle exercises #15, #17, and #18

Student Exercises

1. $\frac{8}{12}$
2. $\frac{6}{4}$ or $1\frac{2}{4}$
3. $\frac{6}{12} + \frac{3}{12} = \frac{9}{12}$
4. $\frac{3}{4} - \frac{1}{4} = \frac{2}{4}$
5. $\frac{2}{3} + \frac{2}{3} = \frac{4}{3}$
6. $3\frac{1}{2} + 2\frac{1}{2} = 5\frac{2}{2}$ or 6
7. 20, 18, 16, 14, 12, 10
8. 30, 27, 24, 21, 18, 15
9. 50, 45, 40, 35, 30, 25
10. 80, 72, 64, 56, 48, 40
11. 100, 90, 80, 70, 60, 50

Answers may vary, possible solutions are given.

12. 12=6+6
13. 12=7+5
14. 12=3x4
15. 12=2x6
16. 18=9+9
17. 18=10+8
18. 18=3x6
19. 18=2x9
20. 155+155=310
21. 173-134=39
22. 25+166=191
23. 87-59=28
24. 89+54=143
25. 165-156=9

Chapter 3—Day 32—Practice

1. 5x5=25
2. 3x6=18
3. 3x8=24
4. 9x7=63
5. 10x10=100
6. 10x10=100
7. 3x6=18
8. 5x5=25
9. 5x6=30
10. 9x8=72
11. 6,420,900
12. 6,420,000
13. 6,400,000
14. 1,854,000
15. 1,850,000
16. 1,800,000
17. $\frac{4}{5}$
18. $\frac{5}{7}$
19. $\frac{10}{6}$
20. $\frac{2}{8}$

21. $\frac{2}{2}$
22. $\frac{0}{5}$
23. $\frac{4}{6}$ → e
24. $\frac{1}{3}$ →c
25. $1\frac{3}{4}$ → b
26. $2\frac{1}{3}$ → a
27. $\frac{1}{6}$ → d

Chapter 3—Day 33

1. 5+3=8
2. 7+4=11
3. 9x4=36
4. 5x11=55
5. 6-3=3
6. 14-5=9
7. 24÷3=8
8. 12÷2=6
9. 2+15=17
10. 5+6=11
11. 81÷9=9
12. 42÷7=6

Pg. 105:

1. $\frac{1}{7}$
2. $\frac{4}{28}$
3. student should circle one apple
4. student should circle 2 pencils

Student Exercises

1. $\frac{1}{2}$
2. $\frac{1}{5}$
3. $\frac{1}{2}$
4. $\frac{1}{3}$
5. 463+616=1,079
6. 548-170=378
7. 561+72=633
8. 415+547=962
9. 895-183=712
10. 123-31=92

Chapter 3—Day 34—Practice

1. 8
2. 2
3. 4
4. 16
5. 4
6. 2
7. 6÷2=3
8. 10÷2=5
9. 6÷3=2
10. 12÷3=4
11. 8÷2=4
12. 12÷2=6
13. 9÷3=3
14. 15÷3=5
15. 42÷6=7
16. 50÷10=5
17. 24÷6=4
18. 24÷4=6
19. 42÷7=6
20. 50÷5=10
21. 24÷8=3
22. 24÷3=8

Chapter 3—Day 35

1. 2+16=18
2. 17+3=20
3. 4x8=32
4. 3x11=33
5. 13-3=10
6. 20-5=15
7. 36÷6=6
8. 16÷4=4
9. 6+11=17
10. 15+4=19
11. 100÷10=10
12. 18÷2=9

1. ≥
2. ≥
3. ≤
4. ≥
5. ≥
6. ≥
7. ≥
8. 4
9. 2

Student Exercises

1. 15, 18, 21, 24, 27, 30, 33, 36
2. 12, 18, 24, 30, 36, 42, 48, 54
3. 0, 9, 18, 27, 36, 45, 54, 63
4. 1, 2, 3, 4, 5, 6
5. 2, 3, 4, 5
6. 3, 5, 6, 11
7. 9
8. none
9. 4, 5

Extra Challenge

10. ≤

11. ≤
12. ≤

Chapter 3—Day 36—Practice

1. 13+4=17
2. 4+12=16
3. 3x8=24
4. 7x10=70
5. 18-3=15
6. 14-5=9
7. 49÷7=7
8. 21÷3=7
9. 5+8=13
10. 7+6=13
11. 50÷0=0
12. 24÷6=4
13. ≥
14. ≤
15. ≤
16. ≥
17. ≥
18. ≤
19. 
20. 
21.
22.
23.
24. 

Chapter 3—Day 37—Activity

Answers will vary based on hymns chosen.

Chapter 3—Day 38

Answers will vary, possible answers are provided.

1. $\frac{2}{9}$, $\frac{1}{3}$
2. $\frac{3}{5}$
3. $\frac{3}{12}$, $\frac{3}{6}$
4. $\frac{6}{7}$
5. $\frac{12}{16}$, $\frac{2}{4}$

Student Exercises

1. 12÷3=4
2. 15÷3=5
3. 9÷3=3
4. 16÷4=4
5. 10÷2=5
6. 20÷4=5
7. 30÷5=6
8. 18÷6=3
9. 49÷7=7
10. 48÷8=6
11. 42÷7=6
12. 64÷8=8
13. 81÷9=9
14. 36÷6=6
15. 100÷10=10
16. 25÷5=5

Chapter 3—Day 39—Practice

1. 7x8=56
2. 8x5=40
3. 5x6=30
4. 8x9=72
5. 7x6=42
6. 3x9=27
7. 10x7=70
8. 6x6=36
9. 525,700
10. 526,000
11. 530,000
12. 7,511,000
13. 7,500,000
14. 8,000,000
15. 111+222+333=666
16. 444+333+222=999
17. 1234+1111=2,345
18. 1000+2000+3000+4000=10,000
19. 1111+2222+3333=6,666
20. 987+111=1,098
21. 11+22+33=66
22. 321+432+543=1,296

Chapter 3—Day 40

1. a. 3 teaspoons
b. 1 ½ teaspoons
c. 4 tablespoons
2-3. Answers will vary.

Chapter 4—Day 41

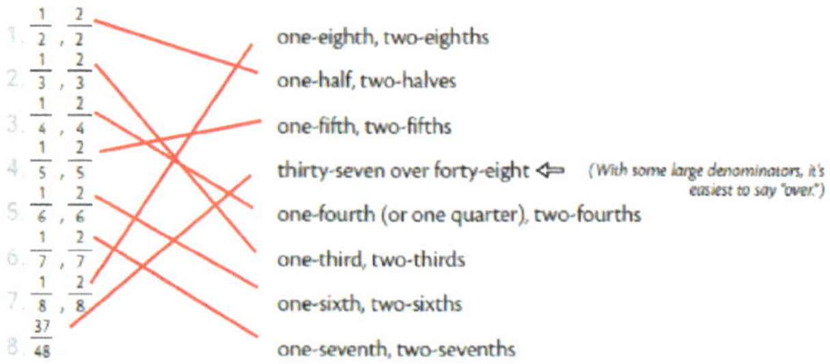

Chapter 4—Day 42

1. 3+8=11
2. 2+9=11
3. 8+3=11
4. 5+5=10
5. 7+8=15
6. 9+9=18
7. 0+9=9
8. 7+7=14
9. 16-7=9
10. 17-9=8
11. 15-7=8
12. 14-3=11
13. 14-4=10
14. 17-8=9
15. 14-6=8

Answers will vary, possible solutions are given.

1. 6=2+4
2. 10=7+3
3. 6=2x3
4. 10=2x5
5. 12=5+7
6. 15=3x5
7. 12=3x4
8. 15=3x5
9. $\frac{2}{5}$
10. $\frac{8}{12}$
11. $1\frac{3}{4}$
12. $\frac{5}{9}$
13. $2\ \frac{7}{12}$
14. $2\frac{1}{12}$
15. Answers will vary, 2.5 in
16. Answers will vary, 4 in
17. Answers will vary, 5 in
18. Answers will vary, 30 degrees
19. Answers will vary, 90 degrees
20. Answers will vary, 40 degrees
21. Answers will vary, 3x1=3

Chapter 4—Day 43

1. 8x2=16
2. 3x3=9
3. 3×4=12
4. 5×4=20
5. 2x6=12
6. 1×12 =12
7. 7×3=21
8. 10×1=10
9. 4×8=32
10. 8×8=64
11. 5×4=20
12. 7×9=63
13. 7×6=42
14. 3×6=18
15. 9×7=63
16. 10×6=60
17. 750, 1,000, 1,500
18. 1,250, 1,340, 1,520
19. 15,900, 16,020, 17,005
20. 145, 451, 541
21. 38+57=95
22. 47-21=26
23. 235-168=67
24. 57+18=75
25. 58-29=29
26. 147+122=269
27. 159-127=32

Chapter 4—Day 44

Pg. 138: $2\frac{2}{5}$

Student Exercises

1. 7 dots over 2 dots, circle 3 sets of 2 on top, 3 12
2. 11 dots over 2 dots, circle 5 sets of 2 on top, 512
3. 17 dots over 3 dots, circle 5 sets of 3 on top, 523
4. 4 dots over 3 dots, circle 1 set of 3 on top, 113
5. 6 dots over 4 dots, circle 1 set of 4 on top, 124
6. 14 dots over 10 dots, circle 1 set of 10 on top, 1410
7. 3 dots over 2 dots, circle 1 set of 2, 112
8. 11 dots over 2 dots, circle 5 sets of 2, 512
9. 8 dots over 3 dots, circle 2 sets of 3, 283
10. 5 dots over 3 dots, circle 1 set of 3, 123
11. 13 dots over 4 dots, circle 3 sets of 4, 314
12. 7 dots over 4 dots, circle 1 set of 3, 134
13. 30, 32, 34, 36, 38, 40
14. 24, 21, 18, 15, 12, 9
15. 50, 55, 60, 65, 70, 75
16. 16, 24, 32, 40, 48, 56
17. 150, 140, 130, 120, 110, 100
18. clock with hands at 12:30
19. clock with hands at 6:00
20. clock with hands at 6:45
21. Answers will vary
22. 
23. 
24.
25. 

26. 

27.

Chapter 4—Day 45—Practice

Answers will vary, possible solutions are given.

1. 4=2+2
2. 4=2×2
3. 9=4+5
4. 9=3×3
5. 12=7+5
6. 12=6+6
7. 12=3×4
8. 12=2×6
9. $1\frac{13}{4}$, 13 dots over 4 dots, circle 3 sets of 4, $3\frac{1}{4}$
10. $1\frac{11}{1}$, 11 dots over 7 dots, circle 1 set of 7, $1\frac{4}{7}$
11. $1\frac{11}{10}$, 11 dots over 10 dots, circle 1 set of 10 dots, $1\frac{1}{10}$
12. 55, 555, 5,555
13. 123, 132, 321
14. 100, 1,000, 10,000
15. 1,040, 1,140, 1,240

Everyday Math

16. Street 1

22+10=22 cookies

Street 2

10+16=26 cookies

Street 3

5+12=17 cookies

Street 4

28+18=46 cookies

17. Saturday 1

22+10+5+28=65 cookies

Saturday 2

10+16+12+18=56 cookies

18. 65-56=9 cookies

Chapter 4—Day 46

1. 10 dots over 3 dots, 3 1/3
2. 10 dots over 2 dots, 5
3. 10 dots over 4 dots, 2 2/4
4. 22 dots over 8 dots, 2 6/8
5. 24 dots over 6 dots, 4
6. 7 dots over 1 dot, 7
7. 7 dots over 5 dots, 1 2/5

Memory Drill

1. 5+6=11
2. 7+8=15
3. 7+5=12
4. 6+7=13
5. 6+8=14
6. 8+5=13
7. 9+6=15
8. 9+8=17
9. 18-8=10
10. 14-6=8
11. 18-6=12
12. 14-5=9

13 16-4=12

14. 17-6=11

15.16-9=7

Optional Exercise

12 in in a foot

$3\frac{4}{12}$

Chapter 4—Day 47—Practice

1. 10÷2=5
2. 14÷2=7
3. 6÷3=2
4. 15÷3=5
5. 12÷4=3
6. 4÷4=1
7. 15÷5=3
8. 20÷5=4
9. 30÷10=3
10. 45÷5=9
11. 32÷8=4
12. 49÷7=7
13. 36÷6=6
14. 56÷8=7
15. 36÷4=9
16. 60÷6=10
17. 50, 48, 46, 44, 42, 40
18. 0, 4, 8, 12, 16, 20
19. 20, 24, 28, 32, 36, 40
20. 6, 12, 18, 24, 30, 36
21. 36, 30, 24, 18, 12, 6
22. 7 dots over 4 dots, circle 1 set of four dots, $1\frac{2}{4}$
23. 10 dots over 3 dots, circle 3 sets of 3 dots, $3\frac{1}{3}$
24. 18 dots over 7 dots, circle 2 sets of 7 dots, $2\frac{4}{7}$
25. 21 dots over 3 dots, circle 7 sets of 3, 7
26. 5 dots over 2 dots, circle 2 sets of 4 dots, $2\frac{1}{2}$
27. 12 dots over 5 dots, circle 2 sets of 5 dots, $2\frac{2}{5}$
28. 8 dots over 1 dot, circle 8 sets of 2 dot, 8
29. 9 dots over 3 dots, circle 3 sets of 3, 3

Chapter 4—Day 48

Pg.151-152:
$1.80
$3.61

Student Exercises

1. $4.96
2. $5.96
3. $7.99
4. $9.98
5. $4.42
6. $9.90
7. Answers will vary, 2.5 inches
8. Answers will vary, 1 inch
9. Answers will vary, 60 degrees
10. Answers will vary, 120 degrees
11. Answers will vary, 2 (1x2=2)
12. Answers will vary, 6 (3x2=6)

Chapter 4—Day 49—Practice

1. 1x8=1=8
2. 3x9=12=27
3. correct
4. 10x4=14=40
5. 6x0=6=0
6. 16÷4=12=4
7. correct
8. 10÷5=50=2
9. 30÷10=3
10. 45÷5=9
11. 32÷8=4
12. 49÷7=7
13. 36÷6=6
14. 56÷8=7
15. 36÷4=9
16. 48÷4=12
17. 9
18. 3
19. 16
20.15
21. 5
22. 2
23. 20
24. 20

Chapter 4—Day 50

1. $62.50
2. $60.25
3. $33.00
4. $24.75
5. $2.75
6. $4.96
7. $6.50
8. $9.99
9. $4.79
10. $30.75
11. $90.75
12. $90.99
13. 1, 2, 3, 4, 5, 6
14. 2, 3, 4, 5
15. 3, 5, 7, 11
16. 9
17. none
18. 4, 5

Chapter 4—Day 51—Practice

1. 15+8=23
2. 4+12=16
3. 14+3=17
4. 5+7=12
5. 13-6=7
6. 14-9=5
7. 20-9=11
8. 11-2=9
9. 6x10=60
10. 5x12=60
11. 3x8=24
12. 4x7=28
13. 48÷6=8
14. 56÷7=8
15. 36÷3=12
16. 6÷6=1
17. $10.00
18. $6.60
19. $9.85
20. $3.20

Everyday Math

21. a. Yes
8.50+15.00=23.50
b. $3.50 leftover
$23.50-$20.00=3.50
22. $22.50 leftover
25.00-2.50=22.50
23. $14.94 total
9.99+4.95=14.95
24. $13.00 total
4.50+1.25+7.25=13.00

Chapter 4—Day 52

Answers will vary.

Chapter 5—Day 53

1. $\frac{1}{2}$
2. $\frac{3}{6}$
3. $\frac{4}{8}$
1. $2\frac{4}{10}$, 2.4
2. $3\frac{7}{10}$, 3.7
3. $5\frac{5}{10}$, 5.5
4. $\frac{2}{10}$, 0.2
1. 1.1
2. 9.2
3. 4.7
4. 5.5
5. 4.7
6. 10.3

Activity

1-3. Answers will vary.

Chapter 5—Day 54—Practice

1. 4, 5, 6, 7
2. x, y, w, z
3. 1, 2, 3, 4
4. 8, 12
5. empty
6. x, y, z
7. 4+9=13
8. 17-9=8
9. 15-7=8
10. 7+11=18
11. 9+9=18
12. 18-4=14
13. 7+8=15
14. 13-11=2
15. 15-7=8
16. 10+4=14
17. 6+9=15
18. 11-9=2
19. 7,638,051
    - 10s: 5
    - 1,000s: 8
    - 10,000s: 3
20. 4,729
    - 1s: 9
    - 100s: 7
    - 1,000s: 4
21. 904,735,234
    - 100s: 2
    - 1,000s: 5
    - 1,000,000s: 4
22. \$1.00+\$1.00+\$1.49=\$3.49
23. \$6.50+\$7.25=\$13.75
24. \$3.67+\$5.32=\$8.99

Chapter 5—Day 55

Activity

Answers will vary.

1. one full block colored in, 3 squares in next one
2. 2 full blocks colored in, 5 squares in last one
3. 4 full blocks colored in, 9 squares in last one
4. 40, 38, 36, 34, 32, 30
5. 30, 27, 24, 21, 18, 15
6. 100, 105, 110, 115, 120, 125
7. 0, 7, 14, 21, 28, 35
8. 0, 9, 18, 27, 36, 45
9. tens place: 5
10. tenths place: 9
11. millions place: 1
12. ten thousands places: 0

Chapter 5—Day 56—Practice

Student Exercises

1. 5x2=10
2. 15÷3=5
3. 12÷3=4
4. 8x3=24
5. 2x4=8
6. 20÷2=10
7. 10/5=2
8. 1x13=13
9. 4x4=16
10. 18/2=9

Answers may vary, possible solutions are given.

11. 8=4+4
12. 10=6+4
13. 8=2×4
14. 10=2×5
15. 16=8+8
16. 20=10+10
17. 16=4×4
18. 20=4×5

Everyday Math

19. 2 pieces of pizza with 2 leftover, 14

20. $27.45
$12.95+$14.50=$27.45
21. $27.00
3x9=27
22. 21 days
3x7=21

Chapter 5—Day 57

$\frac{1}{10}$

1. fill one box completely, fill in 76 squares in second box
2. fill two boxes completely, fill in 38 squares in last box
3. fill three boxes completely, fill in 49 in last box

Activity

Answers will vary.

Student Exercises

1.

2.

3.

4.

5.

6.

7. (x)
8. (+)
9. (–)
10. (÷)
11. (+)
12. (÷)
13. (x)
14. (–)
15. $1.20
15.00-13.80=1.20

Chapter 5—Day 58—Practice

1. 763,805.1
   10ths: 1
   1,000s: 3
   10,000s: 6
2. 47.29
   100ths: 9
   1s: 7
   10th: 9
3. 904,735,234
   100s: 2
   1,000s: 5
   1,000,000s: 4
4. 3x10=30
5. 18÷3=6
6. 24÷3=8
7. 7x5=35
8. 42÷7=6
9. 63/7= 9
10. 8x7= 56
11. 6x6=36
12. 54/9= 6

Everyday Math

13. Answers will vary
14. Answers will vary
15. 9x8=72
16. 40÷8=5

Chapter 5—Day 59

1. 3.56+1.24=4.80
2. 8.53-1.62=6.91
3. 4.59+0.25=4.84
4. 3.27+2.33=5.60
5. 8.06+1.73=9.79
6. 9.55+4.65=14.20
7. 2.44+8.26=10.70
8. 7.32+0.89=8.21
9. 4x3=12
10. 16÷4=4
11. 8÷4=2
12. 4x5=20
13. 5x2=10
14. 10÷2=5
15. $\frac{18}{3}$ =6
16. 40x1=40
17. 9x3=27
18. $\frac{40}{1}$ =40
19. $\frac{16}{4}$ =4
20 6x7=42

Chapter 5—Day 60—Practice

1. 34, 32, 30, 28, 26, 24
2. 65, 70, 75, 80, 85, 90

3. 0, 6, 12, 18, 24, 30
4. 35, 42, 49, 56, 63, 70
5. 530, 520, 510, 500, 490, 480
6. 740,835
   10s: 3
   1,000s: 0
   10,000s: 4
7. 703,451.45
   1s: 5
   hundreths: 4
   100,000s: 7
8. 489,005
   100s: 0
   1,000s: 9
   1,000,000s: none

Everyday Math

9. 80 dollars
   10x8=80
10. $6.06 left over
   10+10+.75=20.75
   20.75-14.69=6.06
11. Giraffe-E3
   Elephant-G6
   Alligator-H8
   Pig-10A
   Tiger-C7
   Zebra-I1
   Monkey-E9
   Frog-B2
   Hippo-H4
   Lion-J5

Chapter 5—Day 61

Total summer rainfall in Springfield, Illinois: 11.6 inches

VEGETABLE	WATER NEEDED IN SUMMER	NORMAL SUMMER RAINFALL RECEIVED	ADDITIONAL WATER NEEDED
Asparagus	4.5 inches	11.6 (answer from pg. 188)	0
Beans	18 inches	11.6	6.4
Cabbage	9 inches	11.6	0
Carrots	4.5 inches	11.6	0
Lettuce	13.5 inches	11.6	1.9
Onions	12 inches	11.6	0.4

1. None
2. Asparagus, cabbage, and carrots.

Student Exercises

Answers will vary, possible solutions are given.

1. 7=3+4
2. 11=5+6
3. 7=1x7
4. 11=1x11
5. 20=10+10
6. 12=6+6
7. 20=4x5
8. 12=3x4
9. 9=3x3
10. 334+437=771
11. 153-44=109
12. 123+437=560
13. 537-188=349
14. 631+149=780
15. 345-139=206

Chapter 5—Day 62—Practice

1. 5+7=12
2. 15-8=7
3. 13-6=7
4. 6+8=14
5. 5+5=10
6. 20-10=10
7. 4+9=13
8. 16-9=7
9. $\frac{4}{5} - \frac{1}{5} = \frac{3}{5}$
10. $\frac{3}{7} + \frac{8}{7} = \frac{11}{7}$
11. $\frac{5}{2} + \frac{9}{2} = \frac{14}{2}$
12. $\frac{10}{3} - \frac{3}{3} = \frac{7}{3}$

Everyday Math

13. 9x8=72
   30.00+10.00+3.00=43.00
   72-43=$29 earned
14. 140-14-99=$27
15. 948-379=$569
16. 34.5+56.7=91.2
17. 334+287=621
18. 1306.99+117.52=1424.51

Chapter 5—Day 63

Pg. 194- $52.28

1. $36.10
2. $163.91
3. $80.90
4. $165.75
5. 
6. 
7.
8. 
9. 
10.

11. completed
12. $\frac{8}{7}$ = 1 $\frac{1}{7}$ one circle filled in, plus one slice
13. $\frac{14}{12}$ = 1$\frac{2}{12}$ one circle filled in, plus two slices
14. $\frac{13}{6}$ = 2 $\frac{1}{6}$ two circles filled in, plus one slice
15. $\frac{3}{9}$ = $\frac{3}{9}$ 3 slices filled in
16. $\frac{7}{7}$ = $\frac{7}{7}$ or 1 one circle filled in
17. $\frac{10}{5}$ = 2 2 circle filled in

Everyday Math

18. 

16÷4= $\frac{4}{8}$ (or simplified $\frac{1}{2}$)

19. 

24÷6=6

20. 

$\frac{8}{3}$ = 2 $\frac{2}{3}$

Chapter 5—Day 64—Practice

1. (-)
2. (+)
3. (x)
4. (÷)
5. (÷)
6. (+)
7. a, b, c, d
8. 1, 2, a, b, c
9. 0, 1, 2
10. 9
11. 2, 4, 6
12. 4, 5
13. 7x5=35
14. 14÷7=2
15. 24÷3=8
16. 9x3=27
17. $\frac{8}{2}$ - $\frac{6}{2}$ = $\frac{2}{2}$ or 1
18. $\frac{3}{5}$ + $\frac{4}{5}$ = $\frac{7}{5}$
19. $\frac{10}{8}$ + $\frac{4}{8}$ = $\frac{14}{8}$
20. $\frac{9}{3}$ - $\frac{3}{3}$ = $\frac{6}{3}$

Chapter 5—Day 65

Answers will vary.

Chapter 6—Day 66

1. 7.7 cm
2. 16 total square inches
   1x1=1
   1x1=1
   4x3=12
   1x2=2
   1+1+12+2=16
3. 3x3x3=27
   500 cc's
4. 6 waterpots
5. 20 or 30 gallons
6. 120-180 gallons
   20x6=120, 30x6=180
7. 3
8. 1
9. 2
10. 1

Student Exercises

1. 15+18=33
2. 37+21=58
3. 35-16=19
4. 146+287=433
5. 78-42=36
6. 253-160=93

	Number	hundreds	tens	ones	decimal	tenths	hundredths
7.	345.58	3	4	5	.	5	8
8.	587.83	5	8	7	.	8	3
9.	145.87	1	4	5	.	8	7
10.	999.99	9	9	9	.	9	9
11.	100.01	1	0	0	.	0	1
12.	68.43	0	6	8	.	4	3
13.	45.38	0	4	5	.	3	8
14.	199	1	9	9	.	0	0

15. $55.50
16. $42.25
17. $84.35
18. $68.31
19. Coins: $2.15
Bills: $97
Total: $99.15

Chapter 6—Day 67

1. 7,160,345,104
   10s: 0
   1,000s: 5
   10,000s: 4
2. 8,436.97
   1s: 6
   100s: 4
   Hundredths: 7
3. 4,720,874
   100s: 8
   1,000s: 0

1,000,000s: 4

4. 2x1x1=2
5. 3x2x2=12
6. 2x4x4=32
7. 5x3x3=45
9. 7x8=56
10. 10÷5=2
11. 9÷3=3
12. 6x4=24
13. 3x9=27
14. 12÷2=6
15. $\frac{15}{3}$ = 5
16. 4x3=12
17. 4x9=36
18. $\frac{12}{4}$ = 3
19. 4x12=48
20. 16÷4=4

21-23. Answers will vary.

Chapter 6—Day 68

1. Answers will vary. Addition.
2. Answers will vary. Addition.
3. Answers will vary. Subtraction.
4. Answers will vary. Addition
5. Answers will vary. Subtraction.
6. Answers will vary. Addition.
7. Answers will vary.

Chapter 6—Day 69—Practice

1. 34+18=52
2. 45-27=18
3. 75-52=23
4. 62+31=93
5. 116+586=702
6. 506-250=256
7. 8, 8+7=15
8. 6, 6+9=15
9. 14, 14-5=9
10. 16, 16-7=9
11. 6, 6×2=12
12. 4, 4×4=16
13. 8, 8÷2=4
14. 21, 21÷3=7
15. $\frac{3}{7} + \frac{5}{7} = 1\frac{1}{7}$

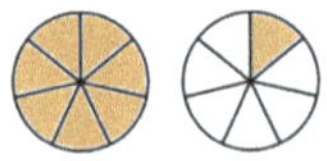

16. 46 + 56 = 136

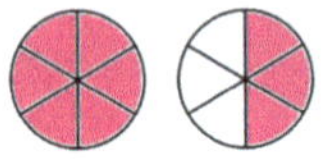

17. 20-13=7

3 p.m.

18. 124x3=372

398-372=$26

19. 56+25=81
20. 60-41=19

36-19=17 more donuts

Chapter 6—Day 70

1. Multiplication
2. Multiplication
3. Addition
4. Addition
5. Multiplication
6. Addition
7. Multiplication
8. <
9. >
10. >
11. <
12. =
13. <
14. >
15. <
16. =
17. >

Chapter 6—Day 71—Practice

1. 5x7=35
2. 25÷5=5
3. 21÷3=7
4. 7x8=56
5. 6x5=30
6. 36÷9=4
7. $\frac{30}{6}$ = 5
8. 6x6=36
9. 5x8=40
10. $\frac{45}{5}$ = 9
11. Answers will vary, 30 degrees.
12. Answers will vary, 90 degrees.
13. Answers will vary, 60 degrees.
14. Answers will vary, 45 degrees.
15. 6,824,100
16. 6,820,000
17. 6,800,000
18. 123,000
19. 120,000
20. 100,000
21. 1x1=1

22. 1x1=1
23. 1x1=1
24. 2x3=6
25. 1x3=3
26. 2x6=12
27. Total area=24 (1+1+1+6+3+12=24)

Chapter 6—Day 72—Practice

1. $\frac{2}{6}$ 
2. $\frac{4}{12}$ 
3. $\frac{4}{5}$ 
4. $\frac{1}{6}$ 
5. 15-8=7
   $\frac{7}{15}$
6. $\frac{6}{5}$ or 1 $\frac{1}{5}$
7. $\frac{3}{6}$
8. fill in clocks

Chapter 6—Day 73

1. per
2. divide
3. per
4. $120
   60+60=120
5. 4 years
   40÷10=4
6. 9 weeks
   50-5=45
   45÷5=9
7. $20
   2.00x10=20

8-9. Answers will vary.

10. 1.50+1.75=3.25
11. 23.45+28.84=52.29
12. 1.89+1.54+3.43
13. 34.45+47.83=82.28
14. 25.75+27.82=53.57
15. 87.23+23.45=110.68

Chapter 6—Day 74—Practice

1. Answers will vary, 120 degrees
2. Answers will vary, 60 degrees
3. Answers will vary, 1 cm.
4. Answers will vary, 1 in.
5. Answers will vary, 2.5 cm.
6. Answers will vary, 5 ¾ in.

Everyday Math

7. $24.10
   12.35+11.75=24.10
8. 16 girls
   28-12=16
9. 84 flowers
   10x6=60
   4x6=24
   60+24=84
10. 13 cards each
   20÷2=10
   6÷2=3
   10+3=13

Speed Test

1. 6×5=30
2. 9×9=81
3. 9×8=72
4. 10×3=30
5. 9×7=63
6. 5×7=35
7. 3×7=21
8. 6x0=0
9. 9×10=90
10. 3×5=15
11. 6×9=54
12. 8×5=40
13. 5×0=0
14. 7×4=28
15. 10×1=10
16. 8×8=64
17. 4×3=12
18. 3×6=18
19. 6×6=36
20. 2×9=18
21. 6×8=48
22. 1×5=5
23. 6×7=42
24. 1×8=8
25. 2×8=16
26. 1×2=2
27. 6×0=0
28. 4×6=24
29. 2×6=12
30. 5×5=25
31. 5×9=45
32. 3×9=27
33. 3x2=6
34. 5×2=10
35. 8×4=32

36. 8×7=56
37. 10×1=10
38. 5×4=20
39. 8×8=64
40. 4×9=36
41. 5×8=40
42. 4×4=16
43. 4×6=24
44. 7×7=49
45. 9×9=81

Chapter 6—Day 75

1. 2x4=8
2. 2x8=16
   1x4=4
3. 32x1=32

4-5. Answers will vary.

Student Exercises

1. 10 boards
   2 boards per foot 5x2=10
   10 boards
   $100
   10x10=100
2. 5x3=15
3. 15x2=30
4. 64+80+64+80=288
   8x8=64, 8x10=80

	Number	hundreds	tens	ones	tenths	hundredths
5.	123.45	1	2	3	4	5
6.	345.56	3	4	5	5	6
7.	567.89	5	6	7	8	9
8.	90.09	0	9	0	0	9
9.	909.09	9	0	9	0	9
10.	999.9	9	9	9	9	0

11. fill in clocks

Chapter 6—Day 76 Practice

Answers may vary, possible solutions are given.

1. 13+6=19
2. 2+11=13
3. 18-10=8
4. 13-4=9
5. 3×5=15
6. 3×3=9
7. 13÷13=1
8. 18÷3=6
9. 167,290
10. 170,000
11. 200,000
12. 32,486,000
13. 32,000,000
14. 32,500,000
15. 15 notebooks
    5x3=15
16. $69.97
    25.99+18.00+12.99+12.99=69.97
17. 4:30 pm
18. 7x9=63
19. 10x4=40
20. 31-16=15
21. 106-28=78
22. 93 - 73 = 23
23. 14 + 14 = 24
24. 27 + 67 = 87 or 117
25. 89 - 39 = 59

Chapter 6—Day 77

1-3. Answers will vary.

Chapter 7—Day 78

Activity

Answers will vary.

1. >
2. >
3. >
4. <
5. <
6. <

7.347

8.437

9.734

Chapter 7—Day 79—Practice

1. 37+18=55
2. 43-29=14
3. 132-8=124
4. 534+885=1419
5. 5.2+3.4=8.6
6. 5.9-4.7=1.2
7. 145,542,436+0=145,542,436
8. 1,234,345-0=1,234,345
9. 0, 4, 8, 12, 16, 20
10. 40, 36, 32, 28, 24, 20

11. 0, 6, 12, 18, 24, 30
12. 60, 54, 48, 42, 36, 30
13. 0, 8, 16, 24, 32, 40
14. 0, 11, 22, 33, 44, 55

Activity

Answers will vary.

Chapter 7—Day 80

Student Exercises

1. 6x5=30
2. 36÷6=6
3. 60÷6=10
4. 6x7=42
5. 8x5=40
6. 40÷8=5
7. 808 =10
8. 9x8=72
9. 4x7=28
10. 497 =7

Everyday Math

11. $14.59

1x5=5

2x3=6

3.59x1=3.59

5.00+6.00+3.59=14.59

12. $168.00

8x6=48

8x8=64

8x7=56

48.00+64.00+56.00=168.00

Chapter 7—Day 81—Practice

1. completed
2. 3+10=13
3. 16-8=8
4. 15-7=8
5. 4x3=12
6. 5x3=15
7. 8÷2=4
8. 12÷2=6
9. 7x8=64
10. 5x6=30
11. 7x6=42
12. 10x7=70
13. 8x5=40
14. 8x9=72
15. Answers will vary, 90 degrees
16. Answers will vary, 55 degrees
17. Answers will vary, 35 degrees
18. Answers will vary, 5 inches
19. Answers will vary, 4 inches
20. Answers will vary, 3 inches
21. Answers will vary, area=70 (7x10=70)
22. Answers will vary, area=15 (5x3=15)
23. 17 C°
24. 40 C°
25. 37 C°
26. 8 C°

Chapter 7—Day 82

	⚀	⚁	⚂	⚃	⚄	⚅
⚀	1, 1	1, 2	1, 3	1, 4	1, 5	1, 6
⚁	2,1	2, 2	2, 3	2, 4	2, 5	2, 6
⚂	3, 1	3, 2	3, 3	3, 4	3, 5	3, 6
⚃	4, 1	4, 2	4, 3	4, 4	4, 5	4, 6
⚄	5, 1	5, 2	5, 3	5, 4	5, 5	5, 6
⚅	6, 1	6, 2	6, 3	6, 4	6, 5	6,6

1. 36
2. 36
3.

	1	2	3	4	5	6
1	2	3	4	5	6	7
2	3	4	5	6	7	8
3	4	5	6	7	8	9
4	5	6	7	8	9	10
5	6	7	8	9	10	11
6	7	8	9	10	11	12

4. 12

Student Exercises

1. 1
2. 2
3. 3

4. 4
5. 5
6. 6
7. 5
8. 4
9. 3
10. 2
11. 1
12. 6,791,000
13. 6,790,000
14. 6,800,000
15. 64,080
16. 64,100
17. 64,000
18. Answers will vary.

Chapter 7—Day 83—Practice

Pg. 271: 12, 24, 36, 48, 60, 72, 84, 96, 108, 120, 132, 144

1. ÷
2. –
3. ÷
4. +
5. x
6. +
7. x
8. –
9. >
10. <
11. >
12. <
13. 375
14. 537
15. 573

Answers will vary, possible solution options:

32	5	6	30	4	5	20	2	6
4	12	10	32	18	9	2	6	9
8	3	8	18	5	1	36	10	54
16	4	4	45	2	9	18	2	0
2	5	5	25	3	9	6	20	24
8	20	18	2	4	8	4	5	4
24	6	4	5	3	15	24	4	6
3	2	5	10	7	5	35	12	3
2	16	20	4	40	6	6	36	2
1	2	2	4	7	28	49	9	6
10	8	10	16	8	2	3	4	12

Chapter 7—Day 84

1. 12
2. 6
3. $\frac{2}{8}$

Student Exercises

1. 3+4=7
2. 8-7=1
3. 15-9=6
4. 4+7=11
5. 8+6=14
6. 15-7=8
7. 9+9=18
8. 18-11=7
9. 14+12=26
10. 22+11=33
11. 22-11=11
12. 14-12=2
13. $\frac{0}{3}$ or 0
14. $\frac{2}{2}$ or 1
15. $\frac{7}{4}$
16. $\frac{3}{9}$
17. $\frac{7}{8}$
18. $\frac{5}{10}$

Chapter 7—Day 85—Practice

1. 60, 57, 54, 51, 48, 45
2. 24, 30, 36, 42, 48, 54
3. 40, 48, 56, 64, 72, 80
4. 670, 660, 650, 640, 630, 620
5. 132, 121, 110, 99, 88, 77
6. 24, 36, 48, 60, 72, 84
7. 2x3=6
8. 10÷5=2
9. 9÷3=3
10. 6x2=12
11. 3x12=36
12. 12÷2=6
13. $\frac{15}{3} = 5$
14. 4x3=12
15. 2x2=4
16. $\frac{12}{4} = 3$

Everyday Math

17. $\frac{1}{3}$
18. 108 cookies, 9 dozen
    450-342=108
    98÷12=9
19. $250
    346-96=250

20. 4 muffins
48÷12=4

Extra Challenge

21. $\frac{2}{4}$ or $\frac{1}{2}$ 

Chapter 7—Day 86

85%

% % % % %

1. It probably will not rain tomorrow.
2. It probably will snow tomorrow.
3. You probably will not roll snake eyes.
4. 100%
5. 10%
6. 50%

Activity

Answers will vary.

Student Exercises

1. 10+50=60
2. 7x8=56
3. 56÷8=7
4. 6+54=60
5. 9x6=54
6. 54÷6=9
7. 60-6=54
8. 1+1=2

Chapter 7—Day 87—Practice

1. Answers will vary, 90 degrees
2. Answers will vary, 45 degrees
3. Answers will vary, 45 degrees
4. Answers will vary, 9.9 cm
5. Answers will vary, 9 cm
6. Answers will vary, 9 cm
7. Answers will vary, area=72 (8x9=72)
8. Answers will vary, area=42 (7x6=42)
9. 87.4
10. 87
11. 90
12. 56,100
13. 56,000
14. 60,000
15. 11, 22, 33, 44, 55, 66, 77, 88, 99, 110, 121
16. 64+78=142
17. 85-47=38
18. 11x5=55
19. 12x8=96
20. $\frac{7}{4}$ or $1\frac{3}{4}$

21. $\frac{1}{6}$

Chapter 7—Day 88

1-2. Answers will vary.

3. Answers will vary, God, God won the battle through David because God is stronger than any man.
4. 6 cubits and a span
5. Answers will vary.

Chapter 8—Day 89

1. 30
2. 36
3. 48

Student Exercises

1. 3+5+3+5=16 cm
2. 2+2+2+2=8 cm
3. 3+5+2+7=17 cm
4. <
5. <
6. <
7. <
8. 189
9. 198
10. 981

Everyday Math

11. 360 ft
    90x4=360
12. Octagon
    240 cm
    30x8=240
13. Octagon
    96 in
    12x8=96
14. Hexagon
    6x4=24
15. Answers will vary.
16. 65+75=140
17. 435+67=502
18. 13,245+2,601=15,846
19. 87-57=30
20. 3.4+4.7=8.1
21. 131-12=119
22. 13.9-1.2=12.7
23. 345-297=48

Chapter 8—Day 90—Practice

1. <u>100</u>, 98, <u>96</u>, 94, <u>92</u>, 90, <u>88</u>, 86

2. 100, 96, 92, 88, 84, 80
3. 70, 63, 56, 49, 42, 35
4. 0, 7, 14, 21, 28, 35
5. 100, 200, 300, 400, 500, 600
6. 4x6=24
7. 24÷4=6
8. 36÷4=9
9. 8x7=56
10. 7x5=35
11. 63÷7=9
12. $\frac{81}{9}$ = 9
13. 9x5=45
14. 9x7=63
15. $\frac{100}{10}$ = 10

Everyday Math

16. 14+4+14+4=36
17. 20+20+20+20+20+20+20+20+20+20=200
18. 3+5+6=14
19. 3x2x1=6
20. 1+10+4=15
21. 2x3x2=12
22. 4+6+3+7=20
23. 3x2x2÷2=6
24. 4x3÷2=6
25. 5+3+1+1+1+1=12

Chapter 8—Day 91

1. 3x3=9
   9x6=54
2. 6x3
   27 cubic inches
3. C
4. A
5. D
6. B

Activity

Answer may vary.

Chapter 8—Day 92—Practice

1. 
2. 
3. 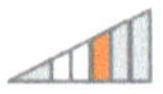
4. 
5. The Law of the Lord
6. two-fourths of the day
7. $\frac{6}{4}$
8. $\frac{7}{3}$
9. $\frac{9}{2}$
10. $2\frac{2}{3}$
11. $1\frac{3}{4}$
12. $\frac{1}{2}$
13. $\frac{2}{4}$
14. $\frac{3}{6}$
15. $\frac{4}{8}$
16. $\frac{5}{10}$
17. $\frac{6}{12}$
18. 4 , 4 + 8 = 12,
19. 7, 7 + 7 = 14
20. 16, 16 - 3 = 13
21. 14,14 - 5 = 9
22. 4, 4 × 4 = 16
23. 0, 0 × 4 = 0
24. 11, 11 × 11 = 121
25. 15, 15 ÷ 1 = 15
26. 9, 9 ÷ 3 = 3
27. 144, 144 ÷ 12 = 12

Chapter 8—Day 93

Pg. 300: Area of rectangle= 2x4=8
   Area of triangle=12÷2=6
   The rectangle piece of candy is bigger.

1. 12÷2
   6 square inches
2. 20÷2
   10 square inches
3. 8÷2
   4 square inches
4. 12÷2
   6 square inches
5. +
6. x
7. ÷
8. +
9. x
10. –
11. ÷
12. –

Chapter 8—Day 94—Practice

1. <
2. <

3. >
4. >
5. 456
6. 564
7. 645
8. 334+159=493
9. 492+213=705
10. 499+519=1,018
11. 363-159=204
12. 125-93=32
13. 505-344=161
14. 40÷2=20
15. 14÷2=7
16. 3x6=18
17. $\frac{1}{8}$
18. $\frac{3}{8}$
19. $\frac{0}{8}$
20. $\frac{2}{8}$
21. $\frac{3}{8}$
22. $\frac{3}{8}$

Chapter 8—Day 95

1. 5x5=25
2. 3x6=18
3. 25÷2=5
   18÷2=9
4. 45

Student Exercises

5. 5+5=10
6. 18-9=9
7. 16-8=8
8. 7+7=14
9. 8+8=16
10. 12-6=6
11. 6+6=12
12. 20-10=10
13. $\frac{2}{4} - \frac{1}{4} = 14$
14. $\frac{2}{3} + \frac{2}{3} = \frac{4}{3}$
15. $\frac{1}{8} + \frac{1}{8} = \frac{2}{8}$
16. $\frac{6}{8} - \frac{1}{8} = \frac{5}{8}$

Everyday Math

1. $\frac{2}{4}$
2. $\frac{24}{8} - \frac{12}{8} - \frac{6}{8} = \frac{6}{8}$
3. 24÷4=6

Chapter 8—Day 96—Practice

1. 90, 87, 84, 81, 78, 75, 72, 69
2. 90, 84, 78, 72, 66, 60
3. 0, 9, 18, 27, 36, 45
4. 90, 81, 72, 63, 54, 45
5. 1,000, 900, 800, 700, 600, 500
6. 4x3x2=24
7. 10÷5=2
8. 9÷1=9
9. 3x2x5=30
10. 8x9=72
11. 12÷6=2
12. $\frac{15}{15} = 1$
13. $\frac{8}{4} = 2$
14. 7x2x1=14
15. 4x2x2=16
16. 11x8=88
17. 60÷12=5
18. completed
19. 6
20. 9
21. 2
22. 4
23. 6

Chapter 8—Day 97

Activity

Answers will vary.

Pg. 313:

18÷6=3

πx2=6

πx5=15

πx10=30

πx12=36

Extra Challenge

20. completed
21. $\frac{1}{6}$
22. $\frac{3}{6}$
23. $\frac{0}{6}$
24. $\frac{6}{6}$
25. $\frac{3}{6}$

Chapter 8—Day 98—Practice

1. Answers will vary, 110 degrees
2. Answers will vary, 130 degrees
3. Answers will vary, 150 degrees
4. Answers will vary, 170 degrees
5. Answers will vary, 7x5=35
6. Answers will vary, 6x4=24
7. Answers will vary, 5x3=15

8. 

9.

10. 

11. 

12. 

13. 

14. 10, (5 + 10 = 15)
15. 10, (3 + 10 = 13)
16. 14, (14 - 5 = 9)
17. 10, (10 - 7 = 3)
18. 1, (1 × 5 = 5)
19. 6, (6 × 3 = 18)
20. 20, (20 ÷ 5 = 4)
21. 6, (6 ÷ 3 = 2)
22. 3x6=18
23. 2x4=8
24. 

Chapter 8—Day 99

pg 323: 27 square inches
3x3x3=27

Student Exercises

1. $\frac{11}{4}$
2. $\frac{3}{2}$
3. $\frac{11}{3}$
4. $2\frac{1}{3}$
5. $2\frac{2}{5}$
6. $4\frac{1}{2}$
7.

Wood	Place
	F5
	C6
	C5
	D5
	F6
	E5
	G5

8. 1x5=5
9. 2x5=10
10. 3x6=18
11. 4x8=32
12. 519-159=360
13. 938+93=1,031
14. 643+59=702
15. 938-93=845
16. 756-178=578
17. 456+456=912

Chapter 8—Day 100—Practice

1. (-)
2. (÷)
3. (+)
4. (+)
5. (x)
6. (x)
7. (÷)
8. (-)
9. 614-244=370
10. 687-139=548
11. 275+348=623
12. 379-299=80
13. 537+196=733
14. 455+125=580
15.498+79=577
16. 399+299=698
17. 799+799=1598
18. 1, 2, 3, 4, 5, 7, 9
19. 3, 5

Chapter 8—Day 101

Check activities and crafts.

Chapter 9—Day 102

Pg. 326: Step 1. 4x5=20
Step 2. 4 x10=40
Step 3. 20+40=60

Student Exercises

1. 3x1=3
   3x10=30
   3+30=33
2. 2x2=4
   2x10=20
   4+20=24
3. 4x3=12

4x10=40
12+40=52
4. {1, 2, 3, a, b, c}
5. {a, d}
6. {x, y, z, w}
7. none
8. {4, 6, 8}
9. {x, y, z}

Everyday Math

10. $9.89
4.35+2.43+3.11=9.89
11. $10.50
1.00+6.00(2x$3.00)+1.50+2.00=10.50
12. $3.30
0.50+0.60+1.00+1.20=3.30
$6.60
9.90-3.30=6.60

Chapter 9—Day 103—Practice

1. 235,784
10s: 8
1,000s: 5
10,000s: 3
2. 37,528.45
1s: 8
Tenths: 4
Hundredths: 5
3. 34,752,034
100s: 0
1,000s: 2
1,000,000s: 7
4. 8+6=14
5. 13-7=6
6. 2+8=10
7. 15+5=20
8. 24-3=21
9. 23 + 0 = 23
10. 14-9=5
11. 8+7=15
12. 16-10=6
13. 45 + 35 = 75
14. 59 + 29 = 79
15. 37 + 17 = 47
16. 
17. 
18. 
19. 

20. 7.00+12.00=$19.00
20.00-19.00=$1.00
21. 13.70+5.30=$19.00
20.00-19.00=$1.00
22. 6.15+7.85=$14.00
20.00-14.00= $6.00

Chapter 9—Day 104

1. 11x4=44
2. 11x5=55
3. 12x4=48
4. 13x2=26
5. 14x2=28

Student Exercises

1. 78,123.5
2. 78,123
3. 78,120
4. 78,100
5. 78,000
6. 80,000

Answers will vary, possible solutions are given.

7. 5+5
8. 6+6
9. 2×5
10. 2×6
11. 10+10
12. 9+11
13. 2×10
14. 5×4
15. 33 passengers
3x11=33
16. 24 puppies
12x2=24
17. 39 eggs
13x3=39
18.

1	4	2	6	3	5
3	6	5	1	2	4
4	1	3	5	6	2
5	2	6	4	1	3
2	5	1	3	4	6
6	3	4	2	5	1

19.

6	4	2	3	5	1
3	5	1	6	4	2
5	6	3	2	1	4
1	2	4	5	3	6
4	3	6	1	2	5
2	1	5	4	6	3

Chapter 9—Day 105

1. 3x7=21
2. 3x10=30
3. 51
4. 68
5. 34
6. 72
7. 72
8. 96

Student Exercises

1. (÷)
2. (−)
3. (+)
4. (÷)
5. (+)
6. (÷)or(x)
7. (−)
8. (x)
9. $\frac{3}{8} + \frac{5}{8} = \frac{8}{8}$ or 1
10. $\frac{7}{9} + \frac{8}{9} = \frac{15}{9}$ or $1\frac{6}{9}$
11. $\frac{7}{9} + \frac{3}{9} = \frac{10}{9}$ or $1\frac{1}{9}$
12. $\frac{8}{5} + \frac{4}{5} = \frac{12}{5}$ or $2\frac{2}{5}$

Chapter 9—Day 106—Practice

1. 39x2=78
2. 78-72=6
3. 4.35+6.12=10.47
4. 1,345+3,915=5,260
5. 37-36=1
6. 35+35=70
7. 7.22-2.23=4.99
8. 74x3=222
9. >
10. =
11. <
12. <
13. =
14. =
15. 45-19=26 kinds
16. 34x4=136 miles
17. 408 miles left
    951-543=408
18.

Item	Place
X	D2
	B4
	D4
	A3
	C2
	G1
	A5

Chapter 9—Day 107

1. 142x2=284
2. 431x2=862
3. 211x3=633
4. 235x2=470
5. 212x4=848
6. 4,345.5
7. 4,346
8. 4,350
9. 4,300
10. 4,000
11. 0
12. $\frac{5}{10} = \frac{50}{100} = 50\%$
13. $\frac{2}{10} = \frac{20}{100} = 20\%$
14. $\frac{9}{10} = \frac{90}{100} = 90\%$
15. $\frac{3}{10} = \frac{30}{100} = 30\%$
16. $\frac{6}{10} = \frac{60}{100} = 60\%$
17. $\frac{7}{10} = \frac{70}{100} = 70\%$
18. $\frac{1}{10} = \frac{10}{100} = 10\%$
19. $\frac{4}{10} = \frac{40}{100} = 40\%$
20. $\frac{1}{6}$
21. $\frac{6}{6}$
22. $\frac{2}{6}$
23. $\frac{5}{6}$
24. $\frac{1}{6}$
25. $\frac{3}{6}$

Chapter 9—Day 108—Practice

1. 47,897,643
   10s: 4
   1,000s: 7
   10,000s: 9
2. 976,432

1s: 2
100s: 4
100,000s: 9

3. 587,753
   100s: 7
   1,000s: 7
   1,000,000s: none
4. 6=3+3
5. 12=3×4
6. 15=10+5
7. 18=3×6
8. 6=2+4
9. 12=2×6
10. 15=7+8
11. 18=2×9

Speed Test

1. 6×5=30
2. 9×9=81
3. 9×8=72
4. 10×3=30
5. 9×7=63
6. 5×7=35
7. 3×7=21
8. 6x0=0
9. 9×10=90
10. 3×5=15
11. 6×9=54
12. 8×5=40
13. 5×0=0
14. 7×4=28
15. 10×1=10
16. 8×8=64
17. 4×3=12
18. 3×6=18
19. 6×6=36
20. 2×9=18
21. 6×8=48
22. 1×5=5
23. 6×7=42
24. 1×8=8
25. 2×8=16
26. 1×2=2
27. 6×0=0
28. 4×6=24
29. 2×6=12
30. 5×5=25
31. 5×9=45
32. 3×9=27
33. 3x2=6
34. 5×2=10
35. 8×4=32
36. 8×7=56
37. 10×1=10
38. 5×4=20
39. 8×8=64
40. 4×9=36
41. 5×8=40
42. 4×4=16
43. 4×6=24
44. 7×7=49
45. 9×9=81

Chapter 9—Day 109

1. 50
2. 500
3. 2,500
4. 35,500
5. 100,000,000
6. 3
7. 3.3
8. 3.33
9. 0.3
10. completed
11. 5400
12. 1,200
13. 54,300
14. 543,500
15. 100,000
16. completed
17. 0.5
18. 1.2
19. 12.0
20. 24.5
21. 245.0

Chapter 9—Day 110

1. 6+6=12
2. 17-7=10
3. 8+8=16
4. 5+5=10
5. $\frac{3}{6} - \frac{2}{6} = \frac{1}{6}$
6. $\frac{3}{7} + \frac{4}{7} = \frac{7}{7}$ or 1
7. 18-9=9
8. 7+7=14
9. 16-4=12
10. 20-11=9
11. 0.5+0.3=0.8

12. 0.7-0.2=0.5
13. 19,928.1
14. 19,928
15. 19,930
16. 19,900
17. 20,000
18. 20,000
19. $\frac{2}{8}$
20. $\frac{3}{8}$
21. $\frac{1}{8}$
22. $\frac{0}{8}$

Everyday Math

23. 143x2=286
24. Answers will vary.
25. 120x6=720
26. 15x6=90
27. 720-90=630

Chapter 9—Day 111

1. 60x2=120
2. Punch:16x8=128 ounces; 1 gallon
Pizza: 16x3=48
Salad: 16x4=64
3. Answers will vary.
4. 25 cm
2.5x10=25 cm
5. 20
6. 200
7. 910
8. 9,100
9. 64.1
10. 641.0
11. 7.5
12. 75.0
13.

Wood	Place
	C3
	F4
	B3
	E6
	A2
	D2
	G2

Chapter 10—Day 112

1. 30 mm
2. 500 mm
3. 92 mm
4. 7 cm x 10 = 70 mm
5. 100 cm
6. 2,200 cm
7. 470 cm
8. 65 m x 100 = 6,500 cm
9. 1,000 km
10. 39,000 m
11. 2,800 m
12. 6 km x 1,000 = 6,000 m
13. 53.0
14. 5.30
15. 0.53
16. 6.4
17. 6.40
18. 6.400
19. 2.4 m
20. 14 cm
21. 1.25 km
22. 0.955 km

Chapter 10—Day 113

1. 65+65=130
2. 150-75=75
3. 86-36=50
4. 32x7=224
5. 3.45+5.05=8.50
6. 17.34-8.16=9.18
7. 1,456+8,544=10,000
8. 110-67=43
9. 22, 24, 26, 28, 30, 32
10. 30, 33, 36, 39, 42, 45
11. 40, 36, 32, 28, 24, 20
12. 150, 145, 140, 135, 130, 125
13. 60, 54, 48, 42, 36, 30
14. 123x3=369
15. 124x3=372
16. 143x3=429
17. 350
18. 9,900
19. 14,000
20. 0.35
21. 0.340
22. 0.330

Chapter 10—Day 114

1. 0.120 g
2. 0.550 g
3. 1.000 g
4. 0.054 kg
5. 1.000 kg
6. 2.400 kg

Student Exercises

1. 4x6=24
2. 8÷4=2
3. 20÷10=2
4. 4x5=20
5. 9x2=18
6. $\frac{15}{15}$ = 1
7. $\frac{22}{2}$ = 11
8. 5x3=15
9. 24÷2=12
10. $\frac{3}{2}$, $\frac{2}{2}$
11. $\frac{4}{3}$, $\frac{3}{3}$, $\frac{2}{3}$

Speed Test

1. 6×5=30
2. 9×9=81
3. 9×8=72
4. 10×3=30
5. 9×7=63
6. 5×7=35
7. 3×7=21
8. 6x0=0
9. 9×10=90
10. 3×5=15
11. 6×9=54
12. 8×5=40
13. 5×0=0
14. 7×4=28
15. 10×1=10
16. 8×8=64
17. 4×3=12
18. 3×6=18
19. 6×6=36
20. 2×9=18
21. 6×8=48
22. 1×5=5
23. 6×7=42
24. 1×8=8
25. 2×8=16
26. 1×2=2
27. 6×0=0
28. 4×6=24
29. 2×6=12
30. 5×5=25
31. 5×9=45
32. 3×9=27
33. 3x2=6
34. 5×2=10
35. 8×4=32
36. 8×7=56
37. 10×1=10
38. 5×4=20
39. 8×8=64
40. 4×9=36
41. 5×8=40
42. 4×4=16
43. 4×6=24
44. 7×7=49
45. 9×9=81

Chapter 10—Day 115—Practice

1. Completed
2. 6 (6+10=16)
3. 12 (12-8=4)
4. 17 (17-10=7)
5. 6 (6x2=12)
6. 3 (3x6=18)
7. 20 (20÷2=10)
8. 20 (20÷5=4)
9. Answers will vary. 4x9=36
10. Answers will vary. 10x6÷2=30
11. Answers will vary. 10x5÷2=25
12. $\frac{5}{4}$
13. $\frac{10}{3}$
14. $\frac{12}{5}$
15. $\frac{6}{2}$ or 3
16. 2 $\frac{2}{3}$
17. 2 $\frac{1}{4}$
18. 240 centimeters long
    2.4 meters × 100 = 240 centimeters
19. 1.304 kg
    Jelly: 510 grams = 510 ÷ 1000 = 0.51 kg
    Peanut butter: 794 grams = 794 ÷ 1000 = 0.794 kg
    Total: 0.51 kg + 0.794 kg = 1.304 kg
    How much more peanut butter than jelly:
    0.284 kg
    794 g - 510 g = 284 g
    Will you run out of peanut butter before jelly?
    No, jelly will run out first.
20. 2.02 feet
    24.236 inches ÷ 12 = 2.0197 feet

Chapter 10—Day 116

1. 0.2500 liters
2. 0.3550 liters

Student Exercises

1. 8 gallons, 12 gallons
2. 72 liters
3. 64 pounds
4. 560 pounds
5. 120 centimeters. Answers will vary.
6. 0.1
7. 0.4
8. 0.5
9. 0.9
10. 1.0
11. 0.4
12. 0.2
13. 0.5
14. 0.9
15. 1.0

16-18. Answers will vary

Chapter 10—Day 117—Practice

1. $\frac{15}{4}$
2. $\frac{5}{2}$
3. $\frac{5}{3}$
4. $1\frac{2}{5}$
5. $5\frac{1}{2}$
6. $3\frac{0}{3}$ or 3
7. Apples:12x$2.00=$24.00
   Eggplant: 4x$3.00=$12.00
   Bananas: 3x$2.00=$6.00
   Pears: 11x$2.00=$22.00
   Lettuce: 2x$2.00=$4.00
   Peppers: 2x$3.00=$6.00
   Melons: 5x$2.00=$10.00
   Total: $84.00
8. 9+8=17
9. 13-7=6
10. 4x4=16
11. 7x5=35
12. 16-5=11
13. 12-4=8
14. 12÷2=6
15. 28÷4=7
16. 2+6=8
17. 13+13=26
18. 12÷4=3
19. 20÷5=4
20. $\frac{1}{8}$
21. $\frac{3}{8}$
22. $\frac{0}{8}$
23. $\frac{3}{8}$
24. $\frac{3}{8}$
25. $\frac{3}{8}$

Chapter 10—Day 118

1. 1 cubit = 20 inches (50 cm)
   1 cubit = 20 inches (50 cm)
   2 cubits = 40 inches (100 cm)
2. 1 cubit = 20 inches (50 cm)
   2 cubits = 40 inches (100 cm)
   1½ cubits = 30 inches (75 cm)
3. 1½ cubits = 30 inches (75 cm)
   2½ cubits = 50 inches (125 cm)
   1½ cubits = 30 inches (75 cm)
4. 5 cubits = 100 inches (250 cm)
   5 cubits = 100 inches (250 cm)
   3 cubits = 60 inches (150 cm)
5. Diameter: 10 cubits
   How many inches (or cm is this)? 200 inches (or 500 cm)
   Height: 5 cubits
   How many inches (or cm is this)? 100 inches (or 250 cm)
   Circumference: 30 cubits
   How many inches (or cm is this)? 600 inches (or 1,500 cm)
   Capacity: 2,000 baths
   Total gallons: 2,000 baths × 6 gallons = 12,000 gallons

Student Exercises

1. 0x6=0
2. 3x8=24
3. correct
4. 12x7=84
5. 3x9=27
6. correct
7. correct
8. 9x7=63
9. 321+123+46=490
10. 658-177=481
11. 432+392+144=968
12. 970-543=427
13. 111+333+456=900
14. 493-388=105
15. Answers will vary. 10x6÷2=30 cm
16. Answers will vary. 10x5÷2=25 cm
17. Answers will vary. 5x4=22 cm
18. Answers will vary, A. 60 degrees B. 30 degrees C. 90 degrees
19. 12, 24, 36, 48, 60, 72
20. 66, 77, 88, 99, 110, 121

21. 110, 105, 100, 95, 90, 85
22. 70, 63, 56, 49, 42, 35
23. 0, 9, 18, 27, 36, 45

Chapter 10—119—Practice

1. $0.9 = \frac{9}{10} = \frac{90}{100} = 90\%$
2. $0.2 = \frac{2}{10} = \frac{20}{100} = 20\%$
3. $0.3 = \frac{3}{10} = \frac{30}{100} = 30\%$
4. $0.4 = \frac{4}{10} = \frac{40}{100} = 40\%$
5. $0.7 = \frac{7}{10} = \frac{70}{100} = 70\%$
6. $\frac{1}{4} = 0.25 = \frac{25}{100} = 25\%$
7. $1.50 = \frac{15}{10} = \frac{150}{100} = 150\%$

8-13. Answers will vary.

14. Check work.
15. $\frac{1}{6}$
16. $\frac{5}{6}$
17. $\frac{2}{6}$
18. $\frac{0}{6}$ or 0
19. $\frac{4}{6}$
20. $\frac{6}{6}$ or 1

Chapter 10—Day 120

Answers will vary, check work.

Chapter 10—Day 121—Practice

1. >
2. >
3. >
4. <
5. >
6. >
7. 4x$200=$800
   2x$1,250=$2,500
   $2,500-$800=$1,700
8. Answers may vary, possible solutions given.
   Oranges: $1.00x8=$8.00
   Apples: $2.00x10=$20.00
   Bananas: $1.00x7=$7.00
   Pineapple: $1.00x9=$9.00
   Kiwi: $3.00x6=$18.00
   Grapes: $1.00x10=$10.00
   Total 50 lbs= $72.00
9. Answers will vary, check work.
10. Optional activity, check student's work

Chapter 10—Day 122

1-4. Answers will vary.

5. 3x4=$12
6. 5x4=$20
7. 1x4=$4
8. 12x4=$48
9. 12x4=$48
10. 4x4=$16
11. 12+20+4+48+48+16=$148

Chapter 11—Day 123

Student Exercises

1. 4   16 = 4
2. 5   25 = 5
3. 6   36 = 6
4. <
5. >
6. <
7. >
8. 3+3=6
9. 16-8=8
10. 10+10=20
11. 12+12=24
12. 14-9=5
13. 8+7=15
14. 16-10=6
15. 24-3=21
16. $\frac{5}{9} + \frac{2}{9} = \frac{7}{9}$
17. $\frac{3}{7} + \frac{1}{7} = \frac{4}{7}$
18. $\frac{3}{9}$
19. $\frac{2}{7}$

Chapter 11—Day 124—Practice

1. 9x4=36
2. 45÷9=5
3. 9x9=81
4. $\frac{40}{8} = 5$
5. 7x7=49
6. 54÷9=6
7. 9x3=27
8. 64÷8=8
9. 8x6=48
10. $\frac{49}{7} = 7$
11. 4   24 = 6
12. 5   30 = 6
13. 6   42 = 7
14. 5x6=30
    5x8=40
    5x5=25
    5x7=35

5x9=45
5x3=15
5x2=10
5x10=50
5x4=20

Chapter 11—Day 125

Pg. 399- 123 marbles
Pg. 401- 14x3=42
Pg. 403- 12x3=36

Student Exercises

1. $2\overline{)24} = 12$
2. $4\overline{)44} = 11$
3. $5\overline{)55} = 11$
1. 9 (9+4=13)
2. 3 (3+10=13)
3. 16 (16-8=8)
4. 15 (15-7=8)
5. 5 (5×2=10)
6. 2 (2×4=8)
7. 20 (20÷2=10)
8. 20 (20÷5=4)
9. 16
10. 4
11. 2
12. 8
13. 2
14. 4

Chapter 11—Day 126—Practice

1. 3,748,600
2. 3,750,000
3. 3,700,000
4. 56,000
5. 60,000
6. 100,000
7. $6\overline{)66} = 11$
8. $7\overline{)84} = 12$
9. $8\overline{)96} = 12$
10. 112x5=560
11. 321x5=1,605
12. 134x5=670

Everyday Math

13. (÷)
14. (+)
15. (x)
16. (−)
17. (+)
18. (÷)
19. Room 1=12x8=96 sq ft
   Room 2=8x8=64 sq ft
   Room 3= 20x8=160 sq ft
   Total Area=320 sq ft
20. Room 1=10x12=120 sq ft
   Room 2=6x4=24 sq ft
   Room 3= 4x4=16 sq ft
   Room 4= 10x8=80 sq ft
   Total Area=240 sq ft

Chapter 11—Day 127

Pg. 410- 25x5=125

Student Exercises

1. $4\overline{)68} = 17$
2. $5\overline{)145} = 29$
3. $3\overline{)108} = 36$
4. >
5. <
6. <
7. <
8. 50, 45, 40, 35, 30, 25
9. 60, 54, 48, 42, 36, 30
10. 70, 63, 56, 49, 42, 35
11. 10x6=60
12. 8x10=80
13. 10x4=40
14. 2x10=20
15. 10x10=100
16. 10x7=70
17. 5x10=50

Chapter 11—Day 128—Practice

1. completed
2. 13 (13+3=16)
3. 17 (17-7=10)
4. 15 (15-6=9)
5. 4 (4×5=20)
6. 4 (4×3=12)
7. 10 (10÷1=10)
8. 15 (15÷3=5)
9. $\frac{4}{5} + \frac{2}{5} = \frac{6}{5}$ or $1\frac{1}{5}$
10. $1\frac{2}{5}$ colored parts
11. $\frac{4}{9} + \frac{7}{9} = \frac{11}{9}$ or $1\frac{2}{9}$
12. $1\frac{2}{9}$ colored parts
13. $\frac{7}{6} + \frac{5}{6} = \frac{12}{6}$ or 2
14. 2 whole colored circles
15. $3\overline{)69} = 23$

16. $5\sqrt{55} = 11$
17. 521x6=3,126
18. 359x2=718
19. 418x6=2,508
20. 77x3=231
21. 3.6 m
22. 1.55 km
23. 3 ft
24. 48 in

Chapter 11—Day 129

1.

$$\begin{array}{r} 92 \\ -\ 80 \\ \hline 12 \end{array} \qquad \begin{array}{r} 12 \\ -\ 12 \\ \hline 0 \end{array}$$

20 + 3 = 23

2.

$$\begin{array}{r} 95 \\ -\ 50 \\ \hline 45 \end{array} \qquad \begin{array}{r} 45 \\ -\ 45 \\ \hline 0 \end{array}$$

10 + 9 = 19

Student Exercises

1. 560÷10=56
2. 6.5÷10=0.65
3. 54.÷10=5.4
4. 15+18=33
5. 78-42=36
6. 4.5+5.6=10.1
7. 146+287=433
8. 35-16=19
9. 37+21=58
10. 2.4-2.3=0.1
11. 253-160=93
12. 6x3=18
13. 15÷15=1
14. 4x4=16
15. $\frac{8}{2} = 4$
16. 23.4÷10=2.34
17. 10÷2=5
18. 12x2=24
19. 12÷2=6
20. 23.4x10=2.34
21. $\frac{3}{3} = 1$

Chapter 11—Day 130—Practice

1. 36, 34, 32, 30, 28, 26
2. 0.3, 0.4, 0.5, 0.6, 0.7, 0.8
3. 75, 70, 65, 60, 55, 50
4. 45, 50, 55, 60, 65, 70
5. 450, 460, 470, 480, 490, 500
6. 320, 310, 300, 290, 280, 270
7. >
8. <
9. <
10. <
11. 3x7=21

$8x21=$168

12.

January	$40	÷	10	=	4
February	$60	÷	10	=	6
March	$80	÷	10	=	8
April	$120	÷	10	=	12
TOTAL					30

13.

14. $37.85

Speed Test

1. 6×5=30
2. 9×9=81
3. 9×8=72
4. 10×3=30
5. 9×7=63
6. 3×7=35
7. 3×7=21
8. 6x0=0
9. 9×10=90
10. 3×5=15
11. 6×9=54
12. 8×5=40
13. 5×0=0
14. 7×4=28
15. 10×1=10
16. 8×8=64
17. 4×3=12
18. 3×6=18
19. 6×6=36
20. 2×9=18
21. 6×8=48
22. 1×5=5

23. 6×7=42
24. 1×8=8
25. 2×8=16
26. 1×2=2
27. 6×0=0
28. 4×6=24
29. 2×6=12
30. 5×5=25
31. 5×9=45
32. 3×9=27
33. 3x2=6
34. 5×2=10
35. 8×4=32
36. 8×7=56
37. 10×1=10
38. 5×4=20
39. 8×8=64
40. 4×9=36
41. 5×8=40
42. 4×4=16
43. 4×6=24
44. 7×7=49
45. 9×9=81

Chapter 11—Day 131—Practice

1. completed
2. 5 (5+8=13)
3. 16 (16-5=11)
4. 7 (7-2=5)
5. 3 (3×6=18)
6. 1 (1×9=9)
7. 16 (16÷4=4)
8. 18 (18÷6=3)
9. dividend, divisor, quotient
10. quotient (on top), divisor (left), dividend (right)
11. $\frac{6}{7} + \frac{5}{7} = \frac{11}{7}$
12. $\frac{3}{9} + \frac{6}{9} = \frac{9}{9}$ or 1
13. $\frac{13}{17} + \frac{2}{17} = \frac{15}{17}$
14. $\frac{6}{7} - \frac{2}{7} = \frac{4}{7}$
15. $\frac{14}{19} - \frac{5}{19} = \frac{9}{19}$
16. $\frac{5}{9} - \frac{5}{9} = \frac{0}{9}$ or 0
17. 145+569+242=956
18. 123+431+567=1,121
19. 521+379+299=1,199
20. 452-333=119
21. 187-99=88
22. 487-298=189
23. 42-37=5 brownies
24. 48÷6=8 feet long
25. $\frac{3}{4} - \frac{1}{4} = \frac{2}{4}$ more snow

Chapter 11—Day 132

1-5. Answers will vary, check work.

6.

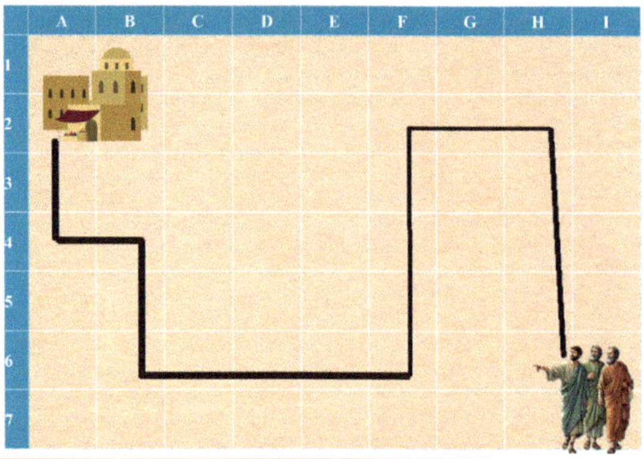

Chapter 12—Day 132

1. 8.00+7.00=15.00
   8.45+7.34=15.79
2. 12.00+22.00=34.00
   12.34+22.22=34.56
3. 30.00+40.00=70.00
   30.45+40.42=70.87
4. 123.456+111.111=234.567
5. 654.321+222.222=876.543
6. 345,567.876+111,111.012=456,678.888

   Pg. 433:
   A-5
   B-4
   C-1
   D-3
   E-6
   F-2

Speed Test

1. 6×5=30
2. 9×9=81
3. 9×8=72
4. 10×3=30
5. 9×7=63
6. 5×7=35
7. 3×7=21
8. 6×0=0
9. 9×10=90
10. 3×5=15
11. 6×9=54
12. 8×5=40
13. 5×0=0
14. 7×4=28

15. 10×1=10
16. 8×8=64
17. 4×3=12
18. 3x6=18
19. 6x6=36
20. 2x9=18
21. 6x8=48
22. 1x5=5
23. 6x7=42
24. 1x8=8
25. 2x8=16
26. 1x2=2
27. 6x0=0
28. 4x6=24
29. 2x6=12
30. 5x5=25
31. 5x9=45
32. 3x9=27
33. 3x2=6
34. 5x2=10
35. 8x4=32
36. 8x7=56
37. 10x1=10
38. 5x4=20
39. 8x8=64
40. 4x9=36
41. 5x8=40
42. 4x4=16
43. 4x6=24
44. 7x7=49
45. 9x9=81

Chapter 12—Day 134—Practice

1. <
2. >
3. >
4. <
5. $\frac{4}{5}+\frac{2}{5}=\frac{6}{5}$
6. $\frac{3}{7}+\frac{1}{7}=\frac{4}{7}$
7. 0.6+0.2=0.8
8. $\frac{6}{7}+\frac{3}{7}=\frac{9}{7}$
9. 50+30=80
10. 200+100=300
11. 300+300+600
12. 600+300=900
13. $3\sqrt{39}=13$
14. $3\sqrt{66}=22$
15. $3\sqrt{99}=33$
16. (4x2)+(6x1)=8+6
   Area=14
17. (3x2)+(5x2)=6+10
   Area=16
18. (2x5)+(2x3)+(4x1)=10+6+4
   Area=20
19. (2x2)+(3x3)=4+9
   Area=13
20. (3x4)+(2x1)=12+2
   Area=14
21. (3x2)+(5x2)+(4x1)=6+10+4
   Area=20

Chapter 12—Day 135

Pg. 437: 492.15-385.56=106.59

Student Exercises

1. 339.94+394.55=734.49
   Approximate: 340+395=735
2. 144.61+437.59=582.20
   Approximate: 145+438=583
3. 834.53-325.61=508.92
   Approximate: 835–326=509

Everyday Math

4. 91.7-49.8=41.9
5. 33.2-9.9=23.3
6. 85.8-77.9=7.9
7. 29.9-25.5=4.4
8.

9. 120
10. -10
11. 110
12. 20

Chapter 12—Day 136—Practice

1. Line #1- Answers will vary. 3 cm
2. Line #2- Answers will vary. 6 cm
3. Line #3- Answers will vary. 10 cm
4. Angle #1- Answers will vary. 60 degrees
5. Angle #2- Answers will vary. 90 degrees
6. Angle #3- Answers will vary. 30 degrees

Answers may vary, possible solutions are given.

7. 6=3+3
8. 6=4+2

9. 12=4x3
10. 12=6x2
11. 15=10+5
12. 15=8+7
13. 18=9x2
14. 18=6x3

Everyday Math

15. 72.75-36.75=36.00
36÷2=18
16.

Product	Cost per item	Selling price per item	Profit per item	Number of items sold	Total profit
Stuffed Animals	$5	$8	$3	10	$30
Yoyos	$10	$20	**$10**	6	**$60**
Shoes	$15	$25	**$10**	5	**$50**

17. $30+$60+$50=$140
Pg. 444: check work

Chapter 12—Day 137

1. 95
2. 63
3. 20
4. 30
5. 30
6. 40
7. 120
8. 122
9. 20.00
10. 13.00
11. 15.00
12. 35.00
13. 83
14. 83.41

Student Exercises

1. Answer may vary, 17
2. Answer may vary, 97
3. Answer may vary, 521
4. Answer may vary, 74
5. 134,410
6. 1,412.231
7. 78,197
8. 8,541
9. 71

Everyday Math

10. 90 people
15x6=90
11. 7 groups
28÷4=7
12. 12, 24, 36, 48, 60, 72, 84, 96, 108, 120, 132, 144
144x2=288

Chapter 12—Day 138—Practice

1-4. Answers will vary, check answers.
5. 111x3=333
6. 222x3=666
7. 333x3=999
8. 241x3=723
9. 4x10=40
10. 56x10=560
11. 23x10=230
12. 177x10=1,770
13. $\frac{20}{10} = 2$
14. $\frac{350}{10} = 35$
15. $\frac{1500}{10} = 150$
16. $\frac{236}{10} = 23.6$
17. $\frac{1}{14}$, completed
18. $\frac{3}{2}$
19. $\frac{11}{3}$
20. $2\frac{1}{3}$
21. $2\frac{2}{5}$
22. $4\frac{1}{2}$
Pg. 450: WITH GOD
ALL THINGS
ARE POSSIBLE
MATTHEW 19:26

Chapter 12—Day 139

1. $104
Pg. 454:

Chapter 12—Day 140

Answers may vary, possible solutions are given.
1. 6=3+3
2. 6=2+4
3. 18=9×2
4. 18=6×3
5. 16=8+8
6. 16=4+12
7. 24=12×2
8. 24=6×4
9. 133x2=266

10. 321x4=1,284
11. 112x5=560
12. 113x7=791
13. 338+245+457=1,040
14. 439+197+376=1,012
15. 151+364+345=860
16. sum(c1:c10)
17. sum(d1:d10)
18. >
19. <
20. <
21. 78 (−) 10 = 68
22. 55 (−) 5 = 50
23. 54 (+) 5 = 59
24. 60 (+) 20 = 80
25. 97 (−) 7 = 90
26. 31 (+) 8 = 39
27. 82 (−) 8 = 90
28. 45 (+) 5 = 50
29. 74 (−) 10 = 64
30. 21 (+) 9 = 30

Chapter 12—Day 141

1-5. Answers will vary, check work.

Chapter 12—Day 142—Practice

1. 4×8=32
2. 24÷2=12
3. 40÷4=10
4. 6×5=30
5. 10+12=22
6. 6+12=18
7. 18-4=14
8. 6÷2=3
9. 14+12=26
10. 16÷4=4
11. 4 64 =16
12. 3 33 =11
13. 4$\sqrt{44}$=11
14. 4$\sqrt{84}$=21

Everyday Math

15. $3,000
    100x30=3,000
16. 7 windows. $6 leftover
    8x7=56, 56-50=6
17. $19.50
    26.50-7.00=19.50
18. $660
    165x4=660
19. 3 weeks, 2 days remain
    23÷7=3.2

Chapter 12—Day 143

Student Exercises

1. 535+376=911
2. 649+393=1,042
3. 642-347=295
4. 624-554=70
5.

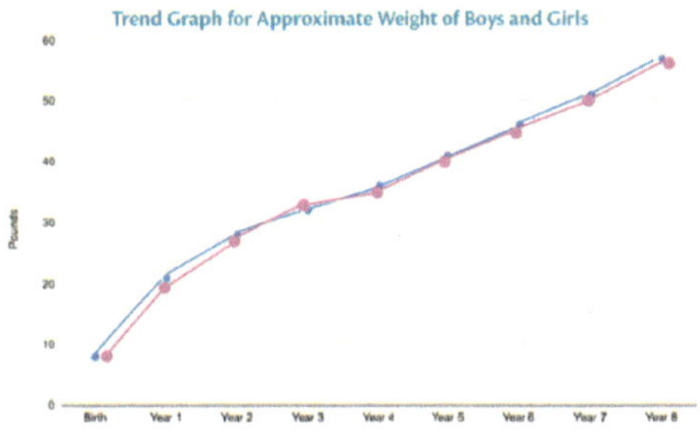

Between what two ages does a child gain the most weight? Birth and 1 year

At what ages are boys usually heavier than girls? 1, 2, 3, 4, 5, 6, 7

At what ages do they weigh about the same? Birth and 8

6. $\frac{1}{4}$
7. $\frac{3}{4}$, $\frac{2}{4}$

Activity

1. 11+4=15
2. 12+7=19
3. 18+6=24
4. 19+9=28
5. 21+9=30
6. 25+8=33
7. 26+5=31
8. 28+7=35
9. 33+7=40
10. 34+9=43
11. 35+3=38
12. 37+7=44
13. 41+9=50
14. 43+6=49
15. 46+9=55
16. 48+5=53
17. 54+9=63
18. 51+8=59
19. 61+5=66
20. 64+5=69

Chapter 12—Day 144—Practice

1-3. Answers will vary, check work.

Multiplication & Division Chart

X	1	2	3	4	5	6	7	8	9	10	11	12
1	1	2	3	4	5	6	7	8	9	10	11	12
2	2	4	6	8	10	12	14	16	18	20	22	24
3	3	6	9	12	15	18	21	24	27	30	33	36
4	4	8	12	16	20	24	28	32	36	40	44	48
5	5	10	15	20	25	30	35	40	45	50	55	60
6	6	12	18	24	30	36	42	48	54	60	66	72
7	7	14	21	28	35	42	49	56	63	70	77	84
8	8	16	24	32	40	48	56	64	72	80	88	96
9	9	18	27	36	45	54	63	72	81	90	99	108
10	10	20	30	40	50	60	70	80	90	100	110	120
11	11	22	33	44	55	66	66	88	99	110	121	132
12	12	24	36	48	60	72	84	96	108	120	132	144